MIND OF THE RAVEN

ALSO BY BERND HEINRICH

The Trees in My Forest
A Year in the Maine Woods
Ravens in Winter
One Man's Owl
Bumblebee Economics

Mind of the Raven

INVESTIGATIONS AND ADVENTURES WITH WOLF-BIRDS

Bernd Heinrich

ecco

An Imprint of HarperCollins Publishers

NEW YORK • LONDON • TORONTO • SYDNEY

ecco

An Imprint of HarperCollins*Publishers*

A hardcover edition of this book was published in 1999 by Cliff Street Books, an imprint of HarperCollins Publishers.

P.S.™ is a trademark of HarperCollins Publishers.

HarperCollins books may be purchased for educational, business, or sales promotional use. For information, please e-mail the Special Markets Department at SPsales@harpercollins.com.

First paperback published 2000.

First Ecco paperback published 2002.

First Harper Perennial edition published 2006.

Designed by Laura Lindgren and Celia Fuller

The Library of Congress has catalogued the hardcover edition as follows:
Heinrich, Bernd.
    Mind of the raven : investigations and adventures with wolf-birds /
Bernd Heinrich.— 1st ed.
      p.   cm.
    Includes bibliographical references.
    ISBN 0-06-017447-1
    1. Ravens. 2. Ravens—Anecdotes. I. Title.
  QL696.P2367H445   1999
  598.8'64—dc21   99-18129

ISBN: 978-0-06-113605-4 (pbk.)
ISBN-10: 0-06-113605-0 (pbk.)

23 24 25 26 27 LBC 31 30 29 28 27

To Raven characters I have known,
especially to Matt, Munster, Goliath,
Whitefeather, Fuzz, Houdi, and Hook

CONTENTS

Acknowledgments

This book grew largely out of the generosity of people who have shared their raven experiences and news, offered help and encouragement, and even aided in the field studies. All have helped to greatly enlarge our understanding and appreciation of this magnificent bird. This book is, then, a tribute to the generous input of: Aaron Adams, Bill Adams, Andy Adderman, Phil Angle, Amy Arnett, David Barash, Mark Bekoff, Cindy Bellinger, Trond Berg, Peter Bergstrom, Bill Boarman, Diane Boyd, Jim Brandenberg, George Brady, Cathy Bricker, Neil Buckley, Thomas Bugnyar, Eric Busch, Duane Callahan, David Campbell, Linda Campbell, Geoff Carroll, Rachel Carter, Doug Chadwick, William Chester, Gary Clowers, Craig Comstock, Eileen Connor, Bob Crabtree, Dorothy Crumb, Moria Daley, Mark Damon, Josina Davis, Richard Donley, Bill Drury, Micha Dudek, Randy Durand, Steve Emslie, Adam Farrington, Anell Farris, Gerald Fitz, Lori Friedman, Herbert Fuchs, Jean Craighead George, Ted Gaine, Ron Gerrish, Terry Goodhue, Donald Griffin, Thomas Grünkorn, Tim Hall, Forrest Hammond, Hilmar Hansen, Fred Harrington, Rolf Hauri, Stuart Heinrich, Kay Hensler, Monika Hilker, Carsten Hinnerichs, Garth Holman, Richard Hoppe, Stuart and Mary Houston, Beat Huber, Wendy Howe, Jim Hunter, Inooqi Irguittuq, Fran James, Delia Kaye, Paula Kelly, Bill Kilpatrick, Don Kilpela, Ted Knight, Catherine Koehler, Kurt Kotrchal, Bob Landis, Rachel Lawler, Bob Lawrence, Gale Lawrence, Don Lego, Ted Levin, Matt Libby, David Lidstone, Scott Lindsay, Volker Looft, Barry Lopez, Valerie Lownes, Marcy Mahr,

Mary Majka, Marvis Mark, Hans-Dieter Martens, John and Colleen Marzluff, Declan McCabe, Sarah McCracken, John McDonald, Terry McEneaney, Lorin McKay, David Mech, Brad Meiklejohn, Larry Melcher, Randolf Menzel, Gail Mihocko, Michael Miller, John Moran, Klaus Morkramer, Kim Most, Dick Nelson, John and Bob Nicholson, Janet Nook, John and Thomas Nutaraviaq, Abe Okpik, Kristian Omland, Nikita Orsyanikov, Linda Osborne, Tim Osborne, Jane Packard, Mike Palmer, Jack Parriott, Ray Paunovich, Doug Peacock, Hill Penfold, Lyn Peplinski, John Pepper, Mike Peterson, Rolf Peterson, Diane Pickard, Raymond Pierotti, Noah Piugaattuq, Andrea Ramsden, Natalie Rapp, Derek Ratcliffe, Jörg Reimers, Barbara Reif, Cindy Riegel, John Robertson, Ethan Rochmis, Michael Romero, Emanuel Rosen, Barry Rothfuss, Lorenzo Russo, Jenny Ryan, Bob Sam, Akaka Sataa, John Sawyer, George Schaller, Joseph Schall, Doug Schamel, Kristin Schaumburg, Charlie Sewall, Paul Sherman, Lorrell Shields, Phil Silverson, Rick Sinnett, Doug Smith, Roger Smith, John Snell, Ron Spiegel, Dan Stahler, Joanne and Neil Stinneford, Theo Stein, Carl Striedieck, Guy Stevens, Todd Sweberg, Jan Tinbergen, William Townsend, Jeff Turner, Charlie Uttak, Bill Valleau, Johanna Vienneau, Tinker Vitelli, Julia Voge, Wolfe Wagman, Dieter Wallenschläger, Chris Walsh, Mike and Ina Wesno, Steve Wheeler, John Williams, Mary Willson, Lesly Woodroffe, August Wright, Brent Ybarrondo, and Ann Yezerski, and all of those whose names now elude me.

It is impossible for me to single out each for their unique contributions. However, I give thanks to Ted Knight, Delia Kaye, Kristin Schaumburg, and Eileen Connor for freely volunteering months of dedicated labor on an often frustrating radio-tracking project. John and Colleen Marzluff invested three years of hard work, and many great ideas, that were invaluable in securing the recruitment hypothesis. I thank my agent, Sandra Dijkstra, and my editor, Diane Reverand, whose instincts on what constitutes a book have been an invaluable guide. My wife, Rachel Smolker, who gracefully endured my sometimes prolonged "absences" to be mentally if not physically with the ravens. Rachel's fine biological insights and keen editorial comments helped me say what I wanted to convey. Kimberly Layfield transformed my often inscrutable raven-scratch handwriting to type, and always did it with good cheer, and with speed and efficiency. The various institutional, state, and federal agencies expeditiously eased the permit process, and I thank them for thereby acknowledging the value of our gaining intimate contact with our fellow creatures.

Preface

I HAVE LIVED AND BREATHED RAVENS since a date I will remember: October 29, 1984. On the afternoon of that day, I was drawn to the commotion of a group of ravens at a moose carcass. If even I could be attracted to a carcass from over a mile away, then

*My last batch of youngsters, including Red,
Blue, Yellow, White, Orange, Green, and Eliot.*

ravens should be ever more so. The birds seemed to be advertising their find, which meant they would have to share it. I was deeply puzzled by the mystery posed by this phenomenon, and I wrote a book about my efforts to understand it. I was then not aware that the study, and the subsequent ones covered in this book, had been presaged more than two years earlier in a journal entry of mine from February 21, 1981, that described a dream from the night before: "I was walking in a dark and very mysterious forest and heard the croaking of ravens, one of the most awesome sounds I know. The ravens' calls told me that their nest was near. The ravens' calls were full of promise. I felt I was close to something new and exciting, and would find it." Then I awoke. I had always wanted to find the nest of a raven. Was that the focus of its intimate life?

Why would anyone dream about ravens? The common raven, *Corvus corax,* is the world's largest crow, usually exceeding goshawks and red-tailed hawks in size. (I will here not use the ornithological convention of capitalizing bird names, adhering to literary usage instead.) It was extremely rare here in northeastern North America fifty years ago, whereas the smaller American crow, *Corvus brachyrhynchos,* which resembles it superficially, was and still is extremely abundant. Cornell ornithologist and bird artist George Miksch Sutton wrote in 1936 that the raven is "wary and solitary in nature." He described it as a bird of the wilderness that few people ever saw, and like most people who know it, he singled it out for being special.

All animals solve the same age-old problems relating to food, partnerships, sex, shelter, home, and caring for their young. Yet ravens have throughout history commonly been singled out to be most like man. Why? What is special about ravens that invites such a comparison?

Humans usually have considered themselves to be different and apart from other animals. Perhaps, as lion researcher Craig Packer points out in his book *Into Africa,* that is because "we make it all up as we go along," whereas an ant has "every small instruction laid out in advance." Ravens, like humans and unlike perhaps most other birds, probably do not have instructions to all of life's problems laid out in advance, or they would not likely have been considered highly intelligent, and mythologized as creators, destroyers, prophets, playful clowns, and tricksters. I have often been startled by their enigmatic and seemingly contradictory responses. But the poetry of biology resides hidden in opposing tensions, and the often arduous fun comes from trying to reveal it.

To begin to see how the raven's mind might be organized requires finding the focal points of its natural history, necessitating an openness to all the input of information one can muster. Since I wrote *Ravens in Winter,* I have intensified and greatly enlarged my original study. I have tried to see ravens from as many perspectives as possible that might reflect not only on their adaptive responses but also on the motivation of those responses. Hundreds of people have shared with me their experiences of seeing ravens' behavior. They have been like extra eyes, probing where and when I could not reach. I have tried to sift carefully through these reports, but I'm able to offer here only a small minority that were appropriate for the context. I have tried to keep in mind that anecdotes can easily become interpretations, and that facts expand in minds when they are not opposed by knowledge. On the other hand, to reject all anecdotes is also to reject facts. With ravens, the line between interpretation and fact is commonly a thin one, but as Mark Pavelka, who studied ravens for the United States Fish and Wildlife Service, said, "With other animals you can usually throw out 90 percent of the stories you hear about them as exaggerations. With ravens, it's the opposite. No matter how strange or amazing the story, chances are pretty good that at least some raven somewhere actually did that." That is because ravens are individuals. Ants aren't.

It may be impossible to prove in a literal or absolute sense that any one particular animal has or does not have emotions, consciousness, or capacity for insight. These subjective, individual, and hard-to-define qualities of mind are found in separate independent evolutionary lines, with the highest end-points reached in some species of primates, cetaceans, and perhaps corvids and parrots. Trying to pin any of the many aspects of mind down to one particular amount in any one particular species, however, is like trying to specify the exact moment in a continuum of time when a child can or could talk. She is *not* talking when she makes gurgling sounds at the age of a few weeks, but she is doing so later when she says, "Mommy, please read me the story about the hungry caterpillar one more time." Some of the varied vocalizations she makes in between the first sounds and the first sentence seem arbitrary, because they are. So it is with all aspects of mind, only we cannot measure them as precisely as we can sounds. And some, like the smell of a rose or the feel of a spring day, no matter how objectively and concretely they can be proven to affect behavior, will forever remain private.

In the last several decades, there has been a blossoming in our awareness of animal minds, despite the large methodological problems of investigating them. Studies have shown many often unprepossessing animals to have amazing sensory and mental (probably unconcious) powers that were unsuspected, and that seem almost beyond belief. Many rival our own much-vaunted powers. Accounts of these amazing capabilities fill textbooks of animal behavior, and they startle and surprise our collective imagination. I know I have been, as are most students of biology, mesmerized by them. To mention only a tiny few almost at random out of the myriad to choose from, I've marveled at Harvard and Rockefeller Universities' Donald Griffin's discovery of how bats can detect and intercept flying moths in total darkness, and Ken Roeder's findings of moths' reflexes that foil bats. I've been awed by Karl von Frisch's elucidation of how bees can perceive and learn to seek food associated with specific scents, colors, and geometrical patterns, and how they communicate to hivemates the distance and direction of food and/or potential home sites for departing swarms. I've admired Cornell's Tom Seely's extension of this work, showing the hive's complex decision-making process, which proceeds seamlessly yet without insight, thinking, and intelligence as we normally define them for ourselves. In contrast, it is therefore all the more intriguing that the detailed studies by Dorothy Cheney and Robert Seyforth of wild vervet monkeys in Africa revealed that some of these animals' calls contain semantic meanings, much like our words do, and the animals apparently know what some of their utterances mean, because they sometimes use them to deceive others for their own advantage. The work of Wolfgang Köhler in Germany, Frans de Waal at Emory University, Jane Goodall in Tanzania, and others has shown that some apes are also capable of self-recognition and deception of others. I was thrilled by Katie and Roger Payne's discovery that humpback whales have elaborate, often hour-long songs whose precise sequence of sounds changes seasonally over the ocean basin, with all individual whales singing a similar song in any one year in either the Atlantic or the Pacific Ocean. I was surprised by the discovery that some dolphins swim around carrying sponges with them, probably to use as a foraging tool. Birds are just as amazing. Even pigeons, as Richard Herrenstein of Harvard proved, can recognize "trees" as a category in photographs, and can even differentiate one artist's paintings from another's. Gustav Kramer in Germany made startling discoveries through clever experiments showing that starlings (like honeybees also) use the sun as a compass to tell them the direction. But the sun is not a stationary beacon. It moves 15 degrees per hour, and the animals compensate, calculate in this movement, as it moves daily across the sky, by consulting an internal clock. He showed that warblers migrate at night, navigating by using the north star as a beacon; and through elegant experiments that will forever be classic in biology, the father and son team of John and Stephen Emlen built on this work to decipher the intricate combination of innate *and* learned components of the amazing star-orientation behavior in indigo buntings. A variety of birds and other animals can use the earth's magnetic fields for orientation. Birds routinely navigate to and from pinpoints on the globe separated by continents, presumably using largely innate, fairly inflexible navigation scripts. But abstract concept formation is not precluded. Just recently, Irene Pepperberg of University of Arizona, Tucson, has shown us through her patient and detailed work that even a parrot may learn to converse with a human, using a vocabulary of about seventy words. Only a year ago, Gavin Hunt of New Zealand revealed how an Australian crow makes two kinds of tools. In all these amazing studies, there is first the sheer surprise and the wonder of what the animal can do. There is then the question of meaning.

Whatever else the behavior might be, it is a product of sensors and the animal's mind. The primary value of the discoveries, however, is not that they might prove any particular of a variety of abstractions such as "innateness," "consciousness," "insight," or "intelligence," or whatever label anyone may try to give them. They are all manifestations of mind, but their marvelousness is independent of the specific microanatomical complexity of neural wiring that allows them to be.

I had my first tame ravens—two of them in my apartment in Westwood—while I was still a graduate student at UCLA in the late 1960s. I would have liked to study them then, had I seen a problem to study. I didn't see any, but in any case, we graduate students were perhaps jokingly advised not to study an animal smarter than ourselves. I worked first with protozoa, in whose simplicity I saw many problems. I choose next to concentrate on a caterpillar, only gradually working my way up to moths. For my first project, I tried to figure out how the large caterpillars of the moth, *Manduca sexta,* which gardeners routinely pick off of their tomato or tobacco plants, "solved" a simple problem: how to consume a huge leaf entirely while it was attached to the plant and the catapillar to it, and regardless of leaf shape. How could an animal do the equivalent of hanging on at the tip of a branch in the top of a tree and eat the branch too? The way the caterpillars did it *looked* like they planned ahead, but they didn't. Experiments showed that blind and simple programmed responses explained their seemingly purposive actions quite nicely. That work later led me to look at behavior internally. I eavesdropped on neural messages that underlie rhythmic stereotyped behavior that was externally visible (flight) or invisible (shivering) in bumblebees. In all of this research, I felt, and still feel, in agreement with most other researchers, that insects' behavior was largely prewired. It was physiology. Bumblebees' foraging in the field showed sophisticated learning, blurring the physiology I felt comfortable with into what could more appropriately be called behavior. Birds seemed behaviorally extremely interesting, but I felt that all their overt behavior that could be known already was, and all that was internal never could be. How wrong I was on both counts!

Years later, when I finally had tenure and ventured to try to solve a small question of behavior in ravens, I had occasion to hear reports of ravens behaving in ways that seemed both intelligent and strange, and although much of that could be dismissed as hearsay, there was never-

theless a sampling of the observations published in the scientific literature by respected and disciplined observers that hinted otherwise.
The reports included ravens hanging by their feet (Elliot, 1977), sliding in snow (Moffett, 1984), snow-bathing (Hooper, 1986; Hopkins,
1987; Bailey, 1993), aerial bathing (Jaeger, 1963), flying upside down
(Evershed, 1930, Täning, 1931), doing barrel-rolls (Connor et al.,
1973; Van Vuren, 1984), social flying (Henson, 1957), using objects
to displace gulls from nests (Montevecchi, 1978), using rocks in nest
defense (Janes, 1976). Manual flexibility included carrying food in the
foot rather than the bill (Owen, 1950), foot-paddling (Ewins, 1989),
and rolling on the ground to avoid a peregrine (Barnes, 1986). Other
strangely flexible behavior involved covering their eggs (Davis, 1975),
poking holes in the bottom of their nest on a hot day (Gwinner,
1965), carrying their nestlings (Stoj, 1989), bonding to a crow (Jefferson, 1991), catching doves in midair (Elkins, 1964), and attacking
reindeer (Ostbye, 1969). The reports in aggregate hinted that studying ravens would be not only interesting but also challenging.

Having now lived on intimate terms with ravens for many years,
I have also seen amazing behavior that I had not read about in the
more than 1,400 research reports and articles on ravens in the scientific literature, and that I never could have dreamed were possible. I
have become skeptical that the interpretations of all ravens' behavior
can be shoehorned into the same programmed and learned responses—
categories as those of bees. Something else is involved, and I wanted to
make some sense of it. My concern with imponderables, however, has
usually been secondary to the quest to find out what they *do,* which to
me is more important than deciding how to label it. Ultimately,
knowing all that goes on in their brains is, like infinity, an unreach-
able destination. The interesting part is the journey.

My goal here is not to be authoritative. Instead, I sketch the world
of a magnificent bird that, as we shall see, has been associated with
humankind from prehistoric times when we became hunters. I focus
largely on unpublished observations, experiments, and experiences
that I hope will engage you to participate in the quest of exploring
another mind.

MIND OF THE RAVEN

ONE

Becoming a Raven Father

THE FIRST PREREQUISITE TO STUDY-
ing any animal is to get and to stay close. You must be able to observe
the fine details of its behavior for long periods of time without the animal seeing or feeling your presence. That's a tall order with wild
ravens in northeastern America. In my part of the country, where food
is sparsely distributed, ravens may range over a hundred square miles a
day, and they fly away at the mere sight of a human. Ravens are shyer
and more alert, and have keener vision than any other wild animal I
know, making it even more difficult to watch their natural undis-
turbed behavior.

Given these difficulties, I felt I needed to try to obtain young and
to be a surrogate parent to them. It was perhaps the only way to learn
about many aspects of their intimate social behavior. Obtaining and
living with young ravens has its inconveniences, not the least of which
is making the hazardous ascents to get them from the nest. The trees
that ravens like to nest in are not the ones I like to climb.

Houdi at ease and confident, showing small hint of ear feathers.

The last few patches of winter snow were left in the shady places under fir trees. The ice had just melted off Hills Pond, and the first warblers were back. But in Maine at the end of April 1993, still a month before the maples would leaf out, that year's baby ravens would already have a coat of black feathers. I was on my way to two different nests. I would take two young of the clutch of four to six I expected to find in each nest. I would then have to attend to every need of the young birds.

It was snowing, and the great pine tree with the first ravens' nest was swaying in the north wind coming across the lake. I wanted to run, but I held myself to a walk, trying to conserve as much energy as possible for the climb ahead. More than once before, I had been frightened when I found myself hanging on to a thick limbless pine trunk as strength ebbed from my arms. The void above the tops of the fir trees seemed to expand as my grip grew less secure.

White feces were spattered on the ground below the nest, a sign that the young were already beyond the pinfeather stage. At this nest, which I had visited often in previous years, only the adult male scolded me; the female always left when I came near. At other nests, both members of the pair may scold, both leave, or one or both remain at some distance.

After carefully putting on climbing spurs and adjusting my backpack, I put my arms around the tree and started. Go slowly, I kept telling myself, one step at a time. I tried to keep looking up, not down. I got very tired just as the solid limbs were getting closer. Fortunately, having trained over the winter to do chin-ups, I was able to sustain the effort. As I hauled myself up onto the first solid limbs, I felt elated. I had once again escaped the fate of some other ornithologists in similar situations. George Miksch Sutton fell from a cliff while climbing up to a raven nest, but was saved by falling on a ledge before hitting the bottom of the cliff. Fellow raven researcher Thomas Grünkorn once fell eighty feet out of the top of a beech, breaking his back in two places. Miraculously, he lived and did climb again (see Chapter 7). Gustav Kramer, an ornithologist studying wild pigeons, died when rocks came loose as he was climbing up to a cliff nest. I am quite frightened of cliffs, by comparison feeling almost safe with tree limbs to hang onto.

The tree was swaying mightily in the gusts, but it had not been blown over in worse gales, and it would not fall or break now. Besides, there was nothing I could do about it. So no worries.

Four fully feathered young hunkered down in a very soggy nest. It had rained steadily for the last two days, and the heavy, waterlogged nest was badly tilting because one of the supporting branches was too thin. The four young were very chunky, clumsy, and cute. When I lifted two out to put them into my knapsack, I noticed their huge bare bulging bellies. They didn't struggle or complain, and the climb back down was easy.

With four young finally resting in the bottom of the knapsack, I hiked home to put my new charges into their new nest—a basket packed with dead grass and leaves almost all the way to the top, so that they could defecate out over the edge. I talked to them in low soft tones, and they immediately broke their silence and answered in raspy raven baby talk. They were almost feathered out, and looked at me with bright blue eyes (which would turn gray near fledging and brown by winter). They raised pinfeathery heads and opened their big pink mouths (mouth linings and tongue turn black only after one to three or more years, depending on the birds' social status). They were begging to be fed! Such trust, especially after just being taken from their nest, is unique for a bird already at least a month old. It is especially surprising given that ravens are innately shy of anything new. As adults, ravens in Maine are among the most shy of all birds. These young had long been exposed only to their parents and siblings, yet they responded to me unabashed. Did they somehow hear something in my voice that put them at ease?

Their sounds tugged at my heart, and I sprang to action. We quickly established a rapport and I chopped up whatever meat I could find, usually roadkills, and fed the ravens bite-sized chunks at about hourly intervals, just as raven parents do in the wild. Baby ravens and crows need a pharmacopeia of proteins, minerals, and vitamins. I fed mine minced mice, grubs, eggs, fish, and chopped frogs. I've seen people try to raise baby crows as they would raise their own babies, on milk and bread. If the young birds didn't die, they suffered from rickets or some other nutritional debilitation that left them crippled.

Parenting in animals has been honed by natural selection over millions of generations, and greatly diverse parenting skills have evolved. Details are critical, and anything different from the species' norm is likely to be harmful. Young passerine birds need a high-protein diet that contains all the vital nutrients in proper proportions. During their critical growth period, muscle, nerve, and bone are built up so fast that body weight may increase 50 percent or more per day. Feedings therefore must be frequent.

On the first day, I fed the four birds six mice, four hen's eggs, two six-ounce cans of cat food, ten ounces of puppy chow, and a couple of mouthfuls of beans I had prechewed for them. They had each gained 600 grams of body weight after they had eaten 8,100 grams of food.

To give you some idea of the food a pair of ravens must provide for their family, I offer here a list of what I fed to a group I parented some years later, six nestlings at about five weeks of age.

Day One: One woodchuck and one snowshoe hare (roadkills that I froze and then chopped up—skin, bones, guts, and all—into bite-sized chunks and thawed before feeding).

Day Two: Three red squirrels, one chipmunk, six frogs, eight chicken eggs (crunched up shells and all).

Day Three: Two gray squirrels, five frogs, six eggs, six mice.

Day Four: One hindquarter of a Holstein calf.

In a few more days, as their appetites picked up, each could swallow six woodfrogs, two mice, one after another in just one meal, and each young was ready for a repeat performance of the same in just one or two hours.

Raising baby ravens has always been a joy to me, but I must explain the difficulties entailed in raven parenthood. Never mind stopping for and chopping up roadkills. Or the fact that a raven will never be house-broken. As for going away for a Saturday afternoon—forget it! Remember, a young raven needs to be fed at least every few hours, and it will require your devoted attention every day. Without that attention, the bird will not bond with you and will become wild and "unpersonable"; it won't give anything back. Basically, you will be stuck with the world's worst roommate. To some, that may be a small price to pay I just wanted to be sure my readers know what the price is.

There is also the bird's side to consider. If integrated into a human family, it will be where it wants to be and it will likely be happy. If it is not bonded to a family member, to which it needs constant access, it will feel imprisoned. Finally, there is also the official point of view to consider. There is no legal qualification for having a human baby, whether or not one is capable of parenthood. But it's not so easy to "have" a raven baby. For that you need both state and federal permits, which require providing a thoughtful rationale.

If you seek contact with a wild bird that will bond, that will "talk" (and even learn to sing tunes), and that will be practical to keep, I recommend adopting a starling. Mozart had one, and evidently he was extremely fond of it. Starlings are black, and when in breeding plumage they glisten even more than a raven does. They are far better voice mimics, and they are easier to keep. Best of all, they require no permit, because they are an exotic "pest" species that competes with our natives. The main annoying thing about them, at least the young ones, is that they will unceasingly yell in your ear and open their bill in it, as if prying under leaves to check for worms. A wild goose won't do that, but it will fly behind you if you leave in a car, as if believing you've joined a flock, so you can't drive in heavy traffic. Each species has its own particular innate responses that have been fine-tuned through millions of years of selection in the natural environment.

Trying to parent wild animals is discouraged these days, perhaps in large part because it is difficult to do and often goes awry in unanticipated ways. When I was a boy, my friends and I raised young crows, jays, robins, sparrows, skunks, raccoons, hawks, owls, geese, and a starling. What we learned not only applied to our pets, but perhaps was also a lesson in diversity, in patience, and in tolerance.

The food that you give baby birds creates even further chores and responsibilities for you. The meat ingested by the young is thoroughly processed into a near-equal volume of liquid waste called "mutes." Raven parents scoop up the mutes in their mouths as they are being ejected, so as not to soil the nest. After the nestlings are a week or two old and the volume of mutes picks up, the parents no longer eat them but carry them off to discard away from the nest.

I lucked out. My four young ravens were already old enough to back up to the edge of their nest. They stuck their rears out over the edge and vigorously wiggled their tails from side to side like a reversible rotor. Only then did they finally let go in a stream that shot out two to three feet.

Despite the young's early bowel control, all that changes as they get older. Eventually, near fledging, they lose it. They go at will, with emphasis on achieving maximum distance. Once they have left the nest, the young become ever more nonchalant about waste elimination. That's why they can't be housebroken.

How can the raven parents swallow mutes, the wastes of not only the digestive but also the urinary system? First, raven mutes normally aren't malodorous, unless the baby bird is sick or overfed and the food is not thoroughly digested. Raven mutes are mostly a white secretion of uric acid crystals from the urinary tract. They aren't at all like chicken manure, which, though it smells horrible, is mixed with sawdust and fed to cattle, so that we eventually eat it indirectly.

All protein metabolism results in toxic, often smelly nitrogen waste products that the body must get rid of by flushing with water. That is why the more proteins we eat, the more water we must drink. Paridoxically ravens live on meat, yet they can get by without drinking, except at high air temperatures when they need water for cooling themselves evaporatively from the mouth and respiratory surfaces. That's where their nonsmelly mutes come in. The white uric acid they contain is non-toxic and relatively nonsoluble in water, unlike the raw sewage, urea, the primary mammalian waste product. Thus, large amounts can be excreted with very little water. It is voided as a white odorless paste rather than a yellow liquid. As a consequence, a raven is preadapted to get by on little water, and the parents' nest-hygiene chores are easier.

Aside from the mutes that can be bothersome, young ravens are vociferous, and not very melodious for a so-called songbird. Ravens are unjustly accused of

Raven parents shove food deep down baby's open gullet.

being poor or uncaring parents because their young beg for food so noisily. Although hunger increases the volume of their begging yells, that volume is itself a product of natural selection. The young may outshout each other in competition to gain parents' attention, but the shouting also attracts predators. Ultimately, the upper volume of noise that evolution allows is capped by predation. If the nest is secure, the cap is high.

In ravens and other birds, there has been tremendous selective pressure to grow fast and develop flying ability soon after birth. A sparrow can grow to full weight in a mere ten days, but in a raven, growth to maximum weight takes about forty days. Young ravens about a week old spend most of their time sleeping while being brooded by the female. They are still naked and unable to regulate their body temperature. They pop up their heads and beg whenever one of their parents gets up and makes short little nasal *"gro"* calls. Later, after the female no longer incubates them, they sleep, waking and begging whenever a parent flies toward the nest.

LEFT: *Recently hatched chick.*
RIGHT: *About a week later, eyes still closed, pinfeathered.*

At about three weeks of age, when they feather out, there is constant stirring and nonstop activity. At any given moment, one young raven may be lying down and sleeping with its bill tucked into the feathers on its back, while another stands, stretching its legs and one wing at the same time, then perhaps reaching with one foot over its back to scratch the back of its head with a toenail. Another may stand on the rim of the nest, vigorously flapping its wings; a fourth may pick and yank at a loose stick, while a fifth is singing. The singer will have a dreamy vacant look about him, frequently fluffing out his head feather and erect "ear" and throat feathers. He'll be acting like an adult male experiencing confidence and/or dominance. He'll cock his head, half close his eyes, and utter queer gurgling, warbling calls that vary crazily in pitch and volume with no detectable rhythm. When a fly buzzes by, all are distracted. Heads shoot up and all watch the insect intently. Seconds later, activities resume. From a few seconds to a minute or so, each individual switches from stretching, preening, wing-flapping, sleeping, playing with sticks, and shaking. There is little indication that the activity of one bird has any influence on the activities of another.

A human baby of the same age of six weeks cannot yet turn himself over, but he can almost hold his head up. If a human baby can't reach the nipple, he gets lots of assistance. The baby ravens must hold their heads up high to beg for food from the day they hatch out of the egg. Those who don't, won't get fed. Baby ravens snatch mosquitos and black flies out of the air in a superb display of eye-bill coordination at only three weeks of age. Even before leaving the nest, they can pick up food and feed themselves, but they won't do it if begging is an option. They can scratch the back of their head with one foot, sleep standing up, and groom themselves extensively.

My four, which I had secured from two different nests, consisted of two males and two females.

Even before they were out of the nest, the males were larger than the females. I called one of them "Goliath." One of the females with a broken toenail on her left foot became "Lefty." The other female was eventually dubbed "Houdi," for the great magician and escape artist Harry Houdini, because she escaped from my aviary several times.

By May 10, as the woods leafed out almost overnight, the ravens were hopping onto the rim of the nest I had made for them in the apple tree next to my cabin. They now constantly "talked" to me in husky, throaty voices.

A week later, they were still unable to fly, but they frequently flapped their wings vigorously. The birds now had velvety black heads and body (contour) feathers, and shiny tail and wing feathers. They had lost the tufts of baby down attached to the ends of some of the head feathers, except for one male who had two tufts of down remaining on the top of his head. He became "Fuzz."

On May 17, I let them all down onto the ground. All four made little *grr* comfort sounds in apparent excitement. They enthusiastically beat their wings and hopped about, picking at leaves, twigs, and grass. They were fearless of a big white husky dog a friend brought. When one of the wild ravens that nest nearby at Hills Pond flew over making *kek-kek-kek* alarm calls at them, they froze, became silent, and made themselves look thin, pitiably shaking in fright as if shivering in a cold breeze, although the sun was shining. I told them, "Relaaax—it's okay," as soothingly as I could, and they did immediately relax. They fluffed out again, stopped shaking, and resumed begging for food.

The young ravens were genuinely endearing, but gave the impression of not being very bright. They appeared to try to cache pieces of meat by tucking them into crevices, but made no attempt to cover them. They often immediately picked up the meat again to repeat the process a couple more times.

When they were only a few days out of the nest, they could fly well, but were very reluctant to do so. Instead, they begged forlornly to be fed when I could not reach them in a tree. Eventually screwing up their courage, they would launch themselves off a branch and generally end up even higher in the next tree. Hopping always onto the branch nearest to their eyes, they kept going higher, still farther away from food.

From the time the birds left the nest, the two males, Goliath and Fuzz, were the most dominant, and the females, Lefty and Houdi, were

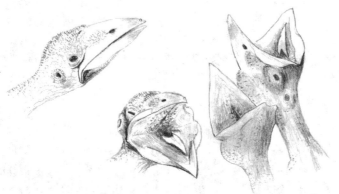

Young ravens, eyes just opening.

considerably less confident. Unlike most dominance hierarchies, this one was not static. In the next few months, there were two main power struggles, one between Goliath and Fuzz, and another between Lefty and Houdi. Goliath remained top bird for the first several months. Suddenly, inexplicably, on August 27, Fuzz assumed top-bird position. Thereafter, he was always the first at most food, and he attacked all his nestmates. At larger or feared food piles or carcasses, he allowed subordinates to go first, then chased them off. Right after Goliath's reversal of status, I wrote in my journal that he "now has funny-looking eyes. It is as if the lower lids are drooping, exposing the whites, giving him a strange bug-eyed look. He has stopped flying onto my arm, when he had been the only one of the four birds to do so. He seems psychologically disturbed and his voice, unlike the others', has now taken on a sharper edge. He also refuses to take food out of my hand. He has changed overnight from being the tamest to becoming the shyest of the four birds."

Goliath was unique in another way. Shortly after he was out of the nest on June 17, he began submissively crouching to me, drooping his wings and vibrating his tail. He did this routinely to me for more than a year, while none of the other birds ever did it at all. I had felt closest to Goliath since he was a fledgling. I even had the impression that he routinely and deliberately tried to make eye contact with me. Perhaps it was reciprocal.

By September 9, Goliath was no longer bug-eyed. He had regained his former status as the top bird, but his reactions to me remained unchanged. At that time, his dominance was expressed in the number of wins during interactions with others (twenty-seven versus nine for Fuzz at one count) and in the position he took on carcasses. Before the status flip-flop, he had seventeen top-bird positions on carcasses versus fifteen for Fuzz (and four for Lefty and one for Houdi). After he became dominant again, he held the top-bird position thirty-four times versus three for Fuzz, none for Lefty, and one for Houdi.

Throughout the winter, Goliath clearly remained the dominant bird, regularly showing erect "ear" feathers and flaring his feather "pants," the trapping of high raven status. Fuzz displayed rarely, and the females never. Most of the fights and chases were between the two females, who had a long dominance battle. In most of these interactions (twenty-one versus three), Lefty prevailed over Houdi.

Goliath's dominance display became even more pronounced by late February. Whenever I brought in a piece of food that he was anxious to have, he immediately flashed his ears, flared his "baggy pants," then slowly swaggered up to the food and took it. The others always hung back with smoothed-down head feathers, then bowed down and fluffed their head feathers whenever they came near him. His dominance was no longer tentative. It was palpable, demonstrable, unquestioned, and unchallenged. Fuzz, in contrast, responded to my bringing food by begging like a baby bird.

At two years of age, Fuzz was still doing the baby bird wing-flutter and uttering high-toned begs whenever I came near him, although the other three birds had not done either since November in their first year. Doting father that I was, I had begun to see them as characters, and I would soon be observing what had never been possible in the field. One of them would end up taking a wild bird as a mate, and nest in the wild by my cabin.

TWO

A Field Experiment

FOR YEARS, I WONDERED IF RAVENS in the wild who had discovered food were instrumental in bringing in, or "recruiting," others to the feast.

In field experiments, you may set up contrived but plausible situations to test responses. This has problems, the main one being that your subjects usually choose not to show up to participate in your plans. Or if they do show up, they are moved by agendas over which you have no control. My usual field approach in the early 1980s was to drag a calf carcass into the woods and then watch from a hiding place, hoping to see something interesting. Eventually, after four years and thousands of hours watching, I determined that various adult ravens lived in pairs near my study, while juveniles seemed vagrant, wandering widely, coming and going. The adults usually defended carcasses I put out, chasing

A group of ravens feeding near one of my observation blinds.
This photo was taken with a 400 mm lens from
a window of my house in Hinesburg, Vermont.

the vagrant juveniles off. At least in late winter, vagrants slept together in large groups and they sometimes recruited their roostmates to bait, arriving in noisy crowds from their sleeping place long before daylight. This crowd then got to feed by overcoming the adult defenders.

In 1988, John Marzluff, who recently had earned a doctoral degree from the University of Arizona at Flagstaff, joined me to help tackle the next problem: how ravens recruited others from the communal roost. Did they perform a dance, as bees do in a hive? Did the birds have specific "follow me" signals? Did the most knowledgeable birds leave the roost early and purposefully provide a cue that the roost-birds follow the first bird out? Did the dominant or the more subordinate juveniles recruit? Who benefited, and why? What were the costs of recruiting?

In the field experiments that John and I were about to do in mid-December 1990, we wanted to test for the effect of a vagrant's status on recruitment, attempting to keep everything else constant. Our idea was to release a bird of known status directly at a calf carcass in the woods and then see which birds, having discovered our offering, would subsequently bring other birds back later.

We had done other versions of this experiment before. First, long-held captive birds, who could not have had knowledge of any food bonanzas in the field, joined crowds at roosts without hesitation after we released them near the roosts in the evening. The next dawn, they showed up at the baits where that particular roost crowd was feeding. Control birds released on the same evening without access to a roost did not show up at the bait. This was definitive proof that the communal roost served as an information center.

This year we would do the reverse experiment of releasing potential recruiters at baits rather than potential recruitees at roost. A huge complication to be expected was that the roost birds were already feeding at other carcasses, and hence uninterested in changing feeding sites. We would try to restrict the number of potentially competing carcasses by removing those we knew about.

To set up this new experiment, we captured twenty wild birds and maintained them for months in our large aviary in the Maine forest. John watched these birds daily to tabulate who made submissive gestures to whom at food. He found that they were all aligned in a dominance hierarchy, starting with the most dominant bird who challenged all and yielded to none, and to whom all others yielded. We could then pick birds from either end of the dominance spectrum for our recruitment trials in the field. We had twenty radios for this experiment, which we would use to track the birds.

The result of the first release of the year, at a food pile near the inlet to Lake Webb, was spectacular! Freed at dusk, our radioed bird did not feed, although she had not eaten for two days. Unlike most of the others, she did not bolt away either. Instead, after we opened the door of the cage, she calmly walked to a puddle near the bait and drank. Then she flew onto a tree above the bait and preened vigorously for half an hour. Next, cawing loudly, she flew north in the direction of a roost. Roosts are noisy at night, and perhaps she heard the din. We knew that the birds at that roost had just finished feeding at another bait. All the conditions that one can never control for in the field were miraculously just right. Better still, our radio signals indicated that the bird entered the roost that night. The result the next dawn was stunning: At first light they came—a string of more than thirty ravens, all flying directly from the roost to the bait that they could have known about only from our radioed bird, who was in or near the lead! Results like that convinced us that ravens can and do recruit from the roost.

For the particular trial I'm about to describe, John picked a bird of low status. Our studies had already shown that the reason only vagrant juveniles recruit to a food bonanza is that it allows them to overpower the territorial adults to gain access to food, although there could be other reasons as well, such as wanting company during feeding at food. We now expected such *subordinate* juveniles as our present subject to recruit more than dominant juveniles, because subordinates should have the most to fear from the adults. On the other hand, in a crowd of juveniles, they experience less aggression at a carcass or bait, since the dominants fight with each other and are also attacked by the defending adults, thus giving the subordinates more chance to get past the socially distracted dominants to get at the meat.

John attached a radio transmitter to the bird's tail. He also attached a large red plastic tag to each wing. Each tag had a large number written on it, enabling us to identify the bird not only by her radio signal but also by sight. He then put her into a pet travel-cage and did not feed her for two days so that she would appreciate the super-bonanza of food we had in store for her. Meanwhile, I drove the two hundred miles from Vermont, in a snowstorm as it happened, to help conduct the experiment. My job was to release this bird at a fresh food bonanza, about 150 pounds of meat scraps, that I had dropped in the woods. The idea was that we would release this bird next to the food; i.e., we would arrange for the bird to discover the food. Would she then find a roost and bring others?

This would be our third release. After seventeen more, we might be wiser. At least, that was the idea. I couldn't wait to get started. We decided to release our bird on the west side of Lake Webb, about a twelve-mile drive from camp. That site was about three miles from a white pine grove where a group of juvenile ravens was then roosting at night. We hoped that our bird would find that roost after she fed. Since ravens are shy, we expected the release to be tricky—the bird could easily flee in panic, paying no attention to the meat and never locating the roost.

I needed to build a secure blind out of spruce and fir boughs from which to watch and release the bird. I drove out to the release site at 8:00 A.M. The snow was about a foot deep, caked in giant cushions all over the red spruces and balsam firs. Near a small opening about 150 yards in from the road, I found a thicket of firs in which I could construct a blind. Every blow of my ax to one of these trees released refreshing cascades of snow onto my head and neck.

After two and a half hours, I had built the blind. I could only see a few pinpoints of light through the back of it, but I could peek out through the interlaced evergreen foliage in the front. It was a deluxe model, tall enough to stand up in. After lugging three garbage bags full of meat from my truck through the snow to the site, I felt relaxed and thoroughly warmed. The ice and snow down the neck was almost welcome.

There was still time to get back to camp once more, so I drove back, jogged up the trail, and made some lunch. After that, I went to John's place to pick up the bird. I saw her large bill and her eyes through the slats of the cage. She looked calm, much calmer than one might expect a captured wild raven to be. But they are never panicky if you handle them gently. They are often even as calm as a sleeping baby in your hands. I checked her on my radio receiver—on frequency 837 megahertz, *her* frequency. Yes—*beep, beep, beep*—she was coming through loud and clear. We would be able to track her movements even in the dark and discover where she went to sleep. We hoped she'd join the nearby roost.

The sky was dark blue when I got back to the woods by the lake. There was no wind. I pulled the tarpaulin off the meat, making it visible for the first time. I made sure the bird had a good view of it from her cage. I unlatched the door, but kept it shut, and gave her fifteen minutes to see the bait from her cage.

I retreated into my blind, settled onto the furry deer hide I'd brought, along with binoculars, radio receiver, and note pad, and peered through the latticework of evergreen boughs.

Within a few minutes, the bird began to hammer the door with her bill. I had attached a fifty-foot-long string to open the cage door, but in three minutes she opened the door herself, without any help from me. Totally silent and with hardly a glance right or left, she walked to the bait and began eating fat. She hacked off gob after gob and swallowed each one, then she walked all over the bait, gently picking here and there as if inspecting details. She picked up a piece of fat and walked directly toward me, stopping at the very edge of my blind to shove the morsel into the snow. Then she covered it with more snow in several back-and-forth bill-swipes. After this first cache, she made another, and another . . . a total of fifteen, all at different locations and within raven walking distance. She seemed to disdain flight and vocalizations, two things I considered major raven avocations. When there is a crowd of birds at a carcass, their behavior is very different; they never walk to make caches. They always fly long distances, and make a lot of noise near the carcass.

Suddenly, her single-minded and silent pursuit of fat-caching was interrupted. She froze in her tracks. I heard the heavy *swoosh, swoosh, swoosh* of wing-beats, followed by the ripping, rasping *quorks* of a newcomer who was decidedly loud, raucous, and forceful. Round and round he flew, then he perched in a red maple tree. I was disappointed. Since the bait had now been discovered by a second bird, our planned experiment was compromised. I was, of course, still interested in what might happen next.

The newcomer loudly snapped his bill while his "ears" were erect and his lanceolate throat hackles glistened and bristled in a perfect rendition of the macho male power display. Next, he gave a few high flute-like calls. Through all of this male dominance display, by a bird I presumed was a local adult resident, our female remained stock-still and silent, like a statue, all the while holding the same piece of fat in her bill. When he finally left a few minutes later, she immediately resumed her feeding and caching.

About twenty minutes later, two birds arrived, flying directly toward and then over the bait. I think one of them was Mr. Big, swinging by with his mate for another look-see. Our bird, fully gorged and having made even more caches, now disappeared. According to her radio signal, though, she did not go far. Just as it was getting dark, she finally flew out over a clearing near the bait, making a rapid series of alarm calls, the kind the birds make when a predator approaches a nest. But there was no predator here as far as I could tell. Finally, she flew down to the edge of Webb Lake about a half mile away, where, according to my radio signals, she settled in for the night.

After I left, I stopped by the spruce tree John climbs every night for a view of the roosts and the flying birds' behavior. That night, John saw more than thirty ravens return to the roost from the direction of Mount Tumbledown. That meant that the birds now sleeping at the nearby roost were already successfully feeding at another site, in a direction where moose are common.

The previous year, all five of the subordinate birds we had released at baits had quickly flown off without touching the food in front of them.

They then had joined the local roosts and headed for those carcasses, a calf and a deer, from which the roost occupants were already feeding, rather than returning to the food we had allowed them to discover.

This time, we didn't think we'd see a crowd the next day, because the roost birds wouldn't leave a productive food site, possibly a moose carcass, until they had finished eating it. If by some chance Number 837 (our subordinate bird) joined this group, she would likely go with them to their feast rather than come alone to our bait.

We had other questions. I previously had found that recruitment normally brings the entire nocturnal roost of thirty to forty or more birds, who can strip a deer carcass in a day or two. Why then isn't a territorial pair more tolerant of a few vagrant juveniles who happen to find the carcass, so that there is no need for recruitment? Why don't they tolerate a couple of strangers at the dinner table rather than excluding them, thereby risking their bringing back a mob who will take everything? (I would later find with Goliath and others that adults at times do indeed tolerate another bird or two.)

I walked up the path to the cabin under an absolutely clear, moonless sky filled by the swath of the Milky Way, with the constellations etched brilliantly against the heavens.

The next morning, as I walked back down in the dark to resume my vigil in the spruce blind, there was a heavy, somber overcast. I had awakened at 5:00 A.M. to the ring of the alarm clock, and built a wood fire to cook my oatmeal and boil a cup of coffee, resisting the temptation to enjoy breakfast at leisure. I sat by the fire for a while before hurrying through the woods, driving to the other end of the lake, and hiking into the woods again, trying to make it to the blind before the first hints of daylight.

A crust of ice had formed on the twigs from an early morning rain. When I settled back into the blind to resume the watch on bird Number 837, I heard the soft tinkling of the iced branches as dawn approached with a slight breeze, punctuated by the soft, dull, rhythmic tapping of a downy woodpecker somewhere behind me.

The radio signal from Number 837 still came from the pine grove by the lake where she had ended up the night before. What a welcome

sound to hear her strong signal and the *swoosh* of her wing-beats at
7:15 A.M.! She arrived alone and silent, hopped from branch to branch,
descended to the meat pile, and as with the day before, immediately
began to feast on fat. She spent the next four hours making food
caches. Unlike the previous day, she usually flew rather than walked to
many of her cache sites, although her radio signal indicated she didn't
fly far. As expected, no crowd of birds came. I started to lose some of
my enthusiasm. The cold was gnawing at my feet.

Things picked up at around 11:00, when a pair (the pair from yesterday?) finally returned. I suspected that their territory, or their main
feeding site at this time, was distant, since they arrived so late. The
male gave a long series of undulating territorial *quork*s, and his mate
gave slow measured knocks in series of three—*knock, knock, knock*—
knock, knock, knock. Number 837 then disappeared from sight for three
and a half hours, but from her radio signal I knew that she was still in
the nearby forest. I heard her beg a number of times—indicating that
she was being aggressively confronted by another bird.

At 2:30 P.M., it began to rain harder. The thick fir boughs of my
blind held out the light, but they did not hold out water. I was thankful when it soon got dark and was time to go home. The adults had
left long ago, but 837 still remained nearby in the woods.

I again stopped by to see John up in his tall spruce tree. He told
me that the birds had again regrouped nearby, after coming from the
same direction as yesterday. Our bird was sleeping separately from the
crowd, at the same place she'd slept last night. We thought she might
recruit the next day, if she found the roost, because she couldn't eat
alone due to the resident pair.

Once in the blind the next dawn, I got out a flashlight, adjusted
the antenna and dials of the radio receiver, and took my first reading.
Our bird was still where she had slept. Good, I had made it up before
she did. By 7:00 A.M., she was gone, and I lost all radio contact.

I waited for what seemed like an hour, then maybe another hour,
but still nothing. I had no idea what would happen, which added a
welcome element of excitement. The bird could come at any second,
or not at all. Perhaps she had joined another group and was now

twenty miles away in one of any number of directions, feeding at a moose carcass on Tumbledown, loitering at the Dryden dump, or at a coyote bait put out by hunters. Maybe the pair would come to feed. Maybe a new bird would discover the bait. This was no longer an experiment, because too many variables had already crept in, and we could not control any of them.

She finally returned, silent and alone, and she fed—fat only, although there was plenty of muscle meat—as if she hadn't eaten in three days. In between snacks, she cached food. The first three caches were within ten feet of the huge bait pile itself.

About an hour later, her food caching came to an abrupt halt when the territorial pair returned. And when they arrived, you knew it! Before, there had been total silence. Now there was constant calling. Number 837 managed to give several "yells," but the pair aggressively flew after her and she fell silent. The pair continued their deep, resonant, rapid-succession short *quork*s that said, "Get the hell away," and they also made the long, undulating territorial *quork*s that said, "I'm here. Take notice." Both birds of the pair gave each call. I was lucky to catch a glimpse of 837 trying to hide. Her red shoulder patches gave her away as she squeezed herself into the thick branches of a spruce some one hundred yards away, but she was not hidden well enough. The territorial birds found her even there, and a long vociferous aerial chase ensued, as their calls receded into the distance. Soon all was quiet again, and the bait was deserted. Our bird was driven away—but not for long.

In twenty minutes, she was back, still alone, sleek, nervous and agitated, pecking on the branches at her feet and scanning nervously in all directions. After ten minutes, she hopped down to the meat, again eating fat as if she hadn't eaten in days. After snacking for a while, she flew off, holding as many pieces of suet as a raven bill can hold. At 9:28 A.M., the territorial pair returned and stayed for forty-two minutes. They still did not feed, all the while giving their long "I'm here" calls. Number 837 stayed away from the bait, but my magic instrument, the radio receiver, told me with beeps that she was still near, apparently hiding. Well, most of the time. Once she gave a yell—the call that recruits. A bad move—as far as I knew, there were no ravens nearby to recruit, except the vicious adults. They immediately pinpointed her, and another vigorous aerial chase ensued.

There were no birds near for a long time after that, and I lost radio contact with our bird. Suddenly, I heard some resonant monotones from the skies far above. Knowing no birds were nearby, I took the risk of stepping out of the blind for a better view and saw what I guessed to be the territorial pair—two specks against a patch of blue sky amongst dark, drifting clouds. The two circled side by side, wing-tip to wing-tip. Occasionally, they dove and turned in formation, then ascended in circles again. Higher and higher they went, thousands of feet up. I followed them through my binoculars for as long as I could until they faded into a hole in the sky among billowing, cushiony clouds, still dancing side by side. This dance moves me more than any human dance performance ever could. This one has been performed for millions of years and will continue for a long time to come.

At 12:24, the pair returned, first making the rasping "Get out" *quork*s and then, for the next eighteen minutes, an almost continual series of the long undulating "I'm here" *quork*s. The calls came from a half mile to the south of the bait, then from the west side. Then they ceased as the pair left. Eight minutes later, Number 837 was back and snacking lightly. She obviously could feed at this bait despite the territorial defense of the adults. She just couldn't feed whenever she wanted to. Presumably, this situation could change if the adults started to use this as their main food supply. So far, they did not seem at all interested in feeding. I was surprised that they went to so much trouble to defend a food site at which they were not feeding, though I thought it might be on their checklist as a possible future feeding site.

Would Number 837 not recruit at all, given that she was able to feed successfully? Our bird's failure to recruit might not seem very exciting, but we were not just raven-watching. We were also hypothesis-hunting. It might turn out, given everything about this bird and this situation, that lack of recruitment would be the most exciting result of all.

At 1:12 P.M., another unanticipated event made us rethink the original hypothesis. Two ravens flew over, then quickly descended to the woods nearby. One of them called loudly, but I knew it could not be one of the pair, because these two did not make any territorial *quorks*. Instead, for the next twenty minutes I heard a completely new spectrum of sounds, a steady singsong of slow rasping *quorks* intermingled with rolling, gurgling calls. There were also many series of rapid, percussion-like knocking calls, each series punctuated with the pop of a bill-snap at the end.

These birds stayed in the woods mostly out of my sight. Once, I managed to glimpse Number 837, conspicuous because of her red wing tags and her radio signal. She was near a second bird, and they were both making gurgling, friendly calls.

One of the new unmarked birds eventually landed on a tree above the meat pile and leaned way below its feet to peer down. It nervously doodled and pecked at branches. Then it again peered down at the meat and made long series of rasping *quorks* that sounded to me just like those our tame birds made when they had an unfamiliar scary animal carcass in their aviary. This bird was obviously afraid of the meat pile.

The new ravens were hyper-alert. They seemed ready to fly off at the slightest disturbance. In my blind, I could sense this, and I rolled over ever so cautiously to get a better look. The fabric of my pants made a tiny rustle—that was enough. Off they went with heavily pounding, fast wing-beats—*swish, swish, swish.* With all the ravens gone, blue jays, totally nonchalant by comparison, flew down to the food. They treated the bait as if it were—well, just dead meat! When they came down, they instantly hopped right onto the meat and started feeding. Just for a test, I loudly whistled "Oh, Suzanna" and athletically jumped around in the blind. They took no notice. I only managed to scare them by coming out of my blind. When I went back in, they returned in a few seconds.

At 2:05 P.M., all was quiet again. I had regained radio contact with 837, who stayed nearby in the woods for about an hour before coming by to feed. The wind had shifted from west to north now, picking up speed all the time. The woods sounded like a giant surf, ebbing and flowing. The temperature kept dropping, and I was soon shaking like a leaf in the wind.

As the afternoon light was fading, I dashed out of the blind, running the half mile to my truck through deep snow to get warmed up again. It had been a long day. I was thankful it wasn't 20 degrees Fahrenheit, as it could have been. I took another radio bearing on the bird—she was down by the lake in the grove of pines, where she had been for the last two nights. She was sleeping alone again.

Then I heard *quork*s above me. Looking up, I saw a fantastic sight—a group of forty or so ravens, gamboling along, flying in pairs, singles, diving, chasing, but generally moving toward the roost that I knew John was observing from his tall spruce. Where did this group come from? Where were they going? Was it a coincidence that they just happened to come directly over the bait?

Down the road three to four miles where John had his truck parked, I saw his silhouette against the rapidly fading western sky. He yelled as he saw me jogging over to his tree, "Did the crowd feed at the bait?" Apparently, he'd seen them.

"No. They kept right on. Did they drop into the roost here?"

"No luck. They flew in a giant circle, covering miles of territory, and then they headed back out again in your direction."

High circling at an old roost site in the evening is something we had noted before on many occasions. It is usually a prelude to the whole roost taking off to a new feeding site. We guessed circling happened when one or several of the birds in the crowd knew of a new food location. Maybe the display brings all the neighboring birds together.

Round and round the growing mob flew, perhaps two or three thousand feet up, as ravens from all around joined, swelling the ranks of the soaring group. Finally, the whole group took off, presumably to settle into a new roost near another food bonanza. I previously had seen the swirling groups of ravens at a roost, only to find the roost empty the next day. I sensed that they represented some kind of mysterious yet unknown social communication.

It is not just the "kettle" flights themselves that fascinate me, but the sensual electricity of the whole event. Garrett Conover, a Maine wilderness guide, once wrote to me: "We camped one night in the lee of the cliffs at Mount Kineo [on an island in Moosehead Lake, in Maine]. Just at dusk hundreds and hundreds of ravens began to arrive and cavort in the calm air in the lee of the cliffs. As dark fell they kept flying, though many landed on the face of the cliff. We lay on our backs looking up through layers and layers of wheeling ravens . . . Later a full moon rose and the birds stayed vocal all night long, quieting only when the moon set just before dawn. At first light we left the warmth of the tent, damped the stove down, and rushed out to see what the ravens were up to. All were gone without a trace or a call. It was like a splendid dream."

Similarly, I once received a letter from Virginia Cotterman, who lives in the Mojave Desert in California. She had observed 1,000 to 1,500 ravens roosting for thirty-six consecutive nights in the desert near her home. One night for the first time, they were vocal long after dark, almost to midnight. Why *that* night? They left before dawn, as usual, but they never returned in the evening.

I wondered if what John and I were seeing here in Maine was one of these coordinated departures. John had seen thirty to forty ravens circling for an hour before heading in the direction of the bait. Number 837 avoided this soaring crowd. She fed and slept by herself without joining a roost. She could not have missed the huge crowd flying over. Had she chosen not to join it because she had her own, almost private large food bonanza? Is the policy of raven vagrants that if you are hungry and have nothing, you join a crowd and follow, and if you are sitting on a good pile of food that you have reasonable access to, then you stay by yourself?

When the chilly dawn came on December 20, I was off as usual, trotting down the trail under a starry sky. After the fifteen-minute drive, I crossed the field to my blind. Thick black clouds were rushing overhead from the west, but a few patches of sky revealed stars. A shadowy figure ran ahead of me over the snow. It stopped and looked back—a fox on its last rounds. I stopped as well, then we both went on, the fox to a den, me into my little hollow surrounded by evergreen twigs.

I had time to pause before entering this morning, because I had arrived early. It was absolutely quiet. No wind. I hoped to hear the ravens come from a distance. In my previous experience, when ravens arrive in the dark on the first few dawns to a new food bonanza, they sound off with all sorts of amazing calls from their repertoire. It's an unforgettable experience to see and hear what seems like an uninhibited display of life and joy. But the stillness was left unbroken. As the sky got lighter, I retreated into my blind.

It was 6:45 A.M., the anticipated magic arrival time. Nothing. 6:50 . . . 6:55 . . . 7:00. Still nothing. It was fully light now, and I was sure the crowd would not appear all day. The birds' decision about where to go had been made last night, or at the latest before daylight. It probably had been just a coincidence that the swirling crowd had flown in the direction of this bait—they had been flying to some other destination, possibly ten to twenty miles farther on. I knew all of this, and it was barely light. I still had a whole day ahead of me of sitting immobile in my blind.

I lay down on my deerskin in the semidarkness and scratched notes on folded sheets of paper, kept dry in my back pocket. After twenty minutes, the cold was already creeping in, due to my inactivity. When the body suffers in this cold, the mind does not work very well either. There was nothing exciting to look forward to. There were no birds to be seen, and there was no assurance that any would come. I had no idea—not the foggiest notion, when I might be the beneficiary of a pleasurable reward. It could be days from now, or weeks, or months.

Indeed, the territorial pair were unpredictable in their arrivals. They came at 7:10 A.M. that day, then stayed in the vicinity much longer than previously. I heard their aggressive interactions with other ravens until about 9:00 A.M. Presumably, these were not only against Number 837, but also yesterday's newcomers. Number 837 did not get to feed until much later than usual. Her descent near 8:00 was her only opportunity to feed. The two other vagrants of yesterday edged close to the food only when she was feeding there. She paid no attention to them, again confirming that she was not defending her find of this food bonanza. They needed each other's company as a shield from the adults who might come at any second.

At 1:20 P.M., a bird I could not see in the woods made sounds that could pass for someone hitting a metal stake into the ground with a metal hammer. The metallic *thud* was repeated several times, then the raven came out of the woods to perch near the bait. Other sounds now: a soft *growl-pop-grr* sequence, repeated at intervals, followed by a rasping series—*rraap, rraap*—to be followed by several soft hooting calls. Then the caller left. What was that all about? What was the bird signaling? And to whom? Why so many different calls in rapid succession?

A little while later, I heard the more familiar rapid series of percussion sounds, each followed by a *pop*. Also, there were regular repeated *quork*s. What strange goings-on I was observing! I felt a certain snugness in my blind. In the woods, a human is almost always seen by the other creatures first. Here, it was the opposite. They could hardly know I existed. Repeatedly, ravens had landed almost on the blind or beside it. They could not see into it, but I could see out.

I liked being invisible among them. What's more, I possessed an almost supernatural power—I could identify some as individuals, and with the radios on, I could find out where they slept and thus find their roosts without actually seeing them at all. Unfortunately, ravens do not form one tight group that can be followed within a certain area. The population wanders over many thousands of square miles. After a bird is marked, it will not stay. Once a bird finishes feeding at a bonanza, it may be in another state the next day. I might complain about this habit of ravens, but to some extent, it is why I study them. What is easy is exhausted quickly.

When I came back down the road to meet John in the evening, I climbed up into the spruce as usual to join him until it got fully dark and all the ravens had returned to their roosts. This evening, there were no ravens at the usual roost. John had indeed read the display correctly. Here the roosts were temporary. They really were "traveling information centers," as we titled a joint research paper in the journal *Animal Behaviour*. This roost had traveled on.

I had long been racking my brain how it could be possible that dozens, hundreds, even thousands (as one report from California indicated) of ravens can use a roost for days, or weeks, and after wandering independently all day, suddenly one evening, *not one bird comes back!* It would be easy enough to explain if the birds always traveled in a group, but they don't. In the early morning, they might travel together to one or several food sites. After feeding, they are more or less independent for the rest of the day, only gathering up from all directions at the roost at night. Even when all feed on *one* carcass, they still come and go from it independently throughout the day. Some may feed only for a few minutes in the morning, and then spend the rest of the day flying many tens of miles over the countryside (to look for a possible next food bonanza?) primarily alone. Was there some agreed-upon signal the night before, like the social soaring that informs them of the next stop?

As I dozed off to sleep that night, I mulled over what to do the next day. We knew a raven crowd would come eventually. Here in Maine they always do, sometime—maybe tomorrow, December 21, the next day, or two months from then if the meat lasted.

Our experimental protocol called for keeping tabs on everything until the crowd arrived, but there invariably comes a time when you say: "Enough. It's time to do something else." It seemed that time had come. It was time to release the next bird. After that, we would have only sixteen more to go. I hoped with more releases, we would begin to see a pattern.

The site chosen for our next release was about six miles (as the raven flies) from the previous one. I built another blind and lugged in five fifty-pound garbage bags full of meat and suet, just to make sure there was enough for a raven to share.

It was 3:00 P.M. John was positioned up in his usual spruce, and I had been settled in my blind for half an hour. Our new bird, a dominant one, radio frequency 843, should be calm and ready to be released. Slowly, ever so slowly, I pulled the string on the door. Would the hungry bird welcome its freedom by rushing out to partake of the feast in front of him?

Not at all. From inside, he pecked at the fully open door. He leaned out and shoveled snow with his bill just in front of the cage entrance for three-quarters of a hour! Will he ever leave his cage? I wondered, as my left leg fell asleep after holding stock-still for so long. Finally he walked out, shook vigorously as if after a long bath, and continued to peck at the snow, still ignoring the meat. He walked thirty feet west, returned, and paraded right in front of my blind. Then he flew up above me, perched in a tree, and preened for another half hour. He remained silent. Finally, he disappeared into the dense fog of the forest as it was getting dark and starting to rain hard. Our bird stayed close by for the night.

It poured all night. As I lay in a warm bed under a watertight roof back at my camp, I savored my warmth and dryness, and I thought of the raven I had released that morning. He had not eaten anything for three days, and must now be burning off calories at a tremendous rate just to keep warm.

As I settled into the spruce blind the next morning in the pitch-black, I was even more uncomfortable than usual. The driving rain had stopped, but a drizzle persisted in the heavy fog. Water settled on the branches, eventually causing a steady dripping. Lying down in the blind on my raincoat to avoid soaking up ice water from the snow, I experienced a new torture, custom-made for raven maniacs—drips of ice water hitting my face at random intervals in random places (right in the eye is the worst). Luckily, John relieved me after four hours, during which time our released bird had flown by only once. Totally unlike the previously released bird, which had alternately fed and then sat tight on some tree in the nearby forest, this one was a mover and seemed interested in joining other birds. First, he flew to visit the twenty birds at our giant aviary on the hill a mile to the north, then he also visited the second aviary with six birds to the west. Often he was out of radio contact. Do males recruit and females not? Was this just an annoying individual variation, which seem so prevalent in these birds, that would necessitate our enlarging our sample size before we would see a pattern emerge?

Next morning was foggy and cold. Eventually, I began to make out the shapes of the territorial raven pair, dueting in low grunting *honks*. Half an hour later, I heard a long series of *knock-knock, knock-*

knock, like a stick hitting a hollow log. The rest of the day brought no surprises.

As on the day before, Number 843 flew over the bait several times, but showed not the slightest intention of landing. Instead, he frequently wandered out of radio contact, possibly visiting other nearby ravens. At least twice, he visited the birds in the aviary a mile away.

Near 10:00 A.M., the drifting fog was swirling through the trees, driven by a steady wind. It was as dark as evening, and then it poured rain. I left the blind. It was a good time to unload the 1,200-pound cow that filled the entire back of my pickup truck. I attached a chain to the cow and a nearby tree, drove forward ten feet, and presto, the raven bait was just where I wanted it. I cut the carcass open, then covered it with brush and snow, hiding it until later, when I would reveal it to the raven world for our next experiment. Before climbing the tree that would be my observation post, I rechecked the two baits. There were tracks at the first bait, and as expected, no tracks at the second bait; but Number 843's beeps sounded close.

Suddenly, a series of quick rasping *quork*s came closer and closer from the direction of the lake, where I had put the first bait. Then a pair of ravens came into view—coal black, with powerfully stroking wings. John had seen them come from the same direction for the past three days. They were probably the pair that had intermittently been harassing my released bird, Number 837. They came by me, flying close over the treetops and steering a straight course to the pine grove to the north, where I suspected they might build their nest next spring.

Ten minutes later, a lone raven flew by, a white shoulder patch shining brightly against each dark wing. It was an adult we'd marked in a previous winter. That was a rare occurrence. We'd seen few of the 463 birds we had marked. With the exception of those residents near the cabin, I would eventually get reports of only eight marked ravens from an area of approximately 240,000 square miles stretching from Quebec, New Brunswick to Nova Scotia, Canada; and northern Maine to near Boston and western New York.

Then I noticed moving black specks against the sky. Hooking one arm around the tree I had climbed, I lifted the binoculars to my eyes and observed ravens gamboling. What a sight! Soaring, diving, climbing, and spiraling down again, pairs and small groups and singles flew in close formation, separated, regrouped, over and over again. Gradually, they drifted over, covering a wide swath of sky in a big semicircle. They flew for miles, making air currents work for them, sailing in a sea. I was thrilled. I could have watched them all day. After about ten minutes, they banked down, folded their wings, and came shooting like so many black falling stars into the pines to the north. Was this the gang that would be led by Number 843 to the so-far-untouched meat pile?

Ultimately, after all our time and effort, we ended up proving what I already deduced from other data from previous years; namely, that ravens who are knowledgeable about food can recruit others from communal roosts. The idea of recruitment is an old one that had been endlessly bandied about in the scientific literature. We were the first to provide sufficient proof that it does indeed occur.

We had ascertained that knowledgeable birds are followed by naive birds, and that both leaders and followers eventually benefit from their behavior. The result is an inordinately simple, beautiful, and elegant system of sharing that relies on mutualism rather then reciprocity.

Our field studies provided a solid and much-needed conceptual framework in a context of behavior ecology. The studies addressed evolved adaptive patterns of behavior. Within that context, the individual variation that is so prevalent in ravens is more of a hindrance than a help in elucidating patterns within the population that we hoped to unravel. These studies could not tell us what was going on in the birds' minds, however, because they were not fine-grained enough. They gave no indication of what was innate, learned, or due to insight. Individual variation might do that. Perhaps the best chance of seeing the involvement of mind would be by embracing individual variation and using it as a tool in future experiments.

THREE

Ravens in the Family

There is something unique about ravens that permits or encourages an uncanny closeness to develop with humans. Many people keep birds as pets, but I've never heard of anyone who has raised a raven to adulthood call the bird their "pet." Instead, they consider it as child, or partner. One family in Maine with whom I recently talked reared a raven, Isaak, who was free on their farm. They described their association with Isaak as "a truly magical experience." They talked endlessly about their "beautiful relationship" with the bird, and said that since he "allowed us into his world," they stayed home summers just to be near him (or her, because the sexes are very difficult to tell apart), whereas before they had traveled. (Isaak, as with most tame ravens, eventually became independent and left.) Another family unabashedly called their raven their "son" and "a true friend" and they said they could "not imagine life without him."

Ravens form powerful pair-bonds.

What is the reason for such attachment? I believe it resides in mutual communication. A raven is expressive, communicates emotions, intentions, and expectations, and acts as though it understands you. This communication is privileged. It occurs when the individual close to the bird is trusted, has *earned* a trust that is not offered lightly. Given that trust, much is revealed that could otherwise never be seen.

I received a letter in December 1993 from Klaus Morkramer, a medical doctor in Oberhausen, Germany, about his raven, Jakob, whom he regularly let "free" in his apartment. There was an opportunity I could not pass up, to get a different perspective on ravens than my usual one, perched in a tree or hiding in a spruce blind in the woods.

First of all, I wondered how fast a raven would disassemble an apartment? I judged it shouldn't take more than about three minutes, maybe five. Did this doctor live in a cave, with the capacity to adapt to a small urban terrorist?

Jakob was born in the spring of 1992, having been orphaned when his nest fell in a storm. He was raised in an animal park at Wolgast in the former East Germany. Klaus had been a fan of corvid birds for a long time, and he considered ravens the *"absolut Spitze"* (absolute peak) of the corvid line. He learned of the raven from one of his patients, and contacted the family who ran the animal park. They sold him the bird for 200 deutsche marks (then about $90). His son Anatol took the train to Wolgast to pick up the raven, bringing it back in a small, darkened cage provisioned with a large sausage for sustenance on the long trip back to Oberhausen.

The sudden arrival of Jakob at the doctor's city apartment in the crowded industrial Essen area necessitated a quick solution to the housing problem. In foresight, a large parrot cage had already been ordered. Initially, it was to be installed on the terrace, where there was a veritable garden of trees, vines, and shrubs. It seemed an ideal place, but at first Klaus put the cage in the house to ease introductions. When it later came time to take the bird out to its allotted place, Morkramer found out that it was too late—he had not taken Jakob's personality into account. Jakob protested to being moved out of the house, and won. "The raven always wins," the Herr Doktor told me.

Jakob had taken his first big step to becoming a full-fledged family member: He took up permanent residence in the living room.

The next obvious move for Jakob would be to leave the cage and roam freely in the apartment itself. The rest is history. Wanting to see the results of this experiment for myself, I flew to Frankfurt, rented a car, and drove to Oberhausen.

Before I took the elevator to the fourth floor apartment, I envisioned a scene not unlike the aftermath of a bull in the proverbial china shop, except that I knew a raven would work with more patience and attention to detail. Imagine my shock when I stepped into the large living room the Morkramer clan shared with Jakob. There were no white streaks on the black upholstered leather furniture. There were no white spots on the oak table. The table had silverware in place, a sugar bowl, a cream pitcher, and several delicate cappuccino cups. All were resting intact and *upright* upon the table. Most surprising of all were the antiques. Klaus has an expensive hobby seemingly incompatible with being a raven-keeper. He is a collector of Roman antiques. Priceless original Roman ceramics sat in alcoves along the side of the room next to large, filled book cases. Invaluable paintings hung on the walls. This was not a cave. It was a museum. In fact, the only rooms in disarray were the kids' rooms and the kitchen. I was told Jakob voluntarily confined himself to the living room, fearing to enter other rooms far from the security of his cage.

When I entered the apartment, Jakob was perched quietly in his four-by-four-by-two-foot cage next to Klaus's favorite black leather chair. The raven seemed tranquil and uninterested in me. Poking his long bill out between the metal bars to Klaus, he held it still for a bill-shake, while gently nibbling fingers. The raven bowed his head sideways and further fluffed out his feathers as Klaus caressed his fuzzy head. "I have to do these greetings with him every day. The raven insists on it," the doctor told me. "Every time I come home from work I have to go through the greeting ceremonies with him. If I'm too brief, he grabs my hand or finger and tries to pull me to him." Klaus's son Anatol is also greeted with soft intimate sounds, but the rest of the family (his wife, another son, a daughter) are greeted with harsh *quork*s.

Jakob finally sidled up to the edge of the cage and thrust his bill out to me. Was this a friendly invitation? I decided it was, and accepted. It was an invitation all right, but not for a love nibble. One bite was enough for me.

Despite his young age, Jakob's tongue and mouth lining were black. Only in those ravens who have learned to be subordinate in the presence of superiors—and in a crowd of ravens, almost all encounter social superiors—does the mouth lining remain pink for several years. Jakob's mouth color alone showed that he had already established himself as the alpha in the household. After completing the greeting ceremonies, we settled into easy chairs around a low table set for coffee.

"Doesn't he want to come out?" I asked.

"Not yet. When he wants to come out he'll let us know."

For the time being, we drank cappuccino. Jakob was preoccupied with the contents of his cage. Klaus told me that whenever he gets mail, Jakob demands to have his fair portion of it. Although he is never denied, he hops around violently, giving loud frustration calls when his keeper comes into the room with a handful of mail and doesn't immediately deliver some to him. As soon as Jakob is handed a few pieces of junk mail, he quiets down and gets busy shredding them into little pieces. This task occupies him for about a half hour. I watched him work hard at it; his chest started to heave and his breathing became heavy.

I could see right off that Jakob's *capacity* for doing damage quickly and efficiently was great. I thought his deeds with the junk mail were admirable, however, although his intentions were not noble. According to Konrad Lorenz, "The capacity of an animal to cause damage is proportional to its intelligence." If this is indeed an adequate IQ test, then Jakob, like many other ravens I've known, was close to genius.

The junk mail having been adequately shredded, Jakob next pulled on the metal gratings of the door of his cage. That was the signal. If Jakob demands, Klaus obeys. Like Grip, the pet raven of Charles Dickens's *Barnaby Rudge,* who eventually "hopped upon the table, and with the air of some old necromancer appeared to be studying a great folio volume that lay open on a desk," Jakob appeared to be biding his time for mischief. He carefully surveyed the room through the open cage door, then hopped down to the floor and flapped his wings violently for about a minute as if revving up before takeoff. After these warm-up exercises, he flew once around the room, then landed on his cage. Anatol brought him a closed cardboard box and set it down on the parlor floor. This drew Jakob's attention, and he hopped off his cage at once and set to hammering holes and ripping off chunks of cardboard. When he had destroyed the box, Anatol offered him a small mail-order catalogue. When finished with that, Jakob fixed his attention on me.

For preliminaries, he looked at me, flared his feather pants, spread his shoulders at the front so that the wings crossed just over the tail, and boldly ambled toward me, stopping only briefly to look me in the eye. He hopped still closer, sideways this time, looked me in the eye again, and drew his head back. Before I knew what was happening, he had delivered a mighty heave into my thigh with his sharp, pointed bill. I jumped back. He advanced again. I was told that he wanted the ballpoint pen with which I was taking notes. Oh! Fearing more blackmail from *Corvus triumphanus,* I surrendered it readily. He soon seemed satisfied, settling onto an arm of a leather chair. He did not move from the spot for more than an hour while we humans chatted. I noticed him watching us with his lively brown eyes.

Jakob had glistening black feathers that were clean and well kept. "Once a week, winter or summer, he gets a bath with the garden hose," Klaus said. "We empty the cat litter from his cage—there is

Relaxed raven.

never any smell—and then we take the cage out onto the terrace and direct the garden hose in one corner of it. First he puts his bill and head into the water stream, then his chest and even his back. As a socially liberated bird, *he* determines the bath's duration. If the hose is shut off too early, he hollers loudly. When sufficiently bathed, he looks like a plucked chicken. When he starts his feather-care, we take him in. He drips and preens till dry."

The garden hose routine was, like many others, a compromise worked out from experience. A big bowl might *seem* more convenient for bathing, but unfortunately a bath is not the first thing on a raven's mind when you give him a big bowl full of water. The very first thing a raven does is tip it over. As Klaus pointed out, "To live with a raven in the house requires a certain capacity for compromise." No kidding. I'd just had my first lesson.

Family mealtime is, naturally, of special interest to Jakob. As soon as he hears the first dishes rattling, he hops onto the topmost perch in his cage, because that perch affords a better view into bowls and dishes as they are carried past into the dining room. He expects to partake in all of the offerings that come by. When the first bowl goes by, he begins to hop impatiently from perch to perch. If it takes too long until he gets his fair share, he hops ever faster and begins to make loud and penetrating *kek-kek-kek* calls. These calls do not at all sound like the plaintive food "yells" birds in the wild give when they are near food but can't yet eat it. Instead, they are the calls ravens make when an intruder such as a hawk or a human comes near their nest. Hearing them, I had little doubt what Jakob was saying. He was saying, "I am frustrated," which, given the context, meant more specifically, "I want some now." Since Klaus understands ravenese, Jakob gets the delicacies pronto. It is only fair that just as the Morkramers obey him, so he "obeys" them, as well, coming instantly to "*Komm*" (come) when he expects a caress or a tidbit in return.

Jakob bravely eats almost everything that is brought to the table, but he has a strong predilection for Chinese cooking, Hessian cheesecake, and raw bird egg (yolk only, unlike Goliath et al.). He likes fruits—figs, tomatoes, strawberries, preferably with cream and sugar, and grapes, but he turns his head away from apples and oranges, shaking his bill violently with disgust. Jakob has become something of a gourmet, yet I suspect he would not pass up good roadkill.

When the food is to his liking, he responds with a short, soft, two-note call that he also gives under other circumstances associated with contentment, such as after a particularly satisfying head-scratch. After a good meal, late in the evening, Jakob often entertains himself, and others, with his "talking"—a rambling, throaty warble.

Living with another creature, you naturally feel closer to it the more activities that can be shared, especially important activities like watching TV. German television is much like American television. It does not have a great deal of interest for raven viewers. So the Morkramers supplement their TV diet with videotapes. During television viewing, Jakob sometimes holds still, watching with one eye from the side of the head, as birds do. But does he see images? It seems so, because one time while watching a show with different mammals, he suddenly became agitated and let loose with alarm calls when a picture of a raccoon came on. He showed no alarm at wolves and deer. Here, it seemed to me, was an excellent opportunity for behavioral test, because it was possible to control what the subject gets to see, which is seldom possible in the wild.

After we sipped coffee late into the night, Klaus lured Jakob back into his cage with a raw egg, but once inside, the bird showed not a hint of getting tired and wanting sleep. Though not participating in our conversation, he seemed alert and interested. Occasionally, he solicited a head-scratch by fluffing his head at the cage bars, or he poked a piece of paper out as if playfully asking someone to pull it.

Why didn't the raven trash the apartment? We can only speculate, but given my subsequent lengthy tests of raven curiosity (Chapter 5), I propose the following explanation. Initially, Jakob had not been free in the apartment. For the first two months, he had remained in the cage. During that time, he had seen most of the apartment, and his interest in its contents had faded. The *new* things Jakob saw, like my pen, were always the objects that people carried about in their hands, so they were interesting enough to warrant further notice. Investigation of the new things could be motivated by curiosity, and it has been the curious birds that through evolution have always found the unexpected and perhaps rare food items that others passed up. The corvid line of birds all share this capacity of curiosity. It is their trademark. One wonders if it is the key that has allowed them to flourish and diversify. They say curiosity killed the cat, but curiosity is also adaptive, provided it is backed up with good judgment.

My next close and personal encounter was with Merlin, an eight-year-old in the family of Californians Duane Callahan and Susan Marfield. It was early August in southern California, at "Camp Pozo" near San Luis Obispo, where I had come to spend the week. Camp Pozo is a ramshackle trailer with its various accoutrements spread out under the live oaks on a hillside. In evidence were several old dead vehicles and various other debris from casual, long-term human occupancy. This little Shangri-la is located at the end of a dirt road on a cattle ranch near the town of Pozo. Pozo itself is little more than a few houses around a saloon under a big cottonwood tree. Duane and Susan have come here for the last eight years to vacation with Merlin and to visit Charles, Duane's brother. Lady, an Australian sheepdog, two horses, and Katche, a cat, reside there as well.

We had come the day before from Santa Cruz in the Callahans' orange Chevy truck with the blue camper top that has been marked up with a few white decorations from Merlin himself. Merlin always travels inside, in his wire screen carry-on cage that is jokingly called his "jail." On camping trips, he treats it as his home base and sanctuary, returning to it eagerly at night, wherever it may be found. At the Callahans' home in the deep, dark redwood groves near Santa Cruz, Merlin's cage sits on a stand of two-by-fours in the living room-kitchen. Merlin is let free in the house daily, and he spends most of his free time perched on Duane's knee. Merlin is calm as a clam most of the time, although once a day he gets animated and flies around the room, negotiating the tight turns around the central fan with no problem. The door and living room window are left open in the summer, but he never tries to go outside, although he has every opportunity to do so.

After we had loaded him into his cage and put that into the camper, we eagerly began the four-hour drive to Pozo. Merlin perched forward, maintaining contact with Duane in the cab. Like a dog eager to come home after a long drive, Merlin became restless, and excitedly hopped about when we got within a mile or two of our destination. Once there, Duane immediately let him out. With few preliminaries, Merlin launched himself high into the air, flying several loops above the chaparral before circling down and alighting on Duane's shoulder.

At camp, he sleeps in his cage just as he does in Santa Cruz, but he wakes up earlier. It is not safe to be out when the great horned owls begin to fly, and each evening, one commonly does fly by Camp Pozo. Merlin appears anxiously to seek out Duane in the evening to be "tucked into bed." In the morning, Duane again lets him out. Even though Merlin is here in the wild, he stays near his "family."

His first calls the morning I was there were typical loud raven calls. Nobody got up. He next tried two series of high-pitched calls that mimicked crow alarm calls. It was the first time I had ever heard a raven sound like a crow. He followed up with a two-note rasping call that I also had not heard before, then he softly uttered a series of "Hi,

Young ravens sleep by bowing head (left) or tucking head into feathers of back (right).

Hi, Hi," and "Merlin, Merlin, Merlin . . ." and some barely audible gurgling noises that I couldn't decipher. He had his own unique vocabulary.

When we arose at last, thirsty for strong hot coffee, Merlin was silent again. Duane said that Merlin used to try to get him up early by making a lot of noise.

Duane, with coffee cup in hand, looked at the cloudless blue sky and declared, "Another hot one. Merlin won't do a lot of flying. He'd rather spend his time on my shoulder in the shade." With that, he walked to the trailer, crawled inside, and exchanged morning greetings with Merlin.

"Merlin—how are you?"

*Merlin and Duane in
mutual greeting ceremony.*

Merlin and Susan.

A few soft grunts came from within.

"Want to come out now?"

"Mm, mm," said Merlin.

After Merlin gave a few more soft *mm*'s and grunts, Duane opened the "jail" door. Merlin, sleek and eager, hopped out. After the exchange of a few more pleasantries, he flew up to Duane's shoulder.

"Want some chicken?"

"Mmmm."

"Tasty?"

"Mmmm."

"Is this special?"

"Mm."

"Want some more?"

"Mm."

"Alright!"

Social amenities and long conversations over, Merlin spun his head back and forth, scanning in all directions. He blinked once or twice, and flew off with strongly beating wings over the clearing and through the oaks. Soon he was high above them. After several circles over Camp Pozo, he banked steeply, and I heard the air being forced through his wings in a continuous rippling sound as he dove and again landed on Duane's shoulder, fluffed out and shook. He was feeling great. Duane continued to sip his coffee, and continued the conversation.

"You are just about the most beautiful thing under the sun," he said, caressing Merlin's head feathers.

"Mm."

Duane and Susan raised Merlin from his pinfeathery fledgling stage, when young ravens have been described as some "grotesque miniature gargoyles." He is their only "child," and he receives considerably more daily attention than most children of even a one-child family.

Merlin ignored me totally, and he would continue to do so for the rest of the week. He pecked lightly at my hand if I intruded it near his face. When I brought my hand near him a second time, he emitted a growl, fluffed his feathers, and pecked much harder. I did not dare to try it a third time. He is bonded most strongly to Duane. If given a choice, he spends his time with no other human. If Duane is gone, he approaches Susan. Other members of the family whom he has known for years are not approached.

His memory for individual people seems to be indelible. When Duane and Susan were away for six months, Merlin stayed with Duane's brother, whom he already knew and accepted. When they came back, his reaction to Duane was instant and strong. "He bolted from Charles to my shoulder instantly, as if I'd never left. Then he stayed on me like a burr the rest of the day. On that day, he also became unusually aggressive, showing his dominance through feather posture. He drove off the magpies, chased the vultures and a Cooper's hawk. Was he trying to reestablish his worthiness as a mate?"

Duane's observations not only attest to the bird's long-term memory, but also address his fidelity. If he distinguishes and remembers individuals of another species as well as he does, it stands to reason that ravens in the wild recognize members of their *own* species at least as well, and bond to them as long and as strongly. How well could *we* distinguish one raven from another and infallibly remember them?

Charles served bacon and eggs, which we ate under the live oak trees. Merlin perched on Duane. He was picky, eating only in small bites. He may already have fed from his staple, the canned dog food always available in his jail. He does not cache surplus food because he rarely needs food for a "rainy day," as do wild ravens. Perhaps that is because there never has been a "rainy day" in his life; food is never an issue—it is always available. No great effort needs to be spent where it is not needed. Nevertheless, crediting him with optimal efficiency in energy allocation may be premature—although Merlin rarely caches food he spends considerable time and effort caching such useless things as wood chips and other trinkets.

After spending a half hour or so on Duane's shoulder, he hopped down to the ground to dig in the soil and to pick at wood chips and other debris. One nondescript four-inch wood chip in particular drew his attention. He tried to shove it into the sandy soil, succeeding only partially, then covered up the rest with debris scraped from the sides. He placed a leaf or two on top. Almost invariably, a small section of the chip still showed. He tried to tamp this part down by pecking it hard. As a result, the whole chip got uncovered, and the whole process was repeated. Then he dug a small trench nearby using alternate side-swipes of his partially open bill. He picked the chip up, laid it into the trench, and scraped the surrounding soil over it with his bill. Finished? No. Within two or three minutes, he was back to dig the chip up, and he then repeated the process in a similar manner with the same or some other chip.

Someone from the appreciatively watching audience offered him a strip of bacon. He flew off with it onto the ground of the nearby hillside, where a flock of about a dozen yellow-billed magpies immediately joined him. He made rasping-growling calls at them, then returned to us. The magpies then dug in the soil all around where he had been, perhaps searching for the cached bacon.

It was barely eleven o'clock, and the sun was blazing hot. Our caffeine levels were up to par, but the heat was already inducing some of us to reach for a cool beer. Merlin, too, got offered a few sips of brew through a tipped flip-top can. A few sips was all he took, although on hot days he has been known to indulge in more than he should. He "gets a little unsteady on his legs and wings," I was told.

We decided to take a half-mile walk through the hills to a spring-fed pond to catch a few largemouth bass for supper. Just as we were ready to leave, Merlin became uncharacteristically loath to follow Duane. Whenever we started to walk, he refused to budge from the roof of a junked car parked under the large live oak by the trailers. He just sat tight, holding up our little expedition as we waited for him. Duane thought Merlin knew we wanted him to come. He told me, "Whenever you want him to do something, he becomes suspicious and doesn't do it. You have to be surreptitious, by acting nonchalant, as if you *don't* want him to do it, before he *will* do it." To get him to come when he didn't want to would be a challenge. Duane wasn't eager to leave him alone, because last year while Merlin was flying near camp, a golden eagle swooped unseen out of the sky from behind and grabbed him in midair. Duane saved him by erupting in a sudden and violent burst of yelling that induced the eagle to drop his intended prey. When Merlin fluttered to the ground he had blood on his feathers and in his mouth.

Duane again tried to coax Merlin onto his shoulder. This time, instead of merely flying away, Merlin growled his agitation calls at his "mate." He was giving Duane a message that even I could read. It said, "Go away—I don't want to go." Duane and Merlin repeated the same maneuvers five minutes later with no different results. Finally, Duane suggested that we just go, and "when we've gone far enough, he'll see we're not trying to put something over on him, and then he might come because *he* wants to." Merlin was never trained with rewards of food. He does what he wants, when he wants, and I suspect that Duane is the one who is trained, not Merlin.

We walked about two hundred yards in the direction of the fish-pond. No Merlin. We found shade and waited ten minutes. When Merlin still didn't appear, Duane went back to do some "negotiating." Sue, Charles, and I continued to wait for another five minutes. "Must be having negotiating troubles," Charles remarked.

A minute or so later, Merlin finally took to the air—not to ride on Duane's shoulder, but to sail over him and fly to us instead, landing on the ground by Susan. When we got up to go farther, Merlin continued to be recalcitrant. He did not follow Duane, but again flew to Susan instead. Duane remarked that he was having some "marital problems" with his bird today. Sue as much as acknowledges that Duane is married to Merlin.

The dusty trail we walked read like an open book of the previous days' and nights' activity. There were tracks of quail and lizards, scats of coyote and bobcat. There were stray feathers of owls and hawks. We flushed cottontail rabbits and jackrabbits from brush along the sides. Merlin paid them no visible attention. He seemed to be interested only in perching in a shady spot.

We came to a white oak where a spring seeps to a trickle of water. The dense foliage above was atwitter with goldfinches and titmice. A succession of hummingbirds hummed down to the trickle to dart back and forth between sips. As many as six of them at a time perched on dry twigs five feet from me.

Merlin stayed near this tree, and the others walked on ahead to the pond. They were soon out of sight. I stayed to observe Merlin watching the birds coming to drink at the spring. He dawdled, pecking bark and softly murmuring to himself in barely audible tones. I approached in order to hear him better. He then produced several renditions of the very loud and high-pitched crow calls. Duane said he often makes them when he is frustrated or upset. As Merlin crow-cawed, he allowed me to get right next to him, acting as if he didn't even see me. When I held out my hand to him, he gave it a sharp peck.

Duane, attracted by the "crow" calls, came back to the spring. "How are you Merlin? . . . Everything okay? . . . You are a pretty bird. . . ." Merlin at first showed his "ears," following up with head fluffing, bowing, and soft murmurings. Duane also bowed, blinked, and made a sound like a yawn. I wasn't sure who was mimicking whom. The bowing ceremony appeared to say, "Look at me—I'm wonderful." Merlin stood tall, puffed his head up some more, lifted his bill, and spread his shoulders. As he bowed, he spread his wings and tail and made a choking noise like a sigh, just like Duane's yawn. "This greeting," Duane said, "is *never* given to strangers." The mutual greetings are also performed every day when Duane comes home from work. If Duane neglects them, Merlin sulks.

"I'd like to make up with you, Merlin. . . ." Merlin tentatively put one foot on Duane's outstretched hand. More pillow talk. He hopped onto the arm, then off, but in a minute or so he hopped back onto Duane's shoulder. The marital rough spot had been smoothed over, and the rest of the fishing trip went smoothly, as Merlin followed us, then rested quietly in the shade while we fished for bass in the pond.

Merlin has emotions. He also remembers faces and events that he associates with them. He has moods. I do not know if he is a thinking being, but there is no doubt he is a feeling one.

Aside from the crow call, Merlin has another unusual call I've never heard before—a distinctive two-note whistle that Duane invented and Merlin copied. Now Merlin appears to use it to alert his "mate" when something strange is nearby. He also uses it to draw attention to himself when he is flying high above Duane. Ravens in the wild also have individually distinctive calls. Perhaps that is one way of recognizing individuals. Certainly, Duane has no problems identifying Merlin when he flies high above him, as long as he calls.

Merlin's communication can also convey more specific information. One day, Duane was sitting down on a hillside and Merlin was exploring nearby. Merlin returned to him, perched on his shoulder, and began to peck him lightly and pull his collar. Then he jumped off Duane's shoulder and flew a loop down the valley. He came right back to *repeat* the same maneuver. Duane said, "I knew unmistakably that he was trying to get my attention, so I went where he directed, and there I found a bobcat on a ground squirrel." Hearing this anecdote, I wondered if raven mates might cooperate, or if ravens form symbiotic relationships with hunters (see p. 193)?

Merlin became animated again as soon as we started our homeward journey from the fishpond. We all separated, but now that he was going where he wanted to go, Merlin at first followed Duane exclusively, then flew the rest of the way home alone.

Temperatures were near 101 degrees in the shade back at Camp Pozo. Duane offered Merlin a large stainless steel bowl full of water. Merlin jumped in and splashed with his wings working like an eggbeater. He emptied the bowl of water three times. He shook and preened on Duane's knee after each bathing bout, then retired onto the trailer roof to finish off his final drying and preening before joining us inside the trailer.

The TV was on, and everyone was slaking their thirst. Merlin was mellow and well-behaved. He spent the whole two hours patiently and contentedly perched on Duane's knees or shoulder, except to politely take the corn chips we handed to him. Duane rolled one corn chip up in a towel. Merlin deftly unrolled the towel by pushing the roll forward with his bill until he reached the enclosed chip. He knew the chip was inside. He paid no attention to Lady and Katche sprawled out on the sofa. He knows them well, and vice versa. They mutually ignored each other even when less than a foot apart, but Merlin teases or attacks strange cats and dogs. He once showed a fright reaction on seeing a picture of a wild cat baring its teeth.

By 4:30 P.M., when temperatures outside again became tolerable, Duane and Charles pulled folding chairs under the oaks and began limbering up their guitars to launch into an impromptu rock concert. Merlin perched as if transfixed on Duane's knee the whole time. Although it was not hot anymore, his bill was open. His shiny blue-black feathers were smooth, and he held his head still. He sometimes blinked rapidly with the white nictitating eye membranes, showing his strong emotion. Amazingly, for another two hours he changed position only once, to hop onto the amplifier that served also as a table for beer cans. I don't know if he appreciates music, but he certainly didn't fidget one tiny bit during the two-hour concert. He was alert and scanning all around throughout. When the medley stopped, Merlin immediately stretched one leg and a wing on one side, then did the same on the other, and shook and preened as if he understood that the performance was over. Becoming lively again, he hopped off the amplifier and started pecking the edges of a guitar case.

Duane thinks that certain sound patterns or frequencies "set him off." For example, Merlin reacts to Andean flute music much as he does to the sound of a vacuum cleaner—with interest. Some might question whether the sound of a vacuum cleaner qualifies as music. Personally, I think music is in the ear of the beholder. My wife, who likes classical music, thinks Merlin's tastes are like mine, which run to Bruce Springsteen and the Animals.

During target shooting with a .22 rifle, Merlin reacts to shots with fright, although a truck motor running full blast doesn't faze him. He reacts to the slightest rustle if he has not yet identified it. In the Santa Cruz apartment, he once came on full alert, craning his neck while Duane and Sue at first didn't hear anything. Then they just barely detected delicate footsteps. A deer walked past the window. On seeing it, Merlin instantly calmed down and again became his old tranquil self. Seeing the deer did not alarm him as much as hearing the very faint but unfamiliar footsteps.

What he sees as dangerous is difficult to fathom. He became excited when he spotted an eagle, even before he was attacked by one. He acted terrified of a real snake, yet treated an almost perfect imitation of one with nonchalance. He was upset for weeks, Duane said, by a frame of two-by-fours that served as a retainer for the stereo set, and he "went bonkers" when Duane put some fresh sage sprigs into his sanctuary. Merlin, like Jakob, was afraid of a tiny toy dinosaur put into his cage, but later he accepted it after he examined it on the floor of the house. He was afraid of a new broom. In general, he is afraid of slight changes in his environment, but he is much more afraid of some things than others. Duane concludes, "I have no clue why some things that you'd think might scare him don't, while other seemingly totally benign things freak him out."

About a half hour before dusk at Camp Pozo, Merlin took off, banked steeply over the trailer, flew over the oak trees, and eventually reached cruising speed at about five hundred feet above us. The sky was clear except for some thin cumulus clouds, and we stood mesmerized, admiring his aeronautic skills as he glided over the hills and then circled back again and again over Camp Pozo. I did not have the presence of mind to look at my watch, but I videotaped for nine minutes. The whole flight must have been considerably longer. When he finally dove back down and on outstretched wings landed gracefully on Duane's shoulder, it was 7:40 P.M. The sun was sinking, and the dry grass shone golden on the eastern hillsides. Duane walked his friend to the trailer, and Merlin hopped off and entered his sanctuary. It had been a most unusual day.

Jakob and Merlin had revealed tantalizing glimpses of ravens that contrasted with my "scientific" field studies requiring large numbers of birds that yield data, which I analyze with statistics. Observing the two birds who had bonded with humans, with an intimacy they normally reserve only for other ravens, gave me a different perspective. Perhaps one could not hope to appreciate the mind of the raven, any more than one could claim to know the sociobiology of a remote tribe, without first living with if not marrying into it, as Klaus and Duane had done.

FOUR

Ringing Necks for Baby Food

GOLIATH AND THE OTHER THREE ravens that I reared from nestlings were one of several groups that I have raised. At times, I had groups of more than twenty ravens caught from the wild as adults and held in my aviary temporarily. Each raven eats a lot of food per day, and as you know by now, feeding them has been a tremendous practical concern. What ravens eat is also of great importance in understanding their behavior, because food is a focus of the bird's lives and has been a major principle in their evolution.

I fed Goliath and all the other young ravens I raised to adulthood on a varied diet consisting mostly of meat, dog chow, boiled eggs, frogs, raw fish, and occasionally insects. They always begged loudly, yet often refused food if they had been fed the same things several days in a row. When only a month or two out of the nest, they yelled loudly and irritatingly whenever they saw me, then pursued me relentlessly. The amount of weight gained for the amount of protein fed seemed to be

Goliath and Whitefeather.

low. Did they get a better diet in the wild? If so, what was that diet? In early May 1995, I decided to try to find out by going directly to a nest.

A pair of loons had returned to Hills Pond on April 22, two days after the ice sheet broke up. The snow in the forest had finally melted, and spring was on its way. The robin, woodcock, phoebe, mourning dove, flicker, red-winged blackbird, cowbird, solitary vireo, sapsucker, white-throated and song sparrows, bluebird, and even tree swallows and kestrels had been back for two weeks. The first wave of wood warblers, Nashville and yellow-rumped, had arrived the previous week. As he does every spring at this time, a woodcock was making his spectacular display in the clearing by my cabin every evening, and the females would soon be sitting on cream and lilac-spotted eggs. None of the songbirds, the Passeriformes, had yet started to build their nests. In stark contrast, the largest songbirds of them all, common ravens, already had nearly full-sized feathered-out young.

Birds are driven by their breeding schedule, which is fine-tuned to their food supply. Woodcocks come when the first soft earth promises earthworms. Kestrels must wait for snow-cleared fields to hunt for mice. The raven is much larger than either, and needs considerably more food. Each member of its rapidly growing brood of young needs huge amounts of protein daily. A raven, if lucky, can catch a mouse or two. It also can find a few stray worms, but it is not equipped to probe for them and routinely get them out of the mud as the woodcock does. Insects are far too small and rare to satisfy the demands of their large bodies. In any case, most of the insects are still in hibernation and unavailable when the young are growing in April. For the local ravens, the growth spurt of the young occurs in the four weeks at the end of April and the beginning of May. In the hundreds of nestings I have seen in New England, there has been only one exception to the relatively precise window of time when the ravens nest (Chapter 29). If something happens and the time slot is missed, or the first nest is destroyed, the birds wait until the next year to breed again. By contrast, in many other regions where ravens are found, they are much more flexible. There are even reports of them nesting in the fall, and it is difficult to characterize these nests as "late" or "early" (see Notes and References).

Ravens usually have four to five young, all of which reach their full adult weight of three to four pounds in about three or four weeks. What food of such vast bulk, I wondered, could the adults possibly find for their young at the time of year to which their breeding cycle is synchronized?

Ravens, like us, are considered omnivores. Given a choice of foods, my captive adults will eat the cholesterol-rich and fatty kind: cheese, grasshoppers, salty peanuts, eggs, butter, potato chips, and hamburgers. Next on the list come mice, birds, deer and moose meat, blueberries, maggots, tomatoes, carrion beetles, fish, oatmeal, and corn. Only a limited selection of this fare is available in the Maine woods in May.

Along the coast of Maine, a most important food for the island-nesting ravens during chick-feeding time is seabirds and their eggs. John Drury, an ornithologist from the island of Vinalhaven, Maine, told me that the ravens there can frequently be seen flying over and landing in raspberry patches where the eider ducks nest. The eider nests in those raspberry patches are soon without eggs, and the nests' down is strewn around. Near one cliff nest of ravens on Seal Island, John found the shells of thirteen eider eggs, and the picked-over skeletons of thirteen black guillemots. The year before, there had been nine eider egg shells and remains of two guillemots at the same raven nest. At a raven nest site on Great Spoon Island, John found shells of eggs of six to seven herring gull, one or two black-backed gull, thirteen eider, two cormorant, plus heads of three gulls, four eiders, and parts of two unidentified bird vertebral columns. Two weeks later, he found eleven more eider egg shells, and the remains of two guillemots and one blackbird. By a raven nest in a spruce tree on Metinic Island, there was evidence of similar fare. In contrast, near Bowdoinham, also on the Maine coast, Tinker Vitelli has since 1988 observed a pair of ravens nesting in a pine grove, and there the young are fed mostly freshwater mussels from a nearby stream. The ravens bring the mussels in whole and carry the empty shells away, never leaving any under the nest tree.

Inland, the Maine forest ravens face a different situation. I suspected these ravens' survival and reproduction were linked to deer and moose. In the fall, ravens invariably show up where hunters leave the entrails of deer or moose. In the winter and early spring, ravens may rely on deer that are killed by the sometimes lethal and tightly inter-related combination of cold, deep snow, starvation, and coyotes. Per-haps the cold storage of winter allows food to accumulate so that ravens can nest before the meat decays in the spring.

Maine has an area of 33,265 square miles, and an annual prehunting season population of about 300,000 deer. In recent years, the annual reported deer kill during the hunting season is approximately 27,000 animals, and the moose kill is about 1,500. The Maine Department of Inland Fisheries and Wildlife estimates that as many animals are killed illegally as are killed legally. Added to the count are the carcasses of beavers, coyotes, and other animals that trappers leave in the woods. This leads to a conservative estimate of about two meat piles per square mile over the year, in addition to the carcasses left from natural die-off.

In the winter of 1992–93, I had picked up fifty-nine raven pellets under roost trees where a vagrant crowd was sleeping nightly while feeding on the calves I had provided. Ten of these consisted mostly of deer hair, and five of mountain ash seeds (see Table 4.1). These findings only suggest *what* ravens feed on. They are not a reflection of *how much*. Eating only one berry would produce seeds in a pellet, but a raven might feed on the viscera of a deer for days and pick up only a few hairs.

Deer carcasses apparently are available, although the plucking of fur off live animals cannot be precluded (p. 225). Of the more than fifty nests I have examined, all have been lined with deer fur. The nest linings also sometimes contain snowshoe hare fur, shredded ash bark, and on rare occasions, moose and bear fur. Once I even found moose, bear, hare, and deer hair all in the same nest; but deer hair is present without fail.

If a pair of ravens with young had a whole deer carcass at their dis-posal, they would have not only insulative nest lining, but also a steady food supply for a several weeks, cold weather permitting. Weather not permitting, they would have fresh maggots rather than fresh meat by spring. My tame birds eat fresh maggots in preference to rotting meat. Having a deer or moose carcass available may often be a matter of luck. Even if the resident pair finds a deer carcass, there is also the problem of keeping it. If a crowd of juveniles takes over the carcass, the food supply will be gone in a few days. In my Maine study area, every one of the hundreds of large animal carcasses I provided was sooner or later taken over by raven crowds.

Although it was only the first week in May, the breeding success of the local ravens had already just about been decided. Unlike other local birds that sometimes raise several clutches per year, or that immediately raise a second clutch when their first is destroyed, I've never seen any of the local ravens renest when their first breeding attempt was aborted. This year, two of the nine local raven pairs had made no breeding attempt. Three had started to build their nests, then stopped; and two nesting attempts failed after the nests had been completed (at least one of them had held eggs). Only two of the nine pairs had carried forward with the raising of a brood. Curiously, these two nests contained not the minimum number of young, which is two or three. Instead, they each contained five, which is near the maximum clutch size, although one year I found a nest that successfully produced seven young. The previous year's nesting results had been similar to this year's, with two successful nests that had brought off eight young together, most of whom were later killed by predators (Chapter 6).

The first nest this year, at Braun's Road, was abandoned shortly after construction. The second, the Weld pair's, is consistently successful each year. The other pair that was so far doing well was the pair in the pines by the Robertson cemetery, which had failed last year very late in the breeding cycle. The very uneven distribution of young among these nine pairs was consistent with either an unreliable local food supply or divergent foraging skills of the pairs. I suspected the latter was the less likely explanation, because the cemetery pair that last year abandoned their nest with starving young had large, healthy young this year, judging from the ample mutes below the nest. Whatever catastrophe had befallen seven of the nine local pairs had not hit this pair. I wondered what food they were succeeding with.

I had just learned of a method to determine what the young were fed without harming them in the least. All it required was for me to get to the nest with some pipe cleaners. The method was pure genius.

It made it possible to get easy and unambiguous results to an otherwise difficult problem, with data that would be direct and pertinent. I thought, this is as good as it gets. I bought pipe cleaners and went directly to the nest tree.

A pipe cleaner is a useful tool. It is a short pliable length of wire with some soft fuzz firmly attached to it. When this cushioned wire is bent around the neck of a young bird, the bird can't swallow large chunks of food, but it does not stop begging and taking in food. The food simply collects in the gullet, stored there until one squeezes it up and out like toothpaste out of a tube.

As I muscled my way up the tree and finally maneuvered round one side of the thinning trunk in the top branches to look into the nest, I saw five healthy young. They were just beginning to feather out, and were above half adult weight already. At that age, the young had no fear of anything near the nest. With blue eyes wide open, they raised their long scrawny necks to me, begging for food. I, in turn, loosely twisted a soft pipe cleaner around the neck of each one. None of the birds showed the slightest sign of having noticed, and each one either continued begging exactly as before or settled deep down into the nest to sleep. Three hours later, after I presumed the adults had made three to four foraging trips to the nest (and after I had adequately recovered from my arduous climb), I returned and climbed the tree again.

As before, the young were calm, and they begged. I noted not one lump in their throats. All the pipe cleaners still were attached. What was happening? Then I saw three loads of meat on the edge of the nest. Each load, which apparently had been regurgitated when it couldn't be swallowed, was a solidly compacted mass, the size of a 35 mm film roll. It was red muscle meat, now almost black from partial drying, and it was leathery, like moistened beef jerky. It was definitely not frog meat, insect meat, snake meat, freshwater mussels, or fish meat. I took it all, and temporarily removed the pipe cleaners long enough to distribute the paper cup full of road-killed ruffed grouse meat I had brought up with me. It seemed more than a fair exchange for this leathery, semidried stuff.

The three regurgitated masses of meat each consisted of numerous pieces of partially dried, dark red meat that had been compacted together. It was not fresh by the smell of it, but neither was it rotten. It was coated with a film of slimy saliva. Most important, there were many compacted deer hairs worked into the crevices of the numerous pieces of meat, indicating the deer hair could not have come from the nest itself.

The meat had more clues. It contained a sprig of a green moss that grows in shady moist places on the ground. It also contained partially dried but green fir needles. Fir needles of that color do not just drop from live twigs. I've never seen fir twigs, branches, or needles in a raven nest. Fir trees hardly ever lose live limbs that would then shed green needles, and needles turn color before being shed by live trees. So the partially dried green fir needles probably came from fallen live fir trees, whose needles dry quickly and detach without first turning color. I suspected the deer meat had come from a cache on the ground among moss under dead fir branches near a lumbering operation. Skidder operators regularly carry rifles in their cabs. Deer in the winter have no fear of skidders, and they aggregate where loggers provide easy paths through the deep snow to felled trees with lots of buds to browse from.

My suspicion that the three feedings were from cached meat was strengthened the next day when I repeated the neck ringing. That time, in four hours the parents again delivered at least three wads of deer meat that sat neatly in place in the birds' gullets. This was fully moist pink meat, not partially dried like yesterday's. No moss and no fir needles were attached, only more deer hair. These meat portions likely came directly from a deer carcass. I was satisfied that regardless of what else the local Maine ravens may feed their young, at least at this nest they fed their young venison.

It might be supposed that the steady stream of calf carcasses I had provided during the preceeding ten winters in Maine might have been beneficial to the local ravens, helping them rear many successful broods. The fact that only four of eighteen potential nests were successful in the previous two years argued against this. I knew of no data from anywhere for such low nesting success by ravens (although I would later learn of very telling exceptions; Chapter 23). My sustained supplemental feeding had probably not helped the breeding pairs. The calves I'd brought attracted crowds of vagrant juveniles. Since no carcass had gone unused by these crowds, none would have been a reliable long-term food source for a breeding raven pair. Given the vagrant crowd, most of the deer carcasses in the woods I didn't know about were probably equally transient. In contrast, in Vermont, where the breeding density of ravens is similar to that in Maine, I had provided only one large animal carcass at a time, and then not as regularly. These carcasses were almost never eaten by crowds, and they lasted months. Nest failures in Vermont still occurred (14 percent out of my total of thirty-five nestings), but not nearly as frequently as at my study site in Maine with regular supplemental feeding, where 48 percent of sixty-eight nests failed. If my food supplementation in Maine had an overall effect of increasing nesting success, then it was for ravens hundreds of miles *distant* from the site, not on the site itself. Given these thoughts, I wanted to get even more food samples from the young, and decided to repeat the observations for one more day.

It was spitting snow on the morning of May 8 as I went back to the cemetery nest. As usual, one member of the raven pair was at the nest. This bird was relatively tame. As she scolded me from the top of a neighboring pine, I noticed that she wore one of my rings on her left leg. The young were stuffed and sleepy. None begged. Immediately after I ringed their necks, two closed their eyes and resumed sleeping.

When I came back to the cabin, the snow had stopped and the sky was clearing. My company, Kim Most and Lori Friedman and their friend Kerry, University of Vermont students and raven helpers all, were baking fresh bread. I drank a cup of coffee and sat on the front step, listening to the energetic song of a winter wren. A solitary vireo called languidly from the mixed hardwood-conifer woods in the back. Then I heard the yells of a raven down toward Alder Stream, in the direction of the clearing a half mile from where I had dropped off six dead cut-open calves two days earlier, and I expected it would be many days before a raven would find them, and many more before they would dare to go to it and start feeding. Looking up into the sky beyond Alder Stream, I saw several ravens dancing high in the sky, then diving straight down and shooting back up. Such an aerial display, and the yelling, never fail to attract me, or other ravens. We took off as a pack of four, to see the ravens and be educated, and be reminded of Aldo Leopold. In *A Sand County Almanac* (1949) he had so aptly written a half century ago: "I once knew an educated lady, banded by Phi Beta Kappa, who told me that she had never heard or seen the geese that twice a year proclaim the revolving seasons to her well-insulated roof. Is education possibly a process of trading awareness for things of less worth?"

The southeast-facing hillside down to the brook had recently been logged. Poplar sprouts some five to six feet tall had grown in the previous year, and deer and moose had come to browse them. Curled fuzzy and brown fern fronds emerging from the ground showed pale green hairy stems below. Down by the brook, the alders were unfurling their catkins, but no green shoots were yet showing through the dry, curled, yellow swale grass that rustled as we walked through it in single file. As John Fowles points out in *The Tree* (1979), nature is, unlike art, created as "an external object with a history . . . but also creating in the present, as we experience it. As we watch, it is so to speak rewriting, reformulating, repainting, rephotographing itself."

We jumped over the eroded beaver paths that had become canals leading into the brook. Along the water where the grass had been matted down, we saw brilliant green sharp tufts of grass rising straight up like phalanxes of short, erect spears. The burgundy stems of the red osier dogwood along the stream lent a striking contrast. The various colors blended pleasingly, because all the colors as they are arranged in nature are "complementary." Nature is the standard for truth and beauty.

We stopped on a cleared slope next to the stream where a large pool reflected bright light like a mirror, making it impossible to see down through the brown water to the black bottom of mud and rocks. Fluffy white clouds drifted from north to south far above the black silhouettes of the pointy firs and spruces ahead of us, beyond which lay the clearing with the calves, where we could hear the ravens. I was feeling the pulse, but I was only beginning to have a sense of what was going on by getting to know their world. I hoped the unfolding of that world would someday create a story where all facts would be meaningful. All data and all observations would be like flecks of paint that would, when seen in total, reveal a masterpiece of evolution.

Ahead of us, just over the spruces, I watched a plume of ravens rising high into the sky, up into the white clouds. The sky seemed black with ravens. I estimated there were about seventy to ninety birds. A brown eagle, possibly an immature bald eagle or a golden eagle, was circling among them. The large raven group split into two major regiments, then smaller squads of a dozen or fewer peeled off. Repeatedly they dived, tumbled, and spun, and rose again on outstretched wings.

We crossed the brook. When we reached the clearing, about a dozen more ravens flew up from the pile of carcasses. Most of the meat from these calves had already been stripped. We quickly built a blind of spruce branches nearby, so we could watch the action from up close later on. Then we returned to the grassy slope of the brook to bask in the sunshine and to float on a log raft in the pool in the middle of the icy stream. A winter wren sang, and a golden-crowned kinglet made its high-pitched contact calls in the fir thicket near the water. The time passed quickly. Time came to leave the raven show and to climb the pine again to retrieve the food from the ringed young and to feed them.

The previous pattern of behavior and feeding at the raven nest held up. As before, one parent was always at the nest. The parents or parent had this time brought six packets of dried meat in five hours with telltale deer hair folded into them, and one packet of fresh deer meat. This pair clearly relied on a deer carcass. Four of the feedings still in the youngs' throats weighed a total of 75 grams and were composed of twenty separate little pieces of meat. Additionally, the parents had fed one of the young some very recently cached food—a piece of sausage along with well-chewed whole grain homemade bread. I knew, because I had left these items on the nest edge after retrieving the birds' food the day before. Just as I was leaving, one bird returned carrying food in its bill. Its voice was muffled, as if it was trying to talk with a full mouth.

That evening at the cabin, we enjoyed a meal of fresh bread and pasta. We built a fire in the pit and roasted marshmallows over the coals. A woodcock displayed on and over the clearing beside us. Before it left to go hunt earthworms for the night, we also heard a few agitated raven squabbles from the pines nearby. That meant there was a temporary roost there, probably of some of the ravens that had been recruited to the carcasses. I felt at peace and slept soundly.

I woke at about 4:30 A.M. as the eastern horizon was beginning to lighten up. Without bothering to make myself food or a cup of coffee, I stepped outside. It seemed warmer, and there was not a hint of cloud in the sky anywhere, nor was there a breath of wind. All the sounds were magnified, and a dawn bird chorus was starting—the fluting hermit thrush, the peripatetic winter wren, the *dee-dahs* of chickadees, the nasal twangs of the red-breasted nuthatches, and the long sad whistles of the white-throated sparrows. The woodcock must have had good foraging in the night, for he was strengthened enough to perform his athletic display, repeatedly rising hundreds of feet into the sky like a giant hummingbird and then whistling and diving again in the dim morning light, just as he had done earlier at dusk.

The ravens at the nearby roost were still silent, and I had time to get into my newly built spruce blind across the brook by the calves. By 4:50, I heard the first raven calls, and soon after that they started arriving by ones, twos, and threes. I expected this, as the birds had already fed well here, and indeed most of the meat had been taken in two days. Those that would come now did not need to follow crowds of others. They could come at their own discretion. As each raven or small group arrived, they settled into the bare branches of a nearby poplar tree, and I saw them as dark silhouettes against a gradually lightening sky. They preened, occasionally shook themselves, and sometimes made soft cooing sounds. There would be no yells today, unlike a day or two earlier, before they first began feeding here when they were still afraid.

By 5:25 A.M., more than twenty birds had gathered in the poplar, and ten were perched in another tree. A group of four or five swooped down to the calves near me. Within five seconds, all the others except two came down, and the pile of calves was then quickly enveloped by ravens. Even before the sun was over the horizon, light reflected off their backs, with the yellowing sky as a backdrop above the silhou-ettes of trees on a low distant ridge. Occasionally, still another bird came in over the trees, set its wings, and glided silently, low over the ground near the calves. Others left at intervals.

Except for occasional bickering, the ravens were quiet, as I'd expect this late in the feeding cycle. The most conspicuous sounds were the other birds of the forest—the peculiar cadence of the sap-sucker's drumming on a dry branch, a "song" where the beat changes but the tune stays the same, and the low thumping drumroll of a ruffed grouse starting strong and slow with hollow-sounding wing-beats ending in a blur of sound. Purple finches, and the first ovenbird so far this spring, called.

About thirty ravens eventually fed at the calves or loitered around them in the near-darkness. None of the new birds we had captured or marked this year were among them. Only one marked bird was in the crowd, a female with a yellow wing marker with the letter *W* on it. I had sighted her a number of times during the last several years.

Daylight arrived. Gently, ever so slowly, I moved a small spruce spriglet from my line of vision. All went smoothly until one bird noticed something amiss—perhaps my slowly moving hand through the curtain of fir twigs. As always happens when one of the group leaves alarmed, the whole group left the bait in a wild clatter that was so sudden, so violent, that their contagious fear was almost palpable. Not one bird uttered a sound. For many seconds, the birds flew wildly all around, not knowing what or where the threat was. They made sev-eral passes around the clearing, then left. I then also left to check again on what the youngsters were being fed.

I climbed up to the nest four more times, to monitor what the young ravens had been fed for an additional 15.5 hours. During that time, they were fed thirteen loads of food. On May 12, three loads were again semidried deer meat. The other load was a white chunk, possi-bly of suet (from someone's bird feeder? see p. 333), along with two crushed robin's eggs. On the next day, I recovered six food loads. All were fresh deer meat.

May 14 would be the last day I could get data, because I had to leave for a week. The young were by then already feathered out and would soon be cantankerous and uncooperative. When I got to the nest, I pushed the lumps in their throats up with thumb and forefinger, as I had done before to deposit the regurgitate into a plastic bag. From two birds, I recovered chunky semidried deer meat, much as I had been collecting before. A third bird again held a chunk of what looked like white suet. Again, the suet had something blue with it—the sky blue color of robin's egg shells. Only this time, something was different. A fully intact robin's egg emerged from the young raven's red mouth.

It was amazing that the delicate egg had not been crushed in the young bird's mouth and throat. Even more miraculous, the egg had survived being taken out of the robin's nest by the adult raven. As I delicately cradled the fragile egg, trying hard not to break it, I realized that the raven must have done the same. Whenever I've given a mouse to any raven, I've heard the crunch of breaking skull or vertebrae as the bird took it in its thick bill. There is also an audible crunch when ravens take an insect. They also squeeze down hard on a piece of meat. Most objects are also pecked. The bird that raided the robin's nest had to reach down into the deep nest cup and grasp the egg in its thick, hard bill. If the egg didn't break, it could only be because the bird had carefully exercised restraint. Why? Did it know it would break the egg if it were handled like other objects? Did it know that breaking the egg would cause the contents to leak and be lost? After dropping the whole egg into the begging young's open throat, the egg would have been crushed had the young swallowed.

My first impulse was to drop the egg into the mouth of one of the young birds, because there seemed little else to be done with it. As nourishment, the tiny egg had minuscule value; as an object of beauty, much. I hesitated.

I rolled the egg in my fingers, astonished by the purity of the light blue color and the symmetry of its shape. I found myself handling it delicately, as though afraid it might crack at my touch. This was more than just a beautiful object. Unlike all the other parts of

Raven carrying three robin's eggs (one in throat pouch).

birds, mammals, frogs, or snakes these young ravens might have eaten, this morsel still had a possible future. It could become a living bird. It had the potential to become a robin with a red breast who sings a beautifully melodious song at dawn. This egg was like the underdog kid who has beaten all the odds. The audience is cheering for him or her to continue, because he or she represents everyone's hope. I gently placed the egg into my mouth, cradled it on my tongue, and took it down with me to find a robin's nest and foster parents. Meanwhile, I was pleased with the results of my pipe cleaner study. It had given me one more little detail that would help to tie the raven's sharing behavior that I had worked on for so many years into a more coherent picture.

Note: In 1998 I mastered the ascent to the cliff nest by my home in Vermont, and I repeated the ringing experiment on the three pinfeathered young in that nest. On the first three ringing episodes, spanning 16 hours, I retrieved four boluses of moistened white bread (40 grams), one bolus of pink (calf?) meat (20 grams), and two boluses of not-so-very-fresh liver (45 grams). After that I retrieved nothing in three days (total of 17 hours) that the young continued to be ringed with the pipe cleaners. Nevertheless, they had been fed; they had gained normal weight and never showed signs of hunger. I presume that the parents caught on to what was happening. They had probably devised a new way of feeding (tiny bits rather than boluses?), and I conclude that my graduate advisors had been right; perhaps one should indeed not try to study an animal cleverer than oneself.

Table 4.1 The Main Components of
229 Raven Pellets at Seven Different Locations

Under DEW tower raven nest
Barrow, AK, June 1993

41 primary lemmings
16 primary birds
23 lemmings plus birds

—
80

Under a cliff nest
Mt. Denali, June 1992

1 egg shells
3 birds
3 rodents
9 caribou fur
1 hare fur
6 unidentified mammals

—
23

At a big feeding spot by my
calf carcasses Weld, Maine, 1995

1 flying squirrel
1 hare
7 mountain ash berries
4 mountain ash and animal hair
1 with 15 *Viburnum opulus*
    seeds, hair, and bone

—
14

At communal roost
Brück, Germany, July 1997

5 transparent plastic bag bits
2 dissolved paper
1 pieces twine (red, blue, green)
1 candy wrapper
1 paper, eggshells, 10 beetle elytra
1 hankerchief (exclusive!)
1 cloth (exclusive!)
1 fur and bones (rat?)

—
13

Under a raven nest Schleswig-
Holstein, Germany, April 1994

11 grain and other plant remains
12 small bones
2 hare remains
3 rodents

—
28

Under winter communal roost on my
hill, Maine, Jan. 1993

10 deer hair
5 berries
44 pebbles, cattle hair, vegetable

—
59

Under sleeping tree
Weld, Maine, August 1995

1 with 39 chokeberry pits
2 almost exclusively white quartz
    pebbles (25 and 12)
1 plastic bag bit
2 beetle elytra, *Maianthermum* seeds
1 with 21 *Actia rubra* seeds
1 with 29 chokeberry pits and many
    blueberry seeds
4 animal fur and bones

—
12

FIVE

Education

Roa, Konrad Lorenz's raven, raided clotheslines to steal ladies' underwear. Roa had been exploring a neighbor's laundry hung on the line just when he was called. He came, taking a small transportable item with him, a pair of panties. When he got a reward of tasty food, he made the association of panties and food. Henceforth, as expected according to classical conditioning theory, he brought these items on his own to redeem them for savory snacks.

It would be difficult to overestimate the importance of learning in the life of ravens. Even so, much of learning involves expression of behavior that is inherited to varying degrees, and released only by specific stimuli at specific times. And as every parent knows, an important aspect of the learning process is simply gaining exposure to what is important. Exposure determines what will, as opposed to what could be, learned. Perhaps the largest part of our educational process, and perhaps also the ravens', involves mechanisms for gaining exposure to appropriate stimuli. These mechanisms may have several com-

Play-fighting in the snow.

ponents. In ravens, youngsters gain the specific experience appropriate for the species' lifestyle by following their parents. Moreover, curiosity allows them to take advantage of this experience and enhances encounters with relevant objects or things.

I got my first glimpse into the process of how young wild ravens "learn the ropes" by watching the Hills Pond family near my Maine cabin. In early May, when the young birds first left the nest but still remained near it, the adults always came to feed them. Gradually, the young flew to meet the parents when they brought food. Then they followed them. In early June, both parents and their young came near the cabin to feed on carcasses I had left out and to dig for maggots and other insect larvae around them. At that time, the young still begged loudly. The parents picked up food in front of them and transferred the food into their open, yelling mouths. After a few weeks, the parents seemed less attentive to the yelling, and the young sometimes tore off their own meat, or dug with their bills and picked up their own larvae at the decomposed carcass.

It is in the interest of the young to stay with their parents as long as possible, but it is in the parents' best interest for their offspring to become independent. Conflict is inevitable. In the Hills Pond family, I saw the first signs of such conflict in mid-June. The young flew closely behind their parents, yelling loudly, while the parents erupted in the agitated *kek-kek-kek* calls they make when they are irritated, as when a hawk is close or a predator approaches their nest. These parents did not act pleased, and the young would soon be on their own.

On June 15, 1993, the parents sounded downright oppressed. I saw them coming up the hill once again, accompanied by their noisy youngsters. They descended on the skeletal remains of a calf carcass, but there was no meat left. For a half hour, the family dug in the leaf litter, searching for sylphid beetles and their larvae that were consuming the hide. Later that day, I saw one of the adults flying down along the Wilton-Weld road. This time, two young ravens were chasing along, yelling-begging loudly close to the parent's tail; but instead of continuing down the road as I expected, the adult bird veered off, dove into the forest, and continued to fly fast through the trees. The two young stayed close behind. The parent (probably the female) came back out of the trees with them, and landed on the tip-top of a pine, the spot farthest away from any other foothold. Her offspring perched on the first available branch far below and continued their clamor. She made agitation calls, snapping her head nervously in all directions as if searching for a place to escape, then took off down the road again. The two youngsters again screamed loudly right behind. This time, she flew *up* and started to circle. As she gained altitude, she continued to make her agitation calls. Finally, she took off north in a straight line far over Adams Hill, still closely pursued. She seemed to be doing her best to escape her raucous followers, but she did not succeed for long; all three were back at the remains of the calf in the evening.

The offspring, sometimes with one parent and sometimes with both, continued to come up to my clearing at least once a day. They sounded desperate and sometimes pitiful, and the adults acted harassed and agitated. A week later, the two young sometimes came alone. The male parent, who had a leg ring, came several times to feed on a dead woodchuck I had provided. As his offspring begged from him, occasionally he fed them. Sometimes he refused them and cached the meat nearby in the grass instead to let them feed on their own. He had become an indirect provider. The young had been led to food and thereby learned what it looked like. Presumably, they would now find their own woodchuck or some other carcass that they had explored with their parents and recognize it as food at some point.

The family, or parts of it, stayed together most of the time until the end of June. By July 15, they had split up permanently. Single juveniles still came near the cabin in late July, attracted by the begging, when I fed my current young tame birds, who were then still free to roam.

My young ravens behaved with me as I had seen the young of the Hills Pond brood react to their parents. They followed me and showed great interest in anything I contacted. In June and early July, their tendency to follow me was so strong that I had great difficulty escaping them to seek peace and privacy. Even I was starting to make agitation calls! I could not hope to maneuver through the forest like an adult raven to try to outdistance them, so I resorted to trickery. My usual ploy was to go in the front door of the cabin, and as they gathered at that door, to exit quietly from the rear door and slink away.

Konrad Lorenz observed that young *precocial* birds (those that leave the nest hours after hatching, like ducks and geese), unlike *altricial* birds (which are nearly embryonic at hatching, like ravens), follow their parents soon after they hatch from the egg. Lorenz determined experimentally that they followed whatever they saw at the nest, normally their parents, during a critical, posthatching interval. By this early exposure, they learned their species' identity, which would later be critical when they identified potential mates.

Is following-behavior like that of the young ravens proof of imprinting? Did the ravens really confuse me with one of their kind? Would they later want to mate with humans as a result? The evidence indicates that they know their own kind very well, despite having followed me. My hand-reared ravens always had as a model not just me, but also others of their own kind. They routinely responded with rapt attention to a raven flying silently in the distance, yet they totally ignored a flying crow, which humans have difficulty distinguishing from a raven. They also responded with the same rapt attention to the ravens' calls.

I suspect that the exposure young ravens get when they are young probably influences but does not determine sexual preferences in them any more than it does in humans. We wouldn't, for example, expect a human child raised with a pet gorilla to be sexually attracted to gorillas as an adult. Our preferences can probably be modified, but they cannot be stretched arbitrarily in any and all directions, because we and ravens probably both already have approximate innate concepts, or "templates," against which we match how an ideal mate looks and acts.

Other things that we and ravens learn are probably less rigidly preprogrammed. Food, for example. There cannot be one innate template of what food should look like, because there is almost no limit to the appearance of what ravens eat. Are there rules for identifying possible food, similar to the "rule" that a duckling follows the first moving thing it sees after hatching? How do ravens gain exposure to what is relevant to them? If ravens learn about food, how can they possibly begin to gain sufficient exposure to hidden and diverse food, given the almost infinite variety of objects in any environment? Can they afford to pass up anything? If so, what? My four young birds, Fuzz, Goliath, Houdi, and Lefty, were the perfect subjects for a test to find out.

For one month after they left the nest, I led the four birds at least once and sometimes several times a day on thirty-minute walks. During these walks, I wrote down everything in their environment they pecked at. In the first sessions, I tried to be teacher. I touched specific objects—sticks, moss, rocks—and nothing that I touched remained untouched by them. They came to investigate what I had investigated, leading me to assume that young birds are aided in learning to identify food from the parents' example. They also, however, contacted almost everything else that lay directly in their own paths. They soon became more independent by taking their own routes near mine. Even while walking along on their own, they pulled at leaves, grass stems, flowers, bark, pine needles, seeds, cones, clods of earth, and other objects they encountered. I wrote all this down, converting it to numbers. After they were thoroughly familiar with the background objects in these woods and started to ignore them, I seeded the path we would later walk together with objects they had never before encountered. Some of these were conspicuous food items: raspberries, dead mealworm beetles, and cooked corn kernels. Others were conspicuous and inedible: pebbles, glass chips, red winterberries. Still others were such highly cryptic foods as encased caddisfly larvae and moth cocoons. The results were dramatic.

The four young birds on our daily walks contacted all new objects preferentially. They picked them out at a rate of up to tens of thousands of times greater than background or previously contacted objects. The main initial criterion for pecking or picking anything up was its novelty. In subsequent trials, when the previously novel items were edible, they became preferred and the inedible objects became "background" items, just like the leaves, grass, and pebbles, even if they were highly conspicuous. These experiments showed that ravens' *curiosity* ensures exposure to all or almost all items in the environment. I never saw them pass up *anything* that was new to them.

In the field, food items are locally distributed: Cranberries grow on the tops of mountains and in bogs, raspberries in clearcuts, wild strawberries in fields, and so on. Therefore, in the context of their own curiosity, the following-behavior (taking advantage of their parents' experience) ensured that odd and perhaps rare sorts of food would always be found. This elegant mechanism would result in food-finding in any environment, with any kind of food, and it explains another long-standing enigma—the ravens' well-known strong attraction to jewelry and other rare baubles.

George Miksch Sutton, the famous bird artist and Cornell ornithologist, said half a century ago that his pet raven "was an aesthete. His love of baubles was not wholly carnal" because he did not eat them. But Goliath and the others treated even pine cones like baubles for the first day or two out of the nest. They soon got their measure of pine cones, and forever ignored them. Having determined they were not food, they soon got bored. So it went with dandelion flowers, maple samaras, cigarette butts, pebbles, shiny bits of glass, inedible winterberries. Blueberries? *Those* "baubles" tasted good, and were *not* soon forgotten. Rings and coins? The birds' immediate attraction to such items was undoubtedly aesthetic, but in an ultimate evolutionary sense the question is, *why* were they aesthetic to them? And the answer is, because of the carnal reward; they *represent* new things, some of which *can* be good food and so they are examined in play. If the birds were attracted to objects merely because they were already *known* to be good food, then what they already knew would be known even better, but nothing would be discovered.

The same "carnal" tendency of ravens also sheds some light on Lorenz's anecdote about the ladies' underwear. Since young ravens are attracted to anything that stands out within their environment, laundry hanging on the line is obviously a prime attraction. When I was still a graduate student at UCLA, my first pair of ravens made themselves unpopular with my mother when I brought them home to Maine. They pulled clothespins off the line and either scattered the laundry or left dirty tracks on those clothes that they left hanging. In retrospect, I understand—the birds had shown interest in what had interested *her,* as well as in what was novel to them. Like picking up shiny new pebbles, picking off clothespins is play. If they carried things away, their plunder was generally the lighter, smaller items, like socks, and of course, ladies' underwear.

A simple experiment proved that the ravens' strong curiosity to explore contrasting or novel objects in their environment functions ultimately in food-finding. Having taken a trip to the seashore, where I behaved like a young raven myself by picking up a large collection of strange new trinkets, I brought my four young charges bags full of such different items as beach-worn pebbles, seashells, crab claws, and seaweed. I included only one potentially edible item—small dead sand crabs, something Maine inland ravens never get to see, much less taste. When Fuzz, Goliath, Lefty, and Houdi were confronted with all of these seashore items at once, which I had strewn through a walkway in the aviary in the forest, they became highly animated and sampled them all indiscriminately. Within ten minutes, they were already focusing on the sand crabs and starting to get bored with the rest. After making only several contacts with these munchies, they sought them out. Within thirty minutes, they had found and eaten every one. By the next day, they basically ignored all of the other inedible trinkets. They had become like Jakob, Klaus Morkramer's raven, who could be trusted even with the expensive Roman antiquities because they bored him.

Young ravens make short work of novel objects. On many occasions, after my tame ravens had satisfied their hunger on run-over raccoon or hare, I have provided them with toys for further amusement. A cardboard milk container is no challenge at all. In two minutes, it is reduced to shreds. A translucent plastic cider bottle amuses them somewhat longer. With their great bills, they drum upon this object and then toss, tear, and crumple it into a misshapen mass of twisted plastic. There are countless newspaper accounts of ravens doing damage to roof shingles, parked automobiles, delicate airplane wings. Young ravens, like other youngsters, are open to exploring a variety of stimuli and to learning what they mean. In the wild, this play behavior is a survival mechanism, but in an urban environment their exploratory behavior can be a source of human annoyance.

According to a 1991 Associated Press article, ravens were caught at the airport in Soldotna, Alaska, pecking holes into airplanes covered in polyester fabric. Six planes were riddled with thumb-sized holes. The local raven expert was summoned, and he claimed that since ravens are "highly intelligent birds" they may have thought there was food inside, because they are used to being rewarded with table scraps in plastic bags. I don't think they thought. They were just playing, and animals who play may get lucky and find unexpected bonanzas.

Ravens also raised havoc in Juneau, Alaska, in a 1991 Easter egg hunt sponsored by a youth center. Almost 1,200 hard-boiled eggs were distributed throughout the Adair-Kennedy Memorial Park. Unfortunately, the staff and volunteers had distributed the eggs hours before the event. By the time they were through registering the attending children, ravens had already crashed the event and were making off with colored eggs in all directions. In this case, their play got them an immediate feast, while a lot of kids left empty-handed. The moral of the story, as reported in the *Juneau Empire,* was: "Don't count your eggs before the ravens come."

As to disrupted public events, my all-time favorite was when the Icelandic Golf Championships had to be moved to another course because ravens crashed the tournament, perched in trees around the golf course, and swooped down to grab and fly off with unattended balls. My tame birds are also hugely attracted to light, round objects, golf balls included.

The ravens' preoccupation with little round objects presented a less trivial problem when those objects were the eggs of the endangered least tern in California. A group of researchers (Michael L. Avery, Mark A. Pavelka, David L. Berman, David G. Decker, C. Edward Knittle, and George M. Linz) tried to solve this problem with negative conditioning. They set out Japanese quail eggs, which resemble tern eggs, treated with methocarp, a nonlethal chemical that makes birds ill. The ravens, not to be outsmarted, soon learned to distinguish between treated quail eggs and untreated tern eggs, and continued to eat tern eggs. Not to be outsmarted again, the researchers refined their tactics. They placed the methocarp-treated quail eggs at the tern colony several weeks *before* the terns nested. It worked—the territorial ravens soon learned that "all" the eggs at that site made them ill. When the terns finally laid their eggs, they had a brief respite. The resident ravens, who otherwise would have been shot, had now become an ally in conservation. By chasing away naive and as yet uneducated vagrant ravens who would have eaten the tern eggs, they inadvertently protected the tern colony.

The ravens' behavior rings true to us because we can see so many resemblances to our own behavior. My son Eliot at about one year in age behaved almost identically to the several-month-old ravens. He was strongly attracted to all objects that were new to him—spoons, cans, lids, an egg beater, film vials, bottles, a metal box, a knife. He would examine all items, if given a chance. All would soon be tossed aside, then ignored. The difference between Eliot's and the ravens' behavior is that the birds are only in this stage for a month or two, whereas the human baby's behavior may continue for several years. Parents soon find that their living rooms, halls, garages, barns, and yards become littered with ever larger piles of obsolete plastic junk.

Ravens' curiosity declines with age. By the time they are four months old, they are already becoming shy of most novel stimuli. As they mature, that initial attraction to novel things reverses. They become increasingly *fearful* of novel objects. I once tested a crowd of wild-caught juveniles with all kinds of objects that my young birds found irresistibly fascinating—film canisters, bottles, cans, silverware, and the like. They ignored the items until they got really hungry. The older birds reminded me of my father. He saw a peanut butter and jelly sandwich for the first time when he was fifty-five years old. It may have amused him, but he would not eat one then, or ever. He "knew" butter doesn't come out of peanuts, unless it was very inferior butter, and that was that.

SIX

The Fate of Young Ravens

BLACKJACK WAS A MARKED WILD raven, Number 21, who fledged from a nest by the lake a few miles from my cabin. He was still in his clumsy postnestling stage when he examined a hamburger grill at the Mount Blue State Park beach. His left leg slipped through the metal grill. Perhaps startled, he jumped sideways. The metal didn't yield but his young leg bone did, and he later lost that leg. Blackjack recovered and did fine minus one leg. By late August, he was still a picnic-junkie, and an entertaining one. Once, while rummaging inside a lady's picnic basket near the same beach, he received another fright when the owner returned to reach into her basket, although she had a comparable shock, judging from her screams.

Blackjack was fairly tame in early August, but became shy of people by early September. David Lidstone, a friend and volunteer raven watcher, kept track of him and other marked raven youngsters with a spotting scope. David continued to put food out at his campsite, and

Two pictures of a raven cliff nest in Vermont with young about a week and two weeks old. Note feces (white) in raven's bill due to nest hygiene, and the iridescent plumage.

that was taken primarily by the young ravens whom David had helped John Marzluff and me tag at the local nests. Early one morning, when Blackjack made an appearance at the accustomed food source, he was viciously attacked by three current-year young from the nest just at the other side of the lake. The dispute could have been over the daily food bonanza, or perhaps Blackjack was a target because he was at a disadvantage with only one leg. The wild young ravens never attacked Jack, a vigorous young tame raven at my cabin. Nor did he attack any of them, although he once waddled up behind a Hills Pond nest youngster and pulled its tail. Soon after his scuffle with the neighbors, Blackjack was gone.

*Young raven with a little fuzz
still on top of its head.*

I have often wondered what happens to young ravens after they leave their parents and their home area. Where do they go? Do they ever come back? What are their chances of survival? We have little more to go on than statistics derived from studies of marked birds in Europe, and they paint a bleak picture. Typically, only half survive their first year. These juveniles constitute a "floater" population without a fixed home. They are largely expendable as far as the nesting population is concerned, because they cannot breed unless a resident with a territory dies, thereby opening up a breeding slot. Adults can remain in residence for decades, and unmated floaters as old as seven years have been documented.

Resident ravens in New England start to defend their nest territory in late fall. By mid-February, in the middle of winter, they begin building their huge stick nests lined with fur and shredded bark on some crag or conifer, generally a pine tree. The female raven, like other corvid females, lays a clutch of four to seven heavily speckled, greenish-blue eggs that she incubates for twenty-one days. At birth, the young are pink, blind, and naked, possessing only a few loose tufts of down feathers and weighing only about 25 grams. They cannot yet regulate their body temperature and are continually brooded by the female as they grow rapidly on the food that the male brings. In eight days, they have increased their body weight by about twelve times. In six weeks, they have reached the adult weight of 1,100 to 1,500 grams, when they leave the nest and begin to follow their parents, staying with them for the next six to eight weeks as they learn to forage. For the young birds in my aviaries, the prospects for survival during the first year would be much better than the odds they would face in the wild; I presumed *all* of my birds would live, with my care and protection.

Normally, the young begin to wander away from their home area in late summer and fall. That's when the heavy mortality begins. Since I wanted to keep Goliath and his cohorts, I needed to confine them in an aviary. When the proper time came, I hoped they would nest inside and become residents there. Then I would let them fly free because I hoped they would stay bound to their adopted home, having skipped the dangerous wandering stage when most deaths are likely to occur.

Most of the 463 ravens we have captured temporarily and tagged at feeding aggregations at my study site in Maine have dispersed. We have seen few of them return. We had hoped that by marking them, we would be able to identify all the ravens we would see for many years, and we would thus have the basis for studying their social interactions. It did not turn out as we had hoped. On the contrary, if a large food bonanza attracts one hundred ravens, it is unlikely to contain more than one or two tagged birds, and these are usually adults from local nests. Most of the others, who are unmarked, must be from somewhere else. What has happened to all the birds with white, yellow, red, blue, or green patches on their shoulders? Where have they gone?

Occasionally, someone may see a tagged raven. Very rarely, the person may even hear of the study and report back to me. The range of the eight nonlocal (beyond fifty miles) reports I have received was south to Boston, north to New Brunswick, Canada, east to Nova Scotia, and west to western New York State. The total area covered by

these eight sightings is over two hundred thousand square miles. It is
a wonder that I have received even one report from such a vast area,
which, judging from the few returns, must be a minimum area. The
main conclusion that can be drawn from the effort of capture-mark-
release work is that in northeastern America, the birds do indeed
travel very extensively.

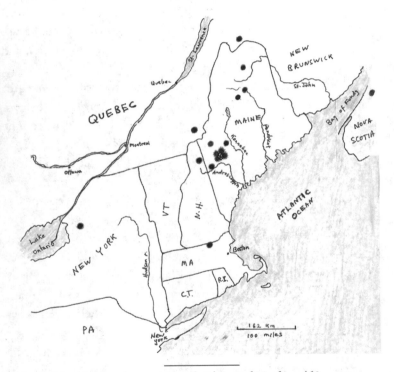

Map showing the recoveries and sightings of 12 of our 463 ravens
that were marked in Weld, Maine. (The four dots near Weld
represent resident adults that were frequently observed at the
same site; all of the others are single recoveries or sightings.)

There is nothing absolute or specific about these distances. In
some areas, the birds presumably travel much more; in others, much
less. How far birds travel is driven by the search for food. The more
sparsely food is distributed, the farther the birds must fly to find it.

But do they travel in sibling groups, alone, or with friends? A set of explicit observations at the time that the young become independent could yield insights.

When I found myself with three recent college graduate volunteers eager to do raven work, and nine radio transmitters, the project was on. The question that seemed feasible and relevant was simply to ask if and/or how much the several young of any one nest stay together and/or intermingle with the young of other nearby nests. Do the young aggregate or overlap in their foraging ranges? To answer these questions, we would have to monitor at least two sets of young during the summer.

I chose my first nest from the white pine grove at the edge of Weld village. This was a pair Marzluff and I had years earlier captured and which had raised young in our aviary (see p. 295). The next-nearest nest was located about a mile and a half from the first. Both nests were in tall, straight, white pines that were limbless for twenty to thirty feet, then had dead, brittle limbs most of the way up to the thick live branches in the crown where the nests were located. As always, it was a white-knuckle experience for me to reach the nests.

The young at the first nest peered down at me over the edge when I got close. They were ready to fledge. And fledge they did the moment I reached the nest edge. I managed to grab two of the five young and put them into my knapsack to take them down to the ground, where I could attach the radios. The other three young landed a bit lower in nearby pine trees on their maiden flights. After descending to the ground and attaching a radio onto the large tail feathers of the first two birds, I climbed up after the others. Again, they flew off when I got near them. With each clumsy flight, they landed lower, and eventually they reached the ground, where I caught them. The process was repeated almost exactly at the second nest, except that this nest held only three young. We used up all but one of our radios. I was jubilant, though exhausted. It had been a great day.

During the next week, I monitored the young by checking on their radio locations, and walked into the woods to locate them visually. On May 31, the members of both groups were perched close together on low branches. They were still within two hundred feet of their nests, but they flew off as I came within fifty feet of them. Three days later, the young were no longer near the ground, but instead perched high in the pine trees, now within a quarter mile of their nests. All eight radio transmitters were working, the birds were healthy and flight-capable, and the experiment was in place. As I left on a long trip, I felt confident that my very eager soon-to-be helpers would be able to carry on the project and we would then see interesting results.

The first volunteer who took over after we checked on the birds during my two-week absence got radio signals from six of the eight transmitters. The notes he took said, "The birds are consistently in the general vicinity of their nests," "They are not yet flying great distances, just making small movements in the nest vicinity," and "They are extremely hard to locate visually." All was as expected, which was why I didn't question the data just then.

Starting June 18, the next volunteer spent five days doing nothing but tracking. Anticipating that the birds would still not have wandered out of radio contact, I asked him to take a compass reading every three to four hours and to write down the directions of the radio signals from specific predetermined spots. In this way, the movements of the birds could be determined more consistently.

He collected a hefty sheaf of data in five days. He had dozens of radio contacts with each of the same six birds the first helper had found, plus one additional bird. Curiously, the additional bird was found only three times.

I eagerly plotted the data. Strangely, one signal, Number 8169, was picked up five times at one station only, but always from the same direction. It was not picked up at all during sixteen other readings at the same station. Number 8510 was picked up every single time the student looked for the birds. He found it from one station only, but from various directions ranging from 70 degrees to 28 degrees from magnetic north. The same applied to Number 9680, but at directions varying from 150 degrees to 350 degrees. In short, the data set said that the birds either had stayed at the nests or had moved only locally for very small distances. But most of the time, they were missing!

As you may have guessed by now, the data didn't make sense. I needed to enter a swamp and try to find the first and most faintly beeping transmitter with a raven still attached. Coming into a cover of evergreens, I heard a golden-crowned kinglet scolding. What would it be excited about, I wondered. It had found a saw-whet owl, I guessed. And so it was. The little owl made itself slim next to the stem on a spruce and peered down at me with its wide yellow eyes that seemed dark in the shadows.

Soon I was in a beaver bog. The radio signal became weak all of a sudden; a hill was blocking the signal. A half hour later, I came to another beaver bog with a big lodge near the edge of the new pond the beavers had created. The lodge had freshly chewed sticks on it. A mourning cloak butterfly fed on oozing sap and spread its wings to the sun. A great blue heron fishing from the beaver lodge lifted off and disappeared around the bend of the valley.

In twenty minutes, I was at the V of the confluence of two brooks. The signal sounded maddeningly close. I searched under a big fir tree and a big birch stump at the water's edge. Three hours later, try as I might, I still had not found the transmitter. A pile of raven feathers lay right at the edge of the water by the roots of a big upright white birch stump under the tall fir tree. So I knew the bird that had carried the transmitter had probably been killed by a raptor. The signal was very strong next to the fir tree trunk, but my search turned up nothing except more feathers. I tried again. I got a very strong signal by the edge of the brook. I checked under the banks. Nothing. Now the signal was strong at a fallen log nearby. I dug. Nothing. Why did the signal strength keep changing? The signal seemed to be ever so slightly stronger when I tipped the antenna up. (You also get a signal by turning the antenna in the direction opposite to its source.) But I had already dug up the leaf litter and the stream-deposited debris all around. Did that mean the transmitter was up in the air rather than down in the ground? At such a short distance, perhaps alignment of antennae is more important than direction in causing slight changes in signal strength. I was not optimistic that I would find it up in the tree. The top was too dense, and not the place a large owl would choose to pluck its prey.

Still, I climbed the tree. Halfway up, as I looked down, I realized that if the raven had been plucked up here, the feathers would *not* be within some five feet of each other below. They'd be scattered all over the place. Besides, the transmitter had a blunt heavy end and no rough spots to catch on limbs. How could it possibly get hung up in branches? The idea that the radio was up there made no sense at all, but I ascended anyway, because I had exhausted my store of logic. Doing stupid things would not, I hoped, prove me stupid. I found nothing, though.

Later, in defiance of all logic, I climbed the tree once more, having again exhausted all other possibilities. When I got to the top this time, having searched all the branches, I happened to look down over the area below. Then I saw it. Next to the fir tree and just below me was the top of a twenty-foot-high dead birch stub. The transmitter lay balanced on top of the stub. The raptor, probably a great horned owl, must have plucked its prey on the stub, as they are wont to do.

The next day, I tuned the receiver to another transmitter, heard its constant clicks, and gave chase at a trot. This time with a helper for support, I tried to set a proper field biologist example by traversing a succession of small valleys and hills, crashing through the underbrush. All of the area we traversed had been logged about ten years earlier, and it was a tangle of poplar saplings, raspberries, blackberries, and fir trees, all sprouting through rotten slash. We sweated and swatted deerflies that burrowed into our wet hair. When we came to muddy brooks, I set a good example and jumped in and waded across. We soon came to old logging roads, where we ran. The signal got stronger as we came over a hill to another big beaver bog. The deerflies may have been bad, but when you're distracted by mosquitos, mud, and tall wet swale grass, you notice them less. After only an hour, we picked the second radio up off the ground. This one was also attached to the remains of a raven. There were still four other radios to track down, so we ran on.

All of these four radios, it eventually turned out, also showed similar signs of having lain still for some time. All were attached to long-decayed raven remains.

In early August, Scott Lindsey, a local schoolteacher, and I climbed to the top of Mount Bald, Mount Tumbledown, and Mount Blue, to do radio surveys of the surrounding countryside. The two radios presumably still at large had at least four more months of life left. But we never got a signal again, even from the mountaintops. Perhaps two out of eight birds were still alive. The statistics that I knew so well were taking on new meaning. These were "my" raven friends and neighbors being killed.

SEVEN

Settling in a Home Territory

Ravens leave home as juveniles, wander, and after extensive exploring, settle down to live by the means that their capacities and experiences dictate. Their area of residence becomes their home, and they remain true to it, as far as we know, for most or all of the rest of their lives. They remain true to home for probably the same reasons we do.

To us, home is an area where we are at ease, because we are familiar with neighbors, with local peculiarities of food and shelter, and we recognize friends and potential enemies. I suspect it is similar with ravens. Alan Gussow (in *A Sense of Place,* 1971) says, "A place is a piece of the whole environment that has been claimed by feelings." Feelings of comfort with familiarity could motivate bird migrants to sustain their long arduous journeys, bringing them back over thousands of miles, year after year, to the very same bush. Do these animals yearn for home as we do? We do not know what they feel. We know only what they do. We also

---

Raven male who has found a territory and is
making his claim through calls that exclude competitors.

know only some of the adaptive reasons for our feelings, which make *us* do what we do. Yet, until shown otherwise, we can assume that the same powerful mechanisms driving behavior in one species apply in another.

Ecologists have generally called the area where a raven settles its "territory," as though ravens behave like other birds, defending boundaries to maintain a specific spacing between themselves and neighbors. Territoriality is considered an adaptive response for monopolizing resources, especially those needed for rearing young. However, there is no reason to suppose that a territorial bird "knows" what it is doing. It just does it, and the behavior, even if unconscious, has served survival and reproduction. Home, or a place an animal stays because it is locally adapted, has a different connotation, but it is not exclusive of territoriality. Nevertheless, nest and home are entirely different. A bird's nest is not its dwelling place. It is a temporary shelter or receptacle for rearing generally only one brood of young.

There are several ways to get insights into how ravens might locate and settle in a home area and whether they use it exclusively. One is to try to follow individual birds, the other is to see the result, determining where the birds end up. We have little information available on the first question, and very little on the second. Thomas Grünkorn, who was doing an extensive marking study of ravens in northern Germany, was trying to get answers by using both approaches.

I had never met him, and we had made no prior arrangement to identify ourselves at the Hamburg airport. After coming through customs, I saw a tall blond man hold a black feather aloft. I went over and smelled it. It had the distinctive musky raven smell, and so we identified ourselves.

Grünkorn had three years earlier finished his *Diplomarbeit*, similar to a U.S. master's thesis, at the University of Kiel on a study of population trends and habitat preferences of ravens over a 7,200-square-kilometer area of Schleswig-Holstein, a northern state abutting Denmark. He had documented 344 of an estimated 400 to 450 active nests for the entire region. A crew of volunteer assistants included Volker Looft, a teacher, Hans-Dieter Martens, a policeman, Jörg Reimers, a computer programmer, and other dedicated unpaid amateurs. They had laid the groundwork over many years, and continued to help on nest surveys and the annual banding of the young.

Volker Looft had for thirty years monitored basically all of the breeding ravens and goshawks in one of the eight survey areas. He had shown Thomas most of the raven territories in his study, and he helped with banding the young ravens in April. Thomas reciprocated in Volker's goshawk project, banding their young in June. Volker's 2,300-square-kilometer area contained fifty to sixty pairs of goshawks. It also contained about the same number of breeding ravens, which usually nested near the hawks (because of similar habitat?). Volker and Thomas had banded close to 2,000 young ravens, and they were ready to band the current 1994 cohort in late April when I stopped by to visit. The study had already revealed that the raven population had rebounded nicely from a big decline in the 1970s.

We immediately made plans for the day following my arrival. Thomas had taken a week's vacation from his job at an environmental assessment project, and planned to be out all day every day, climbing raven nest trees and banding the young. We convened at Thomas's parents' house in the village of Selk in Holstein. Before we left, I got a brief glance at their marked-up Schleswig-Holstein map. It showed a quilt-work of small green patches signifying woodlands, and many of these small patches had black dots on them. The dots represented the locations of the raven nests. The distribution of these nests was fascinating. They were not in a random pattern. If they had been strictly random, some would be close together, others far apart. These, however, were regularly spaced, as though each had some repellent force that radiated in all directions to other nests. The pattern of nests showed that in this part of Germany, the habitat was full of breeding ravens, each pair occupying about 43 square kilometers. If another raven pair were to fit in here or there, it would have to do so where a gap still existed.

The repellent force that created the regular nest spacing was not due to a feature of the landscape, such as the unavailability of nest trees. The probable force was intolerance—pairs of ravens repel one another. Although this was part of the explanation, it could not be the only one. In some other areas, raven nests are also regularly spaced, but they are more densely packed. It was unlikely that ravens of northern Germany were more aggressive and intolerant than other ravens, such as the Italian ravens on the island of Stromboli, where ten pairs nest in an area of only 12 square kilometers. To try to understand what provides the repellent force between nests, and what increases and reduces that force, requires a close look at the behavior of ravens subjected to a variety of conditions.

We hopped into a station wagon and drove over almost entirely flat countryside to the location indicated by the first black dot on the map. Broad fields were either recently plowed or just sprouting fresh new grass or fall-planted winter grain. The fields were broken up by "knicks"—long straight berms of earth and rock discarded from the fields covered with oak shrub, small white-flowering hawthorne and plum trees, hazelnuts, elderberry, ash, and other bushes. Except for a few trees left standing, these knicks were cut back regularly to make the brush sprout more profusely; a tangle of bushes and blackberry brambles then engulfed the berms. Hedgehogs, hares and cottontails, thrushes, nightingales, and other wildlife lived in these knicks and undoubtedly found them convenient corridors between the patches of forest that covered a mere 5.5 percent of the land. Country roads led to and through all of the forest patches, and almost all nests we would see would be within sight of, or a short walking distance to, our car.

We stopped to visit the local head forester, Christian von Buchwaldt, who led us to a recent new raven nest in a stand of mighty 190- to 200-year-old beeches. Buchwaldt told us these would not be cut down. I had never before heard a forester make such a claim for a managed forest. The beeches and oaks, trees some three to five feet thick, towered above us as we walked through the forest on the spring's blooming carpet of white anemones and golden buttercups. In the wetter places grew giant alders, ash, carpinus trees, and birch. Patches of old forest trees alternated with clearings of planted young spruces and low brush, where wild boars had dug in the sod and elk and roe deer had browsed.

The trees had not yet unfurled their leaves, although throughout this warm day the beeches were uncurling their long brown buds, unfolding the tightly packed leaves like pale green butterfly wings. Sunshine dabbled the limbs and trunks of green-algae-tinged beech trees and the brown layer of last year's leaves and beechnut hulls on the ground. For the moment, it was still possible to scan far into the forest for the nests of ravens, mouse buzzards, and goshawks.

Buchwaldt proudly showed us a rare black stork. Its red bill and red face were visible as the bird sat on a large, white-spattered nest in an oak. The black stork was at the time represented by only five breeding pairs in the whole region of Schleswig-Holstein.

Everywhere, the forest rang with the songs of chaffinches, coal and blue tits, nuthatches, winter wren, starlings, and thrushes. The bright, flashing yellow wings of a citron butterfly, flying quickly and erratically, caught my eye.

The sweetest songs and sights were, of course, the ravens. As soon as we came near a nest, the pair rose into the air and flew high above the forest, calling continuously. Unlike other birds under similar circumstances, the pairs flew in tight formation, sometimes with their wing-tips almost touching each other. In New England, the birds at each nest make a great variety of calls, but these ravens made *krrok, krrok* calls and almost nothing else.

The beech and oak trees were giants compared to those in Maine. Smooth round boles reached limbless for at least eighty feet into the sky, with another ten to thirty feet of thick forks and branches into which the seemingly inaccessible raven nests were set. As we readied to climb the first one, I looked up, doubting the tree could be climbed.

Jörg threaded the end of a thin monofilament fishing line through the hole drilled into the end of an arrow. Volker then walked away with the arrow while Thomas unwound the spool of monofilament line. When 100 to 150 feet of line had been unwound, Volker came back while Thomas pulled the line back hand-over-hand to lay it in loose long loops onto a blanket spread out at his feet. With the line in a heap of loose coils on a smooth surface where it would not get entangled, he picked up his bow and shot the arrow with the monofilament line up and over one of the hefty branches in the tree's crown.

The arrow proceeded along the intended trajectory, then fell back toward earth on the other side of the limb, pulling the monofilament line behind it. Thomas motioned for me to remove the arrow and attach a thicker nylon line to the monofilament line. Then he pulled the monofilament line along the path where the arrow had flown, to be replaced with the stronger line that would have been too heavy to shoot across. That line would now be strong enough to pull up a half-inch braided hemp rope. After Jörg pulled the thick rope across, Thomas attached a rope ladder to it, and we pulled the ladder up. Since the sixty-foot ladder was not long enough to reach the limbs, we attached a second rope ladder of equal length onto the bottom of the first. Pulling-up continued until the top of the first ladder reached the limb over which the rope was looped, and Jörg then pulled the rope taut and tied it securely around a tree trunk. The rope ladder in place, either Thomas or Volker climbing up, carrying small bags containing rings and ringing pliers on their belts.

Last year, with a Danish friend, Hans Christensen, Thomas banded 400 raven nestlings in Holstein and neighboring Denmark. Every year he does about 250 trees. Volker had done over 2,000.

After watching the two men climb several trees, I was most impressed. They walked up the ladder like walking up a flight of stairs. I tried as best as I could and failed. All my previous experience made me hang on and hug the tree or pull up with my arms, which had the unnerving effect of flipping the ladder with my feet on it, to the horizontal. With great mental effort (due to the stress) to visualize cause and effect, I overcame some of my previous training effect, and put more of my weight to my feet. I got high into the tree, but when panic reactions overrode my concentration, I again tried to rely on arm strength. Volker and Thomas no longer needed to think about what they were doing. To them it was by now all automatic and unconscious.

Once up the swaying rope ladder, Volker and Thomas would swing up onto the limbs, then almost leap from limb to limb to the nest still farther up. There they straddled the large limbs that held the

giant nests. Sitting at the edge of a nest, as if on some great upholstered chair, they reached in and pulled out feathered birds, then
attached the plastic leg rings to each of the one to five nestlings, and
ambled back down the rope ladder to enter data into a notebook for
later computer logging.

The plastic leg bands that Thomas had made for the 1994 crop of
young were bright orange for easy future identification in the field.
The birds of any one nest were banded with rings bearing the same
number or letter. The individuals of any one nest were distinguished
from one another by a white stripe, solid or broken, that was either at
the top or the bottom of the ring. Thus, every individual was uniquely
tagged and its origin coded. To them, each bird could now be associated with a specific nest, much like to a raven the ability to have (arbitrary) sensation of taste, sight, or smell might tag discrete experiences
that are useful to remember.

While either Volker or Thomas was up in the tree attaching the
leg bands to the baby ravens, Jörg and I wound up the monofilament
line and twine and waited, lying on our backs in the soft dry beech
leaves for ten to fifteen minutes. After the ringer came back down, we
untied the rope from the tree, let down the ladder, and rolled it and
the rope up to be carried to the car for the drive to the next raven tree,
or to search for another nest at a suspected nest site.

It was 8:55 P.M. as Volker stepped off the rope ladder and back
onto the ground from the fourteenth and last tree of the day. It was
getting dark, and the redbreasts sang as the full moon was coming up.
After we had rolled up the ladders for the last time, Volker walked
briskly back to the car in time to listen to the latest soccer reports. By
dawn, he and the crew would be off again to other nests.

With rigorous statistical procedures, Thomas had analyzed the distribution pattern of the raven nests on his map and proved that nests were
indeed dispersed. The raven had always been thought of as highly territorial. These results by themselves confirmed that supposition. They
also suggested that the habitat was "full," and that no new nest sites
could be expected. But the pattern of nest sites had changed over the

years, and had surprisingly revealed something new that ultimately relates to the raven's perception of their world.

Previous data from amateur ornithologists who had tracked the location of raven nests since their recovery in northern Germany in the 1960s showed that the nests were already dispersed, even when there were many fewer birds, well before the present nest density of 2.3 pairs per 100 square kilometers was achieved. Absolute size of territory had changed, but the tendency for nests to be located away from neighbors was preserved. This meant that territory size of ravens is not a given. Ravens do not exclude others at a set distance from the nest. Barring a sampling problem, territory size seemed to have changed through time. Territories were large at first, then in later years they apparently had shrunk. Surely, the ravens hadn't changed. Their reactions to each other had.

Another pattern in the black dots on the map was that very few new ones had appeared in recent years. Each nest was still producing three to five young each spring, more than doubling the whole population of these very long-lived birds every single year. Why did the number of breeders stay constant? Johannes Goethe, working in the Mecklenburg area of northeastern Germany, and Derek Ratcliffe in Britain had found the same patterns of dispersed nest spacing, with at first a steady increase and then a limit or ceiling in the number of breeding pairs. What was happening to all the birds? They seemed to vanish.

Only isolated individuals of the 1,860 nestlings that Thomas and Volker had so far banded found breeding territories within his study area. Thomas had documented five individuals that settled down to breed, but only at ages seven, five, four, four, and four years, even though ravens can breed at the age of three years. These results showed that breeding in ravens is a privilege of the select few, and that privilege is not easily acquired.

Derek Ratcliffe had drawn data from the efforts of hundreds of eager cliff climbers and amateur British ornithologists who had for decades ascended to the nests of ravens. He had at his disposal an impressive data set on the population ecology of ravens in Britain.

These data allowed him to assess how raven territoriality related to different land uses, the available food supply, and to raven persecution past and present. Ratcliffe determined that the key to nest spacing is primarily food supply. Where food is scarce, territories are large, and where there is much food, along seacoasts and in areas of intensive sheep pasturage, territories become smaller. In the Shetland Islands, for example, twelve active nests were found within 5 kilometers of a large garbage dump (Ewins, Dymond, and Marquiss, 1986). Ratcliffe concluded that under conditions of copious food supply, ravens become extremely tolerant of the close proximity of neighbors, and territoriality virtually breaks down.

Many corvids are mutually attracted to one another, and clumped nest spacing and even dense colonies is typical in species such as rooks, jackdaws, and pinion jays. Ravens are almost unique among corvids, and more like raptors in their generally wide nest dispersal and strong territoriality. With many birds, territoriality is considered the result of an inflexible and innate antagonism to neighbors that *functions* in monopolizing an adequate food supply, but is not directly related to it, so that territory size remains constant despite food supply. Is the raven different? Is the raven's intolerance a function of its stomach contents? We don't know. We've seen the result—the black dots on the map, but we've seldom seen the dramas that transpire to achieve the specific nest spacings.

I suspect that neither Grünkorn's nor Ratcliffe's results can be explained by one cause alone. Low food supply probably reduces the level of tolerance and causes nest dispersion, while increasing time of residency (as neighbors get to know one another) may increase tolerance, instead tending to reduce nest dispersion.

We know that the ravens' adult occupancy in their territory is not all-or-nothing. For several years, I have seen a pair spend weeks near a cliff less than a mile from my house in Vermont. They roost there at night, and in the spring start to bring sticks for a nest foundation, then leave for a month or more. Goliath and his eventual mate White-feather, a wild raven, who was a sometime resident of my aviary, were absent from their territory for a two-week stretch after breeding in the summer of 1996, and for months during the following years. After finding a mate, ravens apparently try to settle down, but whether they are successful in becoming permanent residents and consistently raising young is another matter. It may take a long time to get established, and even once a site is chosen, the birds do not necessarily stay there all the time. "Vagrant" and "resident" are not absolute categories. Individual differences, or the birds' finely attuned responses to conditions, could determine just how resident or vagrant they are.

EIGHT

To Catch and Track a Raven

In the literature of ravens as well as in common usage, a group of ravens feeding together is almost invariably referred to as a "flock." Unfortunately, this is an assumption. A flock is a group that has a membership. The term "flock" could apply to a group of geese, starlings, crows, or ravens that flies together toward a common destination or that stays together for some other common purpose. Many birds stay in cohesive flocks for long portions of their lives, feeding together, flying together, and sleeping together. Their behavior contrasts hugely from that of others who may be found in equally large numbers at a certain time in a certain place. For example, hundreds of ravens could be at a carcass because that is the only one around. Do all the individuals come and go independently? After watching marked individual

Two ravens rolling in the snow.

ravens at baits in crowds I have the impression that they are no more flock members than diners in a city restaurant are necessarily members of a cohesive group. Nevertheless, raven individuals might possibly maintain a group membership in some way that was invisible to me. There was only one way to find out: track them.

I had a second reason for wanting to track ravens from one feeding crowd. Early in my raven studies, I noted that many birds in a feeding crowd would remove great quantities of meat from a carcass to hide it at a distance in the snow, as if for later use. Yet, as soon as a carcass was stripped of meat, all the birds seemed to vanish. Did they leave the area because their memory was so poor that they could not remember where they had hidden their stash? Did they stay to feed on hidden caches, but remain hidden so that I missed them? Tracking several birds from a feeding crowd would help answer these questions as well.

Kristin Schaumburg, a student at Sterling College, had come to me earlier in the winter wanting to work on ravens for college credit, since each student at the college was required to do a project "in the field." She was willing to work for free, she had a car, and the thought of radio-tracking ravens appealed to her.

Delia Kaye and Ted Knight, two former University of Vermont students, were also interested in the outdoors and wanted to take a break from making a living. That was enough person-power to do a project. I had secured time off by doubling my teaching duties for a semester, and so our merry crew of four gathered at the cabin. Our plan was to capture and radio-tag ten ravens from one feeding crowd on a carcass. The birds would then be released back into the remainder of the feeding group at the same carcass. About two days later, we would remove the food. Might all the birds then soar off together and stay together as a group?

First, we had to capture the ravens. The event was scheduled for Sunday morning, January 6. All day January 5, we worked to get ready for the big raven roundup. Wolfe Wagman, another friend and former University of Vermont student, worked with us to fix holes in the big chicken-wire raven trap. I tested the door of the trap by pulling the wire from the newly renovated and camouflaged spruce blind nearby. In the previous weeks, I had already lured a big group of

ravens to feed inside the open trap. Four carcasses still remained inside
to be eaten, and we added some fresh pork lung for a final irresistible
inducement. We were ready! We finished the preparations as it was
just getting dark, and went back to the cabin. Ten more student rein-
forcements came that evening from the University of Vermont to join
us. All were ready for a big capture.

The alarm clock rudely clanged next to my ear at 6 A.M. in the
darkness. A heavy rain pummeled the roof, but there was nothing else
to do except hang tough. Four of us rolled out of our beds, stumbled
down the trail, and dove into the dark spruce blind. There, lying on
our backs, we felt the drops of ice-cold, snow-filtered water dripping
off the soaked fir boughs—*splat, splat, splat*—onto our faces. We
lasted for two hours.

Only two birds showed. One perched for a while directly over us,
giving voice to a repertoire of sounds that reminded me of gurgling-
dripping water. Later, another flew over. Its wing-beats were irregular,
not the usual slow, steady, businesslike *swish, swish, swish*. What had
happened to the crowd? The raven crowds' sudden absence continued
for the next week, the week after that, and still two more weeks.

Ravens again came into the trap on February 4 after a fresh snowstorm. In preparation for another roundup attempt, we built a large fire
under the stars, slapped big steaks stuffed with garlic onto the grill,
and put our arms around each other. The snow had come! Everything
was right. This time we would catch all the ravens we wanted, and all
we needed was ten. We would then get answers to our questions.

The capture was not successful until February 10, 1992. We
caught ten birds, and using Superglue and dental floss, attached a radio
to the tail-feathers of each one. We released them at the bait close to
the capture site where a large crowd had continued to feed. We then
removed the bait, and soon had results from the radio-tracking coming
in. No birds were hanging about in the woods near the bait. Nor did
they gather up to leave all together as a group. Two birds apparently
left the county immediately. As we found over the next three months,
none of the ten were regularly sleeping or foraging together. They dispersed all over the place.

We worked in shifts. One of us drove at night, taking radio bearings trying to detect birds at sleeping roosts. Another drove daily along one of three fifty- to sixty-mile road loops, stopping about every two miles to see if we could get any radio signals. Usually, we got none. It was tedious work. To find any birds, we had to enlarge our search far beyond the local area. We were searching over a 1,500-square-mile area, driving more than a hundred miles per day.

One night around 9:00 P.M., Ted came back with word that he'd again got a radio fix on Number 9680 roosting down near the lake. She was not sleeping with any of our other marked birds. It was a beautiful cold clear night, with the moon glow making the snow a milky bluish white—a good night to take a hike to find out if she was with the others or alone. Snow crystals crunched underfoot as we descended the trail from the cabin. We soon entered a thick, dark spruce bog where the wires from the earphones and the rods of the antenna kept snagging on branches. We lost our sense of direction, but whenever we put on the earphones, we could reconnect using the loud metronomic *click, click, click* of the birds' radio. We went ever onward into the swamp, toward a dense stand of white pine trees. We approached the likely spot cautiously, stopped, looked up, and suddenly heard a wild clatter of powerful wing-beats crashing through the branches. More than twenty huge black shapes scattered in all directions, vanishing into the night. We had our answer: Number 9680 did not sleep alone, even though she was not with any of the ten other ravens with which she had been feeding when we caught her.

The next day, February 11, Kristin, who tracked over toward Augusta, got no signals all the way there; but while coming through the Belgrade Lakes region on the way back, she got Number 9680 again—the very same bird we had traced near Webb Lake the previous night. Before breakfast that morning, 9680 had also circled over us at a bait that we inspected some thirty miles from where Kristin later found her. She had traveled all that distance, despite the plentiful food that was locally available. Number 8510 was also near Belgrade along with 9680, then he flew on to the Dryden dump, where Number 8300 had been all day.

It was my turn for the night run that evening. It was snowing when I started out in my four-wheel drive. The snow was falling so thickly that my headlights shone up against an almost solid wall of white. I could just barely make out the white lines—the snowbanks on each side—with my peripheral vision. I tried to take readings from all the high elevations, which meant driving on narrow logging roads at times. They all had solid walls of dark forest on both sides, but they were well plowed to allow for the passage of trucks in the daytime.

My first contact was with bird Number 8510, at McGrath Hill Road near Wilton. We had routinely found this bird at this spot at night. Some birds had roosted in one location only for two or three days, then were gone. This bird came back to the same spot each night for weeks. As we would later discover, it had taken a mate, and it stayed in the area at least until spring but did not nest there. It slept there only with its mate. Going on to East Dixfield and past Carthage, I found nothing. At one spot along the lake near Weld, I picked up a weak signal from Number 9239 that I later localized to Center Hill: I'd located my second bird, and my loop of fifty-nine miles was nearly complete. It had been a good night. I had collected two data points to plug into the graphs, ultimately showing these birds were apparently not organized into groups, at this time, under these conditions.

Two years later, in 1994, I organized the next obvious project with Ted, Delia, and a new recruit, Eileen Connor. We would tackle the challenge of studying the "territorial" residents. I had again taken on a double teaching load the fall semester in order to have the winter semester off for this project. As before, the first priority was to catch the birds—this time the territorial adults. We could not use our walk-in trap because by now the territorial birds were all wise to it. We had to capture individual birds that I knew. The rub was that they probably knew me.

The walk-in trap being useless, we had to use leg-hold traps, a task I didn't relish. From a local trapper, I got a large selection of them. I judged a Number 1 would be small enough not to injure a raven's leg, but to be safe I hacksawed off one of the two springs, halving the power. I weakened the remaining spring by heating the trap in the woodstove among red-hot coals. We then duct-taped pipe insulation around each of the two jaws of the trap. When the trap was not in use, the jaws were held slightly open with a laundry pin to prevent the insulation from gradually depressing and losing its cushioning effect. I tested one of our altered traps by snapping it shut onto my fingers. It caused no undue damage. I hoped the traps would still be strong enough to hold a raven by a leg or foot.

Setting a trap presumes you know how to depress the spring. As you bend the jaws over to engage the trigger on the pan, you've got to bend the metal tag holding the trigger just right so that the added pipe insulation that you've got to press down does not cause too much tension on the trigger. You've got to be able to set it so that a light tread will trip it. Setting your trap this way in the cabin is one thing. Out in the field, it's quite something else. It has to be set in the snow. In the field, the sun causes melt in the day, which then freezes at night. Once ice gets on the trap, it becomes inoperative, so when a trap is set in open snow, as ours would be, the metal had to be kept free of snow. There can be nothing under the pan, a place where snow could lodge easily. To mitigate against that happening, I put wax paper over the pan in a shape that just fit in the opened jaws. I tried to keep the wax paper dry by sprinkling barley chaff over it and the trap. I then sprinkled snow to hide the dry chaff. The chain attaching the trap to a small tree was also hidden under the snow. Setting the trap was still the easiest part. The next step for ensuring trapping success was knowing where to put down the set.

During the previous ten years, we had found seven adjacent nests of ravens. This was an ideal situation, because we could now distinguish, at least in theory, if the birds defended a large territory or only their nest site itself. If the latter was true, then several pairs might show up at a single large food bonanza like a large cow carcass. But providing a large bait could attract hundreds of birds and was not the way to go. We'd catch mostly vagrants, and vagrants in a trap at any one site would tip off residents to stay clear of that site. To increase the chances of catching only territorial birds, we had to discourage vagrants. The best solution might be to use very small baits, and place them with a trap very close to nest sites. Therefore, we had provided no baits all winter long, so that few vagrants would be in the vicinity when we started trying to trap residents.

I set out baits one weekend in late February, just before the nesting season began, when both members of the resident pair should be near the nest site. The female, who after the eggs are laid is fed entirely by the male, might still be captured because she would still be feeding herself. I hoped to get both members of each pair gradually used to feeding at one spot, then set the trap there. But first I had to find an irresistible bait. I had located a dead horse at a local farm and provided chunks of horse meat half-buried in the snow. Then I added peanuts, bits of leftover pancake, and corn chips.

After the first week, there had been raven tracks at only four of the dozen baits I put out. I learned that details matter. At one bait, a raven had fed on a piece of meat sticking out of the snow. I then strewed a handful of peanuts all around. Two days later, there were again raven tracks all around—but the raven no longer went to the meat it had fed on previously. The tracks showed that the bird had paced about a good deal, and it had taken all but two of the peanuts—and the two it had left were those that I had placed onto the meat that it had fed from previously! I obliterated the raven's tracks every day, to see where new tracks might be put down, to determine what the bird might be up to next.

On March 2, six days into the raven-catching project, it was still not yet time to set the first trap. Increasingly, I'd started to consider my study of the ravens to be a miniature military campaign. I was the general, plotting strategy. The ravens were the wily foe. The situation resembled what the Chinese philosopher Lao-tzu in the text of Tao Te Ching described more than 2,500 years ago. He said that the best athlete wants his opponent at his best, and the best general enters the mind of his enemy.

For a practice-capture, I had set a trap just outside the aviary, where a small crowd of ravens had been attracted by my four tame birds at their food. First I covered the trap with a thin sheet of cellophane that fit just inside the opened jaws. I set it on dry wood duff on top of the snow, sprinkled more wood duff all over the trap, then sifted snow over the whole lot to camouflage it all. I placed a small piece of horse meat in the snow—small enough, I thought, that the birds wouldn't get suspicious.

The ravens came to the aviary first thing in the morning, as expected. From a corner of the cabin window, I watched them by the baited trap. Nothing seemed to be happening. After an hour, I knew something was wrong, and I went out to see what it might be. There were many raven tracks by the meat, all right, but none were on the trap. The birds must have noticed that the sifted snow over the trap was different than the rest of the snow, but that was not all that was wrong with this trap set.

Temperatures had been near zero degrees Fahrenheit in the night, and I had not expected any melting and refreezing. Recalling Murphy's Law—if anything can go wrong, it will—I gently poked the top of the snow I had sprinkled over the trap. It was congealed into a crust, so that even if a raven had walked across the trap, it would not have sprung.

The next step was to make false trap sets near the baits to get the residents used to sifted snow and other unfamiliar telltale marks that might make them suspicious. Delia and I set off down the hill under a dark gray morning sky. I didn't like the looks of that sky. If it snowed for only a half hour, we'd have to pull up all our traps. We set them nevertheless. Then it snowed for a half hour, then twenty-four more hours. The blizzard raged for days.

Meanwhile, for three days in a row, one of the traps alongside the aviary had sprung without catching anything. One morning, while I was a half mile away by Alder Stream, I heard a commotion from near the aviary up on the hill. Hearing alarm calls, I rushed back. One of the traps had indeed been sprung. The raven had pulled out. Conclusion? Number 1 traps were too small. I'd have to get all new traps.

As expected, no ravens came by for the next few days, and I feared they would not allow us any more mistakes. While I was practicing with the new traps I had procured, we also felt ever more the pinch of the weather. We had a series of warm, clear days, and that meant the snow was no longer powdery and easy to sprinkle for camouflaging. The soggy snow could be put on only in thick wads, which melted in the sun in the daytime, then froze rock solid at night.

By March 10, we still had not retained a single raven in a trap, not even just for practice near the aviary at the cabin. But the time had come, if not gone, to try to go for the real thing, one of the territorial pairs. In the cool rain, the crust on the deep, soggy snow no longer supported our weight, and we plunged down with each step, even while wearing snowshoes. Nevertheless, Delia and I built two blinds of spruce and fir branches about one hundred yards from each of two bait sites by two nest sites. Here, we'd watch our captures the next day.

We started at 5:00 A.M., ready in the dark for the big event. We had learned a lot in the previous two weeks about what not to do. We figured that we'd finally get it right. By dawn, we were snowshoeing to the first blind, where Delia would take her stand to watch a set of four traps ringed around a piece of meat that had had fresh raven tracks around it the previous afternoon. We put each trap in a scrape in the frost-hardened snow, using dry potting soil as a foundation. Then we sprinkled more dry soil onto the trap itself and put a piece of wax paper over the pan to cover the surface between the open padded jaws. We camouflaged the traps with a dusting of snow dug out from under the crust, and we put snow dusting even where there were no traps. We also hid the chains under snow. We did everything just right.

I wished her good luck and trotted back down the trail to go to my stand at the Weld nest. There, I also found frost-hardened raven tracks from the day before near the piece of meat. A very good sign! The birds had been back, even though we had dug up the meat, disturbed the snow all around, and moved the meat to a small hillock where it was now visible from the blind. The bad sign was that one of the birds flew over making alarm cries just before I finished setting the traps. Knowing the bird would now be suspicious, I aimed to distract it from the traps around the bait by putting two big bright orange cheese puffs onto the snow. When the raven came back, it would see these strange things. Knowing I had come to do something, this raven would associate the orange puffs with me, then avoid them. It then might think I hadn't come to the meat, and would then not be alarmed by the meat, since that had not been changed. Although I then had suspicions how raven might size up a potential meal, I still did not know that their knowledge and expectation of whether friends or enemies had been associated with that food can have a tremendous bearing on whether or not they eat it.

I was in my blind only forty-five minutes when I heard heavy wing-beats. The raven had come back! It made only three or four alarm calls, then quieted down. Five minutes later, it flew down from near the nest in the pines (I could not see the nest itself) to the top of the snow near the bait. It did not go to the bait, though. The raven paced back and forth and all around for some ten yards to either side of the bait. Something, it knew, wasn't quite right here— and it flew off.

A half hour later, near 8:00 A.M., its mate came to the nest, making soft, high-pitched calls. Fifteen minutes later, one of the pair flew down to the snow, and walked up to the bait with little hesitation. My heart began to pound. Really hard. Weeks of frustration, hundreds of miles of driving, days spent on snowshoes walking laboriously through deep, heavy snow, endless planning, endless details—all were coming to fruition. Was *this* the moment? I had watched and baited them for weeks to learn their habits, to get them habituated to minor disturbances, so that they would not notice a hidden trap buried in the snow. And here was a raven, miraculously walking directly to a piece of meat with four traps set all around it.

The bird stopped, looked cautiously all around, and advanced again. Then I saw it pecking—it was feeding at the meat! It fed, and fed some more, and more. Amazing! I had set the four traps, I thought, on hair triggers. It was at least fifty to sixty yards from my blind to the bait, and I had only a very narrow field of vision through all the many trees of the forest through the spruce boughs of my blind. Slowly, ever so slowly, deep within my spruce blind I raised my binoculars—*rrack rrack rrack*—instantly, the bird erupted in alarm calls and flew off. It had likely seen only a glint of light off the binocular lenses. It could not have seen me.

About fifteen minutes later, a raven was back, again feeding at the same spot. After about five minutes, the raven flew up to the nest, possibly to feed its mate. It again made the high-pitched, soft mewing calls. Ten minutes later, one bird left the nest, making soft *gro* calls as it flew off. They always announce both their comings and goings with at least some kind of call. The female, who stayed at the nest, did her knocking call once.

By 9:15, a raven was again back at the bait. It fed leisurely for several minutes, then flew off. I was dumbfounded, and I was shivering from the cold and the excitement. It was the third time the bird had come this morning. The raven *had* to get caught on the third time, I thought. Wrong.

Some minutes after the raven left, presumably satiated, I heard very excited, sharp, high, quick caws. Two crows came flying over. The crows saw meat, turned around, flew back, and immediately landed, making the same excited caws. Would the ravens now chase them off? I waited—in less than two minutes one of the crows already started to feed. It was caught within seconds. The just-trapped crow was by no means silent, but its mate, hopping frantically in the branches twenty to forty feet away, was the more hysterical-sounding individual. I waited a minute or two to see if there might be a response from the ravens. I could see or hear none, so I came out of my blind and released the crow unharmed.

As I left, snowshoeing back out of the woods, the two crows circled over me and continued to caw excitedly. I did not expect the ravens to be back for a while. For once, I was correct. No more tracks were seen near the bait here again, even though the pair nested in the pine tree just above it. I never did figure out how the raven avoided the trap. I never caught either of the pair, only two more crows.

The next day, March 13, I trapped at a new spot where ravens had never seen me at a bait. It was a quarter mile from a nest and it was also a 1.5-mile snowshoe hike from the road, where I had seldom ventured. The trap was set at a carcass. This time, in only one hour I had caught a bird! It was silent, but as with the crows, its mate called loudly and angrily near me as I took the unharmed bird out of the trap.

That same day this bird—the only resident we would catch— flew off with a radio attached to his tail feathers, beaming pulses at a radio frequency of 14,8130 hertz, or "8130" for short, our name for this bird from now on. Our study could now begin, although on a much more modest basis than we had planned. Most of the work had already been done.

Number 8130 seemed calm throughout the whole handling procedure. Since his feathers had become soaked, he was clumsy in flight when we released him up near the cabin at dusk. He would have to dry off before flying back to the nest. At first, he stayed until after dark in the Alder Stream valley just below the cabin. Wondering if he would travel at night, I got up near midnight and made another radio check. He was gone, but there was a very faint signal from the nest direction. So he *had* flown in the night. After this, Ted, Delia, and Eileen tracked him every day from dawn till dark. In the first two weeks of tracking, he spent most of his time alternating between the nest site and a cow carcass that a coyote trapper had dropped in the woods.

I returned to the nest on March 25 to determine if the female was, as I presumed, incubating. It seemed time for her to incubate, because I had seen the two birds approach the nest, with one carrying nest lining, on March 12, the day before I captured him. There were no birds near the nest this time, and none flew off the nest when I banged on the tree to try to flush off the incubating female. Was the female an exceptionally tight sitter? Was the nest abandoned due to my previous presence here, or due to stress associated with my tagging one of the birds? The nest was about ninety feet up in a giant, thick white pine tree that was nearly limbless up to sixty feet. I did not dare risk my neck climbing this very difficult tree to check the nest contents. To find out if the nest was still active, I would have to watch it.

I made myself comfortable about fifty yards from the nest, hiding under a dense young fir tree next to another thick pine. My attention was fully engaged as I tried to detect a sound or a flash of black wings.

Then a miracle happened. The radio signal got louder. 8130 was approaching. I heard the beautiful, clear, bell-like xylophone knocking of a female raven. Within seconds, I heard it again, closer. Then I heard it again and again. There was no doubt, both he and the female were approaching the nest from the northwest. Looking through the fir branches in the direction of the calls and radio signal, I saw black specks approaching. Not one, or two, but *three!*

I could easily conjure up a rationale (however true or false it may be) to account for a third bird in a territory. Perhaps it was a male sneaking in to mate with the female. Perhaps it was a neighbor trying to damage or destroy the nest to enlarge its own territory. But whatever reason I could come up with, it invariably had to do with the third bird being up to no good, and the pair doing their best to evict him or her.

One glance at these birds, however, told me that I was seeing something entirely different. The three birds flew calmly, wing-tip to wing-tip. As they came closer, I heard the soft *gro* calls that signified trust and friendship. These three birds were friends, and the idea that a territorial pair of ravens had friends outside the pair-bond that they tolerated, if not invited near the nest, seemed extraordinary.

I watched spellbound as the three, now chatting softly among themselves, all flew into the nest tree. Two of the three birds landed directly by the nest, while the third flew near it, then veered off and calmly flew back in the same direction from which they had all come. The pair stayed at the nest for only a minute, chatting constantly in low tones. I heard the begging call that females make when wanting to be fed near the nest. When the two left the nest, they first perched in a nearby pine for another minute, and then flew off in the same direction the third bird had just flown.

I wanted to build a blind right then and there and keep the nest under long-term observation, but other approaches were needed. My staying there would merely increase the chances of disturbing the birds and possibly disrupting their breeding effort. I would have to forgo further surveillance until much later, when disturbance would less likely disrupt the breeding.

We left the nest alone, but we frequently found our bird at a trapper's cow carcass about two miles from the nest. When we covered this carcass up with brush to see where else 8130 might forage, he usually disappeared from radio contact. Judging that incubation must surely be in progress, I visited the nest again on March 29. When I came close, 8130 was in the pines near his nest, and he made alarm calls and left. As before, no bird flew off the nest when I hit the nest tree with a heavy stick. Odd, I thought. Where is his mate?

Again, I hunkered down in my hiding place under the fir tree next to the thick pine. Again, about half an hour later I heard the beautiful, clear, xylophone-like female calls from the northwest, and they were repeated, coming closer every few seconds. As before, *three* ravens came to the nest as a group, flying wing-tip to wing-tip. All three flew directly to the nest or within several feet of it. I again heard a female's food-beg. Within several seconds, one bird left and flew back in the same direction that they had all come. Just as before. The pair remained at the nest for a minute, making soft conversational comfort sounds, then they perched in the pines by the nest. Unfortunately, this time they discovered me, and both made alarm calls and left. When I returned a week later, the nest had been abandoned.

Number 8130 had become our focus. Our very reason for living at the cabin revolved around our day-to-day surveillance of this one bird. Our conversations revolved around him—a bird only I had actually seen, though he revealed himself to all of us via the radio transmitter attached to his tail. Later, throughout April and early May, whenever I saw him he was with only *one* other bird. Who had the third bird been? A dozen tagged birds might have given us a clue. We could only guess, and we inevitably came back to the apparent breakup of the nesting cycle just at the moment when the eggs were about to be laid.

As we loitered in the evening around the cast iron stove while waiting for fresh bread for supper, our conversation and thoughts about the nest failure ran to the fanciful and jokingly anthropomorphic.

"What do you suppose happened?" someone would ask, and we'd launch into imaginative discussions of various scenarios. For example:

Here she is one night, we reasoned, as always expecting him to come to her side by the nest to feed her and sleep beside her. He doesn't show up. She becomes hungry and anxious. Finally, long after midnight he comes back, weak and disheveled, with that strange extra tail feather and a weird shoulder patch. "You been *where?*" she'd ask. He would try to explain. And she'd say, "Yeah—*sure!*" So now he's all weak and lazy the next morning, just when he is supposed to mate because it's now time to lay the eggs. But he isn't interested in sex just yet, nor does he feed her. She eventually sees another male, a neighbor. She displays to him. Feeling acknowledged, he follows her, hopeful. Maybe it is a male from a neighboring pair they both knew already. Maybe it's an unattached male. In either case, it is a bird that her partner knows, too. So he tolerates him, if she does. The funny part is that this scenario could in principle actually be close to the truth.

Some anecdotes mean more than others, because they are decisive and leave less room for interpretation. The three birds at one nest were not an interpretation. During February, I had at three different times seen groups or a group of five ravens flying together. Later in March, I had twice seen three flying together on the other side of the lake. I had always favored the most probable explanation, that these were groups of juvenile vagrants. Now I wasn't so sure. More was going on here: Territorial ravens were not always reflexively aggressive to all others outside their pair. But when and why not? I needed to see details of bonding behavior, something that my tame birds might teach me.

NINE

Partnerships and Social Webs

MY OBSERVATIONS OF THE THREE birds spurred my desire to understand more about bonding behavior. The best and perhaps only thing I could do for the time being was to engage in long-term observations of the birds I knew well, Goliath and his groupmates and subsequent aviary birds. I already knew my birds were highly attentive to, apparently curious about, and at times aggressive toward strange ravens coming near. Another juvenile raven in the vicinity caused them to watch, listen, and call, and the newcomer usually came down to the aviary. All sorts of social behavior transpired, the nature or purpose of which I didn't understand. Perhaps the newcomers were interested in the food in the aviary, and the aviary birds were trying to defend their food; but I'm not convinced that this obvious explanation is the right one. The same behavior occurred when no food was visible. When I put food outside in the snow, the newcomers commonly ignored

Raven friends.

it, yet interacted with the birds through the wire. They postured at each other as though trying to show off and go through vocal repertoires.

Wandering juveniles are a gregarious lot. By the dozens, and sometimes hundreds, they fly high in the sky riding the updrafts, barrel-roll back down, catch air, and spiral back up. They sky-dance in pairs and sometimes triples during these raven flight "parties." I have tried to see if the two are stable pairs, or if third birds cut in to exchange partners, but the action is usually too fast and too far away for me to be able to tell. Similarly, apparent pairs come to baits in the morning from communal roosts of juveniles. Could juveniles form pairs? I was skeptical, even though field data had long suggested the possibility. But Goliath and his mates gave me fresh insight on pairing.

Except for mouth color, ravens a month out of the nest look almost identical to adult ravens. Adults can begin to breed at the age of three years, but sometimes do not until seven. Males and females have unisex garb. We can distinguish them by certain displays and vocalizations, but neither shows up reliably until the birds are well into their second year. Yet in the aviary, distinct pairs form before one year, and paired birds show distinctive displays toward each other (Lorenz, 1940; Gwinner, 1964).

On February 8, at nearly one year of age, Lefty bowed, fluffed, fanned her tail, and did her first very clumsy rendition of the female-specific knocking display to Fuzz. That is how I knew she was a female. Did he? I needed to wonder, because I previously had one pair of ravens, two very large dominant brothers, who bonded and built a perfect nest, then fought viciously when one tried to mate with the other at egg-laying time, i.e., just when the nest was finished. From this I deduced that the condition of the nest, rather than the condition of the female, is likely the cue for mating.

The brother's friendship survived despite numerous vicious fights at attempted matings, and despite availability of females. Lefty, in the next few days, repeated the display numerous times as if practicing or communicating to no one in particular. I never heard Houdi, the other, slightly smaller raven I suspected of being female, give this female-specific call until about a year later.

As I write, in June 1998, I have another group of six juveniles that are a year out of the nest. Of these six, a male, Blue has been paired with a female, Red, since last January, when they were only seven months out of the nest. (The birds are named for the color of the ring that marks them.) They perch next to each other regularly, preen each other, and feed together. Blue is the most dominant bird of the group. He and Red eventually fed amicably side by side, but he attacks all birds except Red. For all appearances, they are a couple, yet if they were in the wild they would be wandering juveniles, and they would not breed for two or more years.

We don't know why young ravens socialize, but by doing so they enter a social web that could have relevance beyond information exchange about food. Socializing reduces aggression; to be familiar to the others at food is to be tolerated. In addition, in order to meet potential partners, the birds have to show themselves, getting exposure to others. Partnerships may lead to mating. In addition, being in a feeding aggregation could also be a large part of making social contacts, leading to alliances. If so, the birds must recognize individuals. It has often been assumed that some birds can do this, but we don't know how. I hoped eventually to observe individual differences in my charges, to maybe get hints if and how they might do so themselves.

When Goliath and my other three young ravens were just weeks out of the nest, they followed me when I went for daily rambles in the woods. By the end of July that year, the four were showing signs of independence. I feared that, like other young ravens I had reared, they were nearly self-sufficient and would soon leave me. Since I wanted to study pairing (and more), I retained them in my aviary complex.

By the first fall, some pairings already started to develop. Commonly, one bird perched close to another and bent its head down to solicit preening. Such solicitations indicated who wanted attention from whom, and whether or not they received preening indicated who was interested in providing satisfaction. At first, every bird preened and was preened by every other one. The preening partners gradually became more specific, until they eventually became nearly exclusive. For example, Goliath formed a preening partnership with Lefty, and

Fuzz preened exclusively with Houdi, and vice versa. They were "going steady." (I presume that it was only random chance that Goliath ended up with Lefty, his sibling.) Might these pairs also become nesting partners? To find out, I had to separate each pair into their own aviary.

In January 1995, I let Fuzz and Houdi remain in the complex in Vermont and took Goliath and Lefty to my aviary complex in Maine. The pair were housed for the most part in their own side aviary, separated from the main aviary that was then holding a crowd of wild-captured juveniles. On April 26, I let the pair into the main aviary with the crowd. Would Goliath be beaten up when entering their domain? Not at all! Goliath seemed eager to join them, then immediately attacked *all* the top males, especially G67, the most dominant one there. Within a day, G67 and all others yielded to Goliath. He chased them and did the macho display until all stopped fighting back. He had concentrated his attention on those who resisted. He seldom approached females, with the exception of one bird with a white feather on her left wing. He initially greeted her with the same macho display he used when approaching males. Being a female, she responded not with fighting, fleeing, or submissive gestures, but with her knocking display that says, "I'm a powerful female." He then stepped aside. I could thus sometimes tell the sexes apart by behavior, and maybe that is how they do it as well.

Afterward, I tried to chase Goliath and Lefty back into their own private side aviary. They knew the door, because they had once been free to use the whole aviary, but they both seemed determined to stay in the big aviary. Goliath looked at me pointedly, then pecked branches angrily and erected his macho-pose ears. He had never reacted to me that way before, and I had always shooed him back easily. Lefty, who also refused to go back, found a hole in the wire and escaped the aviary. She sat above in a birch tree for a while, made some *kek-kek* calls, and flew off down the valley toward Alder Stream. While all this was going on, "Whitefeather" flew into Goliath's aviary. Once there, she immediately made long, undulating, territorial calls I'd never heard from her before, and in *his* aviary the calls were especially surprising. Normally, these calls don't attract other birds, but Goliath reacted instantly by flying to her, back into his aviary. I expected him to attack the interloper. Instead, he sidled up to her as she gave a bow-ing, fuzzy head display and repeated a long series of the three repeti-tive knocking female-indicating calls. She meekly stepped away. He then did a little macho display, but without much bluster. Then all was calm. No fight. She later did long series of the quick repetitive knocks. I felt a match had been made, and closed the door to "their" aviary. There was no point in me trying to get Goliath and Lefty together again. Fate had decided otherwise, and I was curious to see what would happen next.

When Lefty came back from her visit to the wild ravens feeding at a calf carcass a mile away, she paid no attention at all to Goliath. She perched above that part of the aviary housing the strangers, totally ignoring the side aviary with Goliath and Whitefeather. So much for what I presumed had been a love match. If Goliath and Lefty previ-ously had a relationship, it was clearly only one of convenience.

It was time for me to release all the wild birds other than Goliath and Whitefeather. I opened the door of the big complex, leaving a raccoon carcass as bait just outside the door. Some of the birds stepped out to feed, but they did not fly away. Instead, they returned to the aviary after their meal. In the evening, others tried repeatedly to get back inside, walking paths in the snow along the wire. It took two days before they stayed out.

Weeks later, Goliath, my tame, hand-reared bird, and Whitefeather, the wild raven, were still paired like a loving couple. They perched next to each other, preened each other, and cooed softly to each other. Lefty had departed for good. I wondered if the Goliath-Whitefeather pairing was also one of convenience and might easily be disrupted when an opportunity arose. There was only one way to find out: pro-vide that opportunity. On July 22, I brought Goliath and Whitefeather to Vermont, to release them into the home aviary with the Fuzz and Houdi pair. I would there keep the four together for four months.

I released Goliath first. Hopping out of his crate into his home of months earlier, he seemed fully relaxed. He shook himself. Fuzz made the *rap-rap-rap* calls, on high alert. He had been subordinate to

Goliath before. In Goliath's absence, he had not been challenged. Now he proceeded to make a macho display, sidling up to Goliath. Houdi stayed up out of the way in the sleeping shed, giving some knocking calls. Fuzz stood tall, bill-snapped, flashed the white nictitating membranes of his eyes, and strutted. Goliath did not respond. Fuzz then attacked, and Goliath assumed the "crouching pineapple" submissive display, which I had never seen him do before. The contest was over, just like that.

A raven pair, alert to a neighbor's calls.

Goliath, previously dominant over Fuzz, was now submissive to him. Established within the first few minutes of the encounter, this relationship was maintained from then on. Their tense social interaction deescalated in the next forty-five minutes, but the result was clear. Houdi had stayed totally out of the strictly male-male encounter. Then I released Whitefeather. Would she be challenged and beaten by the resident female, just as Goliath has been beaten by the resident male? That's what I would have bet, especially since Whitefeather, a wild bird, would now be in a strange new place without the home court advantage.

As Whitefeather flew out of her crate onto a perch, Goliath flew to her and macho displayed. She started knocking. Very soon they had bill-to-bill contact and briefly preened each other. Fuzz made a new ringing-rasping call I had never heard him give before. It was long and oft-repeated. Houdi, hearing Whitefeather's knocking, responded for her own part with paroxysms of knocking. Neither female engaged in body contact like the males. For about ten minutes, both females knocked almost continuously, then their vocal duel tapered off. Whitefeather had the last word, and Houdi became quiet.

After that day, Goliath continued to be subservient to Fuzz, and Houdi, in turn, always yielded to Whitefeather. Houdi, who before had routinely engaged in knocking displays every day, no longer knocked at all. Not once. Indeed, the noisy Whitefeather seemed to inhibit all of her vocalizations. In turn, the previously vociferous Goliath became totally silent as well, and he always yielded physically to the now-expressive Fuzz whenever there was food.

Although the dominance had been reversed, the pairing or preening bonds had not. The females continued to preen only their males. Whitefeather, the dominant female, stayed with Goliath, her now subordinate male. Similarly, Fuzz, the most dominant male, remained true in his preening to Houdi, the now-silent subordinate female. Each day, the members of the respective pairs sat long hours side by side with each other.

Curiously, Goliath, who until then had been the only one to greet me routinely by doing the crouching, tail-quivering, entreating display, no longer gave me this display. Conversely, Fuzz now greeted me with the standard male macho display used to impress potential mates and male rivals.

The dominance hierarchy amongst the birds was most readily seen at contested resources, principally perches, food, and the bathing bowl. When it was time for a bath (in the large wheelbarrow), the most dominant bathed first and the most subordinate last. But there were constant attempts at line-crashing, which were met with much shoving.

July 22, 1995, was a scorching hot day with temperatures near 90 degrees Fahrenheit. After I poured the water into the tub, Fuzz came down and walked right in. He ducked and splashed until all of his feathers were comically matted into tufts, streaming water. Bath done, he flew up to the roost to preen, and only then did Goliath cautiously approach the water. Fuzz would normally preen for a half hour, but seeing Goliath approach the water to enjoy a bath, he interrupted his preening and flew back down, still totally soaked, to take another leisurely bath. When finished, he went back up to his perch to resume his preening. Having barely started, he again stopped to chase Goliath, who had attempted to reach the water once more. This sequence was repeated eight separate times before Fuzz finally allowed Goliath to bathe, or he allowed himself to preen. I could see no practical point to his costly exertions to exclude Goliath from a bath, except maybe to show that he could do it. There are people like that, but I was surprised to see ravens being so unreasonable.

When Goliath was finally allowed to bathe, he was continually interrupted by Whitefeather. She didn't actually get into the tub with him, but came near, obviously intending to join him in the two-foot-diameter wheelbarrow bowl that was surely big enough for two. Goliath invariably interrupted his bath, however, just to chase his partner off. Long-paired partners become progressively more tolerant of each other, bathing and feeding together.

Houdi tried to be next. She was immediately chased off by White-feather even after Whitefeather was done bathing. During the time Whitefeather was trying to prevent Houdi from bathing, she kept giving her female power-call, the knocking. Houdi, in trying to sneak in to take a bath, kept looking all around at the others, as if to make sure they were preoccupied before she dared a try. They always saw her. Eventually, she gave up.

After four months I separated one pair from the other, not only to terminate the fidelity test, but also to permit each pair to eventually build a nest and raise young, since both pairs could not be expected to nest in the same aviary. I kept Fuzz and Houdi in their home aviary in Vermont, and brought Goliath and Whitefeather back to their aviary at my cabin and study site in Maine, where they had originally met.

On November 21, 1995, the day I separated the pairs, Houdi underwent a remarkable change. She regained her voice—and with a vengeance. She did the knocking calls continuously for hours on end. When the pairs had been together, Fuzz had bitten off the end of Goliath's tongue. I had thought that possibly Goliath's silence had been due to that injury. It wasn't. As soon as he was away from Fuzz, he regained the full range of his voice with no change at all in his calls that I could tell, tongue or no tongue. He also resumed his macho displays.

Despite their separation, the stories of Fuzz and Houdi, and Goliath and Whitefeather, later became closely intertwined. By Christmas, Fuzz and Houdi were still the epitome of a loving couple, preening each other almost every minute of the day. He preened her about twice as often as she him. To solicit preening, she pursued him, perched right next to him, and bent her head over as he folded over one feather after another with his bill . After a while, he reached under her throat, and she arched her head up over her back for him to preen her throat, feather by feather. I've never seen a single parasite on either one, although they did remove rare specks of dirt from the feathers. Fuzz did not seem to solicit preening from Houdi. Instead, he did many of his macho displays accompanied by *oo-oo* calls, while seeming to push her. Houdi responded with her knocking display, and occasionally placed her foot on his back. He sometimes sidled up to her, reached out with his nearest foot, and grasped hers firmly for many seconds at a time. I could not tell what the foot-grasping meant, except that sometimes it seemed as if it might be used as a mild restraint.

Houdi was playful. She slid and rolled down a two-foot-high mound of snow on her back. At first, the snow-sliding seemed almost like an afterthought as she simply allowed herself to topple over while Fuzz was preening her on top of the mound. After that, she did it several more times on her own.

On January 28, Houdi brought two sticks into the shed where I hoped they would nest. Fuzz took an immediate interest and followed her closely. When she brought the third stick up into the shed, he picked one up, too, and followed her to the spot where she had deposited hers. Was she telling him something?

In the next two weeks, Fuzz alone carried sticks. He not only carried them to the designated spot for the nest, he also held them fast with his bill and rapidly vibrated his bill with the stick in it. Such vibrations normally function to anchor sticks to each other and to the nest substrate, but their nest-building seemed more like play. At first, as many sticks fell out as were carried up. Houdi accompanied Fuzz and watched him work; she carried no sticks herself. They were almost three years old. Were they too young to seriously build a nest? Perhaps all the neural connections necessary for nesting behavior had not yet been made or activated.

Almost a month later, on February 22, the nest still consisted only of a loose bundle of about ten sticks. Were they missing something? I put an old sheepskin with long tufts of wool into the aviary. Both looked at this material intently. He then grabbed a stick lying nearby and took it to the nest. She took a piece of sheep wool, shredded it, then carried it to the nest as well. Now he got busy! Again, as if he'd suddenly caught on from her cue, he began carrying more and more sticks in, even two at a time. In eighteen minutes, he had carried in eight sticks and she had brought up three loads of fur. Both had also played with more sticks and fur. When I gave them dead grass, she picked it up and carried it in one large load. After that one billful, she picked up a huge ash stick, one that was an inch thick and two and a half feet long. She debarked the stick, held it in one foot, and let it dangle below her before eventually carrying it to the nest. Then she was done. She sat on the nest edge, preened, and did many knocking calls. Did her actions tell him to get busy? It seemed so, because he started carrying in as many as three sticks at a time, although still ignoring the wool and grass.

Only one day later, Fuzz had finished the stick "basket." I gave them piles of inner bark from a dead poplar. This they both carried into the nest, accompanying each other. Both were now almost equally engaged carrying load after load of bark fibers and wool. By the next day, Houdi was making three trips with nest lining to the nest to his one. He still picked up sticks, dropped them, picked them up again, as though forgetting what they were for. It was only the lining that the nest then needed. The next day, February 27, the nest seemed ready. It had a beautiful, deep, soft lining. The whole structure measured thirty inches across and twenty inches tall, and the nest cup was twelve inches across and six inches deep. I expected eggs any day.

On March 8, I saw Houdi sit down in the empty nest, turning around in it several times, but sitting quietly most of the time without moving. After one of her nest-sitting sessions, from 8:19 to 8:35 A.M., I checked the nest, but it was empty. Fuzz still accompanied her every move, perching next to her even when she was in the nest. At 8:54, she again went into the nest to sit. After she sat for six minutes, she hopped out and he took his turn for four minutes. At 9:27 A.M., he hopped out and she sat quietly in the nest again for seven minutes, while he did his macho displays. He then took another turn for four minutes, while she did her knocking displays. She took two more turns of nest-sitting in the still-empty nest for nine minutes and three minutes, respectively. Several times, he laboriously picked hair off a moose hide I'd provided them, then spat it out and discarded it. Meanwhile, several times when she hopped off the nest onto the ground, he walked up to her to hold her foot, as she lay down on her side beside him. Were these preliminaries to sex?

The next morning as a wild raven flew over, Fuzz went into paroxysms of deep angry *quork*ing. Later, he perched in uncharacteristic silence with a partially fuzzy head. At frequent intervals, his whole body shook and vibrated for a few seconds at a time. I had never seen him do that before, and would never see it again. He was not shivering from cold, because it was a warm day. I sensed a great excitement in him. Had he taken a cue, from the finished nest, that it was now time to mate?

Two mornings later, on March 10, 1996, I found the first egg in the nest. Houdi sat on it most of the day. Fuzz perched stolidly nearby for hours, making *oo-oo* calls, bill-snaps, macho displays, and undulating territorial calls.

Already at dawn the next morning, he began making long, deep, rasping, territorial *quorks* telling all neighbors to keep away. She was on the nest. He maintained high alert, looking around in all directions. At 6:30 A.M., she hopped out of the nest, stretched her right wing, and hopped over to him on his perch. He greeted her with his macho bill-clicking display, and she responded with her knocking display. He then waddled sideways to her along the perch and reached out with his right foot to grasp her left. She skooched down, arching her back and rotating her tail, and he hopped onto her back. He lost his balance, and she then flew to the ground. He followed. She crouched again, with rapid tail vibrations, and he responded with a similar crouch and tail vibration with his wings held widely to the side and his bill up at an angle of about forty degrees. He did this for only a second or two, then hopped onto her back again, perching directly on her, while maneuvering backward to make cloacal contact. In two to three seconds, it was all over, and she walked over to the calf carcass to feed. He flew back up to his perch. I quickly took the opportunity to check the nest. No second egg yet.

At 7:10 A.M., she hopped out of the nest and he hopped on, remaining there for forty-three minutes before she came back and perched next to him at the edge of the nest, making soft, throaty calls. He sat deep in the nest mold, unmoving, but later got off the nest while she was down below and feeding. He'd been off for only two minutes when she went up to the nest to resume incubation and/or to lay the second egg.

As I tried to approach the nest at 9:40 A.M. to chase her off to check for a second egg, he got defensive for the first time. Indeed, his anger was downright intimidating. He puffed himself out, made the rapid *kek-kek-kek* alarm calls, and pounded the wood right next to me so hard that splinters flew. She, in turn, remained silent and refused to come off the nest. Was he feeling his testosterone? During breeding time, the testes of male birds increase in size more than thirty-fold. Testosterone supposedly increases aggression. But Houdi, presumably with little testosterone, was soon as aggressive toward me as Fuzz. She was so aggressive, I needed to resort to trickery to find out what was in the nest.

"Want to do an experiment?" I asked a friend who was visiting.
"Sure!"

"Okay—go up to the nest and see if there are two eggs in it."

She did. And there were.

Neither bird tried to repel her. Neither made alarm calls. She had not interacted with the birds before, and should have been more threatening to them than I, who had raised them from chicks and whom they had trusted until that very moment. I wasn't much surprised, through, because ravens are consistently surprising.

I needed more reliable access to the nest in the coming days and weeks, and since visitors were not available every day, I had to improvise. For the most part, I learned that I could hold them off by carrying something bulky in my hand. A paper bag would do. They never took my advances personally. Afterward, without holding a bulky object, I often sat for hours within five feet of them and the nest, and they both ignored me—provided I did not move toward it. My slightest movement towards the nest set them off in vehement threats that would make my heart pound, because I knew I'd get hammered. I expected fully that Houdi would fly in my face and peck hard.

Five eggs were laid, about twenty-five hours apart from each other. I marked each one with a small piece of numbered duct tape to record the sequence in the clutch.

Matings occurred almost precisely at the same time and in the same way at dawn every morning, and stopped on the morning that the fifth egg was laid. It was always the same ritual. Each mating was initiated when the female hopped off the nest to stretch on her first morning break. He sidled up to her and did his macho display. She responded with her knocking display. He then drooped his wings and vibrated his tail. She then went down, crouching in the same display, and he hopped on to mate.

Ravens have always been assumed to be monogamous and keep their mates for life. But as with almost everything about them, it depends strictly on circumstances. There is ample documentation in the literature showing that if one member of the pair gets killed, a replacement may appear within a day. Recently John Marzluff, who has continued working with corvids in the western United States, documented extra-pair copulations in ravens inhabiting the open country in Idaho. He saw four males other than the mate (who was wing-tagged) copulate with a female, and he saw extra-pair copulations at the two nests under observations. John said "the sneakers wait until the exact moment when the territorial males leaves, which is very infrequently. They streamed in as soon as the male left. They would do this only at egg-laying time—exactly when it would result in fertilization." Interestingly, these copulations differed from the ones I saw of the legitimate pairs. These extra-pair copulations occurred at any time, not just at dawn, and they occurred when the female was sitting and remained sitting on the nest, not off it. It seems as though the secondary males know what is going on, not only with regard to the precise reproductive status of a female and/or her nest other than theirs, as well as her mate's mate-guarding behavior.

The eggs were a little more pear-shaped than a hen's egg, and smaller than a "Grade A Large." The first one laid, which was infertile, was so thickly blotched with grayish black that the greenish-blue background color was almost totally obscured, especially at the thick end. Numbers two and three had distinct blotches, and the fourth one had even fewer. The fifth egg was the most distinctly colored. It was light blue-green with only tiny pinpoints of dark markings, mostly at the narrow end.

I removed each egg after it was laid, intending to return them all after egg-laying ceased, to see if the birds would continue to lay as some birds do, until they have a full nest. They stopped after laying five eggs, even though never finding more than two in the nest at once. When I returned all the four eggs I had previously removed, Houdi hopped onto the nest with five eggs, even though a minute before she'd been sitting on just one, without exhibiting a flicker of surprise. From that point on, after all the eggs were laid, only she incubated. When she did leave the nest temporarily, she hopped about frantically and then went right back on. She rarely fed on her own. Instead, Fuzz brought almost all her food, feeding her just outside the nest or in it. He nudged her bill if she hesitated to take his offerings. Sometimes, she begged like a nestling.

The third and fourth eggs hatched on April 4, twenty-one days from laying. Two chicks had hatched two days earlier, which suggested that the eggs I had removed from the nest for three and four days had already been partially incubated on March 16 when they were returned to the nest.

Fuzz and Houdi's behavior toward me changed markedly after the young began to hatch; their aggression increased to new levels. As the pair became ever more defensive, I dreaded retrieving the young to weigh them every two days. At birth, the chicks weighed 25 grams each. By April 13, the four young each weighed 350 to 360 grams. Their eyes started to open, and although they were still naked, black stubble of new feathers was visible under the skin. In fourteen days, they increased their body weight by twenty-four times, to 600 grams. They became feathered out by thirty-two days, but did not leave the nest until the age of forty-eight days.

The just-hatched young were vocal from their day of birth, but they did not reach up their heads to gape and make their high-pitched, hoarse, begging cries unless the parents perching on the nest edge made a soft little *grr* sound to them. At two weeks, they begged in response to almost any disturbance, no longer requiring the parental *grr* calls to get them to gape.

Fuzz was a devoted father. He and Houdi fed meat to their young directly after tearing it from animal carcasses and storing it temporarily in their throats. When they spat the meat up, I often saw saliva coating it. I wondered if the saliva contained digestive enzymes. Both parents ate only after the young would eat no more. Mice were a preferred baby food. The ravens would tear them into small bits, carefully draping the entrails over a nearby limb to discard them or eat them themselves. When Houdi sat on the young and Fuzz came with a throat full of meat, he first gave her some of it as she stood up, then he fed the rest to the young himself. When I gave them several mice at once, they grabbed as many as I had and cached them, retrieving and tearing them up afterward.

The foods they ate were not necessarily the same as those they fed the young. The young got only meat. When available, the pair ate or cached berries and butter, but gave neither of these foods to the young. The parents ate the biggest, roughest pieces of meat with bones and skin after giving the young the juiciest, choicest parts.

Throughout all of incubation and until the young were feathered, when she stopped brooding them, Houdi never once bathed. Her wings, breast, tail, and especially her feet became badly soiled. Fuzz bathed in water when available and in snow. She only preened.

The time had finally come when I felt secure enough to open the aviary, because I did not believe they would leave their young. On April 20, I opened the door. They were not in a rush to leave, but Fuzz eventually did fly out. The neighboring pair of resident ravens that nested a mile down the road were there, as if on cue, to meet him. The neighboring male instantly lit into Fuzz. A huge chase followed, and that was the last I ever saw of Fuzz.

Houdi hid in the nest shed, silent while her mate was being attacked outside. It was only hours later that she ventured back down onto the floor of the aviary to feed, and the next day she went to the calf I had provided outside the aviary.

For the most part, she remained silent. Eberhard Gwinner, who studied ravens in Germany in the 1950s and 1960s, had reported that a raven missing its mate called the equivalent of its name by using a vocalization only the missing bird had used before. Not Houdi. Instead, her most common call was her own name—the knocking that said "*Notice!* It's me. I'm female." Only one time, at around 10:30 the next morning, did she finally go through a repertoire of other calls: rasping honks, undulating, territorial *quork*s, "dog whines," and *rap-rap-rap*s. Three times I saw ravens fly by at great height. She became silent and perched immobile in a pine tree. On another occasion, when one raven came close, she flew into the aviary and hid next to the nest in the covered shed, cautiously peeking out. She did not venture out until the bird had left.

Near 6:00 the next morning, she did the female knocking calls for minutes at a time, and no other calls. To me, these calls made sense if she thought Fuzz was searching for her. However, they attracted not him, but another wild male, who was not unwelcome. Houdi stayed in the trees outside the aviary, next to this new male, who I presume was the same neighbor who had chased her mate off. The male did his macho displays of bowing, blinking, and groaning to her, and she reciprocated by continuing her knocking display. It looked like a typical boy-meets-girl situation.

After a half hour, the male left and Houdi resumed tearing meat from the calf I had provided in the nearby woods. She fed the chicks as though nothing had happened. I sat near the calf and she came right next to me, exhibiting no fear or apprehension. What a joy to see her flying freely over the woods where the poplars were all tasseled out, the willows starting to bloom, and the first warblers, the solitary vireo, and the winter wren were singing! She brought huge mouthfuls of meat, one after another, to the seemingly insatiable young in her nest in the aviary. Often, she dropped a big load of meat near the nest, then made as many as four smaller trips to feed the young from that one load.

Once a pair of ravens suddenly arrived, both making long, nasal honks. Houdi was at the carcass tearing off meat. She instantly aborted her work and silently, without a load of meat, retreated into the aviary to hide in the nest shed. She didn't make a peep and looked totally intimidated. Meanwhile, the pair helped themselves to "her" calf. Undoubtedly, this never would have happened if she had a mate. A mate in this situation would have been in an unstoppable fury. I felt sorry for her, and handed her a hard-boiled egg from my bedroom window that opens directly to the nest shed. She immediately fed it to her young.

Almost twenty minutes after the pair had left, she still remained silent and in hiding. Sometimes, she stood stock-still, head straining down horizontally to peek under the roof overhang. Was she wondering, "Are they still there?" Did she know that this time, her paramour's mate had come along and that the female bird would not be so tolerant toward her as he was?

To encourage her, to let her know the coast was clear, I went to the carcass and called. She would know that no wild raven would be near with me there, and indeed, she instantly flew out to me. She wasn't there more than a few minutes before a lone raven appeared. She stayed put, made a variety of calls, then she flew after it! The two settled next to each other. I presumed it was the male again, alone this time. He displayed his macho poses to her, and she acted coy. The party didn't last long, though, as a second raven appeared within several minutes. Undoubtedly the first bird's mate, who chased after Houdi, but seeing me, then swerved and flew off directly to the Swamp Road nest where I had suspected the pair was from, as they had often fed near my house in winter.

It was 8:00 A.M., and I had to leave just as things were getting to be fun. I could not come back to resume watching until 3:00 P.M., when I saw the neighboring pair arrive again. Only one of the pair gave Houdi chase. The chaser had a fluffed-out fuzzy head, and Houdi depressed her feathers, making herself look thin. The duo went out of sight. There was a long silence. I became worried. Thirty-five minutes later, I heard raven calls and saw two flying over. One came down. It was Houdi. She resumed feeding at the carcass and I felt greatly relieved, being confident now that together we'd raise these young, despite the pesky female neighbor.

After this, the male visited often. Always he and Houdi displayed to each other. He flew after or toward her at times, and sometimes she flew over to meet him. Never once was there any aggression between them.

On May 1, I was almost startled to hear her start the day with a "shout" of vocalizations; fifteen minutes or more of knocking calls, kek-kek-kek calls (no predator near!), dog whines, long, undulating, territorial quorks, upward-inflected, rasping calls, and many others. She was animated as never before, looking in all directions, flying over the fields and forest, then flying back to land on the trees by the aviary and her nest. She was full of force and power, not shy and tentative anymore. Something was up.

The next morning, in contrast, she made no calls at all! It was very unusual, yet I didn't anticipate anything. Too bad, as it turned out.

I left in the afternoon on my long-planned trip to Maine, first leaving her a calf. When I returned on May 5, Houdi was gone! The young were very hungry. A friend told me later that he had seen Houdi in the late afternoon on May 3, flying along the Hinesburg Road (close to where the other pair have their nest). She had been identified by the two missing wing feathers on her right wing. I never saw her again, and I was left with her four young.

• • •

My trip to Maine had been to check on Goliath and Whitefeather. After I had provided them nest material on March 6, they had completed their nest in just two days! On April 23, when their young were ten days old, I watched them as Whitefeather made squeaking noises to Goliath, and they held each other's bills as in a long kiss. For long periods, they sat close enough together to be almost touching. They paid so much attention to each other that I wondered if they had forgotten their young. I tried to remain hidden and watched from a shed through a one-way mirror. In the two hours and fifteen minutes that I watched, he brooded the young for one thirty-seven-minute session, she for four sessions lasting five, three, five, and fourteen minutes. He fed the young three times, each time afterward eating their feces. She fed them only once, and she ate the mice I had brought. He was totally quiet the whole time. She was very noisy, making three long sessions of knocking. At one point, she made a long series of rasping high honks, rapid nasal honks, *caulk-caulk-caulk*s, *rap-rap-rap*s, undulating territorial *quork*s, and ripping, rasping *quork*s. All these calls were apparently directed to distant neighbors to the north and southeast, because I could hear the ravens at three distant nests in those directions respond with their own calls.

As described in more detail elsewhere (Chapter 12), I eventually pawned off the four abandoned Fuzz-Houdi offspring on Goliath and Whitefeather. Then I opened their aviary as well. I expected things to proceed differently. First of all, Whitefeather had been captured from the wild. She was in home territory. She had seen the hills, the forest, and the lay of the land from the air. She might still know the neighbors. She would not get lost or disoriented if chased by them. Goliath had been free in these forests as well for a time after he had fledged. It was with confidence that I tore out one side of the aviary, so that the birds would have easy and direct flight access to their nest, filled now with six young rather than just two.

Within minutes, both adults were out. After tarrying briefly in the large maple trees next to the aviary, they were, like the Vermont pair, also immediately met by a neighboring pair of ravens that magically appeared out of nowhere. Here there were no chases. Instead, all four birds perched together, making mutual dominance displays, but I could not determine the outcome, because all four soon left together as a group, vocalizing energetically. Their voices soon were lost in the distance. Through my binoculars, I saw them disappear, tiny black specks high in the sky to the north. All morning they were gone. I heard not a sound. All afternoon they were gone as well. I fed the young myself, thinking I now had created six orphans. I fretted all night, but at dawn the next day I heard two ravens calling by the aviary. Goliath and Whitefeather were back! Their absence the first day was to be their only one for the next two months.

Some of my greatest contentment that summer came from sitting on a log next to the aviary and watching both birds fly in and out, caring for their young. Goliath readily came up to me, taking tidbits from my hand. Whitefeather, being wild-born, never came that close, but she didn't act alarmed, and came within twenty feet of me. Unlike Fuzz and Houdi, they at no time scolded me, even when I climbed right up to the nest to look in at their young.

When the young fledged, they behaved like young ravens do normally. At first they hung out near the nest, tearing and picking apart everything within reach. Gradually, they expanded their excursions, following one or the other or both parents, sometimes going alone. By July, they were taking independent excursions, and soon after that I did not see them anymore.

Several days after the young left, Goliath and Whitefeather left as well, but following a month's vacation from the heavy duties of child rearing, they were back again. As before, they raucously sounded off with a barrage of calls every dawn. I expected them to adopt this hill as their territory and to settle here for good.

Goliath and Whitefeather, living free around my cabin, had given me a chance to view ravens' interactions with neighbors at close range, and I could routinely watch them. One day after Whitefeather flew from the aviary, I left a chunk of meat on the path in the woods about two hundred yards north of the aviary, where neither she nor her mate immediately found it. I saw a pair of ravens flying at great altitude, and these newcomers tumbled down out of the sky toward the meat, landing in a red maple tree, where they called several times. White-feather left her perch by the aviary in a big hurry and went directly toward them. I then saw all three within a few feet of each other, at least one bird doing macho poses and Whitefeather knocking to them. Seeing me, the two newcomers left, and she returned to the aviary. A few days later, Goliath and Whitefeather were flying with another raven, and she took the lead in what seemed to be escorting it away; she flew closely behind the other raven, who was yelling a lot. Goliath lagged behind. Again and again, the stranger flew back to the clearing by the cabin, and eventually a second one joined it.

It never failed when a raven called from the distance that Goliath and Whitefeather instantly tensed up and responded, trying to "out-shout" strangers with territorial advertisement calls. One spring morning, with Goliath beside me, I again heard raven calls in the nearby forest. Goliath, who usually went ballistic at the sound of strangers, acted unconcerned this time. Strange, I thought, so I went into the pines to check, and sure enough, it was Whitefeather. Later, as both were at the aviary, I saw a raven flying at a distance of perhaps a mile. Both were after it immediately. A few hours later, I was up in a spruce and saw a raven flying silently in the valley heading toward Lake Webb. Whitefeather and Goliath erupted in rasping caws, and Whitefeather took off after it, and the two flew together. Not knowing otherwise, one might well have thought Whitefeather and the stranger were a mated pair. The "pair" flew together for miles—to Gammon Ridge, Mount Blue, and then down to Alder Stream. Goliath stayed near me and the nest and made undulating territorial *quork*s. After five minutes, Whitefeather came back alone, and both she and Goliath gave the undulating territorial *quork*s. An hour or so later, a pair flew over the aviary. First Whitefeather took off after them, then Goliath. This time, there was an aggressive chase. I heard staccato chase calls and the begging of the chased birds. This time, it was not play. They were gone more than five minutes, Goliath return-ing first. Both did deep, rasping caws and *rap-rap-rap* calls.

A pair or pairs came by repeatedly, and each time, they seemed to be chased off with greater vehemence. Goliath took off after them even when Whitefeather was far away foraging, an indication to me that strangers are apparently identified even from a distance.

The strangers that kept coming by were not merely interested in food. I once had a deer carcass plainly visible in the field when two ravens came flying very high in tight formation from the east, the direction of the nearest nesting pair at Hills Pond. They flew over and beyond the carcass and zoomed straight down toward Goliath and Whitefeather's nest instead. Goliath and Whitefeather both took chase. It was one of the most vehement chases I'd ever seen, lasting for at least twenty minutes. I heard honks, knocking, begging, *rap-rap-raps*, undulating *quorks*, *caulks*, rasping *quorks*, agitation calls. All the while, the intruding pair kept trying to come back to the aviary, but Goliath and Whitefeather continued to hold them off. Did the intruders want to destroy the nest and/or young?

After a while, the neighbors apparently realized that the new pair on the hill, Goliath and Whitefeather, could not be driven off. They stopped coming, but shouting matches between them continued daily. Whenever one of a neighboring pair made their territorial advertising calls from their nest area, Goliath and Whitefeather faced in their direction and responded with ear-shattering territorial calls of their own. At no time did a raven come by when these calls were given.

In early June, Goliath and Whitefeather led their young all around the neighborhood, visiting the many different carcasses I had spread out. At that time, the young still begged from their parents. A parent would tear off meat, then feed it directly to a nearby begging youngster. Later, the young started tearing off some of their own meat, and the parents began treating their offspring as competition. Goliath, who had been the main provider until then, sometimes even ferociously pecked at them.

On September 17, 1996, shortly after Goliath and Whitefeather returned after a long absence following their raising of young, I heard a raven calling from the area of Hills Pond. Goliath, who was as usual perched on the tall dead birch stub next to us just above the fire pit, turned attentively to look in that direction. Strangely, he was neither excited nor defensive. Minutes later, both Goliath and Whitefeather took off and began soaring. A raven flew up from where we'd heard the calling, and joined them. All three circled amicably together, far above the clearing. They circled close to each other, freely intermingling, so that it was no longer possible for me to distinguish them. After five minutes of socializing, the newcomer broke off and flew back to where it had come from, and Goliath and Whitefeather returned to our hill. Goliath was in a playful mood. On his descent, he did two complete corkscrew tumbles. These were the first such aerobatics I'd seen him do. Then I heard a rush of air as he came down at 45 degrees, suddenly dove straight down, and did three corkscrew twists before landing gracefully beside his mate.

The next morning, the pair was again soaring over the hill with a third raven. A sharp-shinned hawk rose to meet them, diving at one of them. One raven flipped onto its back and extended its feet, and the hawk flew on. The other raven returned to its territory down by Hills Pond. Another pair flew over at great altitude, which Goliath and Whitefeather ignored. In late afternoon, a group of eight ravens came from the north. Goliath and Whitefeather ignored them, too. Did they have specific friends?

On September 28, Goliath was flying with another bird just above him, the two moving in perfect synchrony, a perfect pair. Whitefeather was near me, seemingly unconcerned. The flying pair stayed circling over the valley and the pine-clad hill near the cabin. The only calls I heard during the fifteen minutes that the two flew together were the *caulk* calls, which to me sound relaxed, friendly, and reassuring. I often hear these calls when Goliath and Whitefeather are together alone. After the flight, the third raven flew north, and Goliath came back to the cabin and his mate.

It rained on October 28, but in early afternoon the rain stopped, the fog lifted, and dark clouds drifted from the northwest. The air was clear, and I could see to the tops of all the surrounding hills. Near 2:20 P.M., I suddenly heard the *rap-rap-rap* calls from Goliath and Whitefeather by the aviary. Had they seen a stranger? I walked out of the cabin and faintly heard a raven in the distance to the northeast making the *rap-rap-rap* calls. Goliath launched himself instantly and flew toward it, but then doubled back to land on a maple near the cabin after making a large circle. He continued to look in the direction of the other raven, standing tall, erecting his "ears" and rubbing his bill vigorously on his perch in an exaggerated display of power. In the distance, the raven called again. Again, Goliath took off instantly after it called, then all was quiet. Five minutes later, I heard the *caulk* calls and saw two ravens flying around the hill by the aviary. Was one Goliath? A moment later, I heard the knocking calls. Then I saw four ravens coming up the hill. There were no chases, no antagonistic interactions. The two lead birds were Goliath and Whitefeather. They swung off and landed near their usual perches on the hill, while the other two birds turned around and went back down the hill, toward the east where Goliath had flown, apparently agitated, a few minutes earlier.

Although the data on distribution of nest sites is congruent with the concept of exclusive defended territories, I suspect now that raven society is far more complex. My observations of Goliath and White-feather suggested that adult birds differentiate and discriminate between friends and foes. Since aggressiveness is a function of food supply as well, I wondered if the conflict between vagrants and territorial residents was really between familiar versus unfamiliar birds rather than juveniles versus adults.

I had not answered my earlier question about why there were three birds at the nest on the other side of the lake, but I was no longer surprised that such things could happen. I had learned that raven partnerships are flexible, with sexual activity and jealousies possibly confined to a very narrow time slot at the time of egg-laying.

Many of the behaviors associated with nesting are primed by hormones, but as in all other birds and also in us, the complexity of response indicates that the birds' minds are driven by more than just hormones. The relationships within and between pairs and with other ravens could suggest that these birds evaluate and make choices.

TEN

Pairs as Cooperative Teams and Sharing

Oᴌᴅ ᴘʀᴇᴊᴜᴅɪᴄᴇs ᴀʀᴇ ʜᴀʀᴅ ᴛᴏ change. I had until then long cherished an image of the raven perched on a lofty crag surveying its domain and reflexly chasing out all others that cross its skies. That image is not necessarily a false one, but it is only one dimension of a multidimensional picture.

Goliath and Whitefeather and my other tame ravens had given me glimpses of another dimension to the raven's social life that I had not suspected. I began to suspect that ravens, aside from being cantankerous loners that were intolerant of others, could also form strong attachments to others. Perhaps their superb flight capacities even allowed them to maintain friendly contact with neighbors in a loose confederation, like that more typical in other corvids.

A pair of courting wild ravens within a crowd. The male is striking his "macho" pose. The female (at right) is doing her knocking display, which includes winking with her white nictating membrane.

Ravens do not usually breed until their third or a later year of life, yet already in their first fall, Goliath and many other ravens I have kept began to form strong partnerships. In the field, such partnerships would be difficult to detect, although in crowds of juveniles performing aerial dances, pairs predominate. The problem is that we have no way of knowing if these pairs last for a moment, an hour, a day, or a lifetime. Ravens, like humans, are exceptional animals that may pair up as friends years before breeding. My observations with aviary birds shows that some of the friendships last and others don't. In the wild, adult ravens probably stay paired up the entire year, not just when rearing young. Raven pairs fly around together in the daytime, sleep close to each other at night, regularly communicate with each other, and mutually preen each other.

Could the pairing result in cooperative hunting partnerships? Economics provides a rationale for raven's gregariousness in communal roosts, since the birds who join roosts are led to carcasses that others of that group have found. Are there also economic reasons for both raven subadults and adults to pair up? Might couples work together in mutually productive partnerships, because a couple could get more food than either of the two could get independently?

Wildlife filmmaker Jeff Turner photographed ravens at sea cliff colonies of murres and kittiwakes at Cape Pierce, Alaska, in the spring of 1997, and described seeing a raven "dive like a hawk and hit a kittiwake in the air. White feathers flew! This raven was unable to grab the kittiwake on this occasion, but I saw ravens plucking fresh kittiwakes every day. Usually only one raven dove and hit a kittiwake again and again, eventually forcing it onto the ground when it and others jumped on it and killed it. The ravens often worked in pairs. We also saw them go up to kittiwakes on the nest, grab a wing and yank them off, then they or a partner rushed in and took eggs." John R. Moran, an ornithologist, saw groups of ravens at the same site attacking and killing large gulls and geese.

Hunting behavior of ravens could be more efficient when two or more work together rather than alone, and numerous anecdotes indicate that ravens indeed hunt effectively in teams. Take squirrel hunting, for example. A squirrel on a tree can easily escape almost any pursuer either by running to another side of the tree, or going up or down the tree, depending on the direction of the pursuer. If a mobile hunting partner could position itself in the path of the escaping squirrel, the retreating animal would be cut off from its escape route. Gary Keene, an observer from Maine, saw something resembling this behavior: one raven chasing a gray squirrel across the road while another awaited on the other side.

The most commonly observed raven teamwork has been of ravens taking prey away from predators. Wildlife biologist George Schaller told me of watching raven pairs in Mongolia cooperate in snatching rats from feeding raptors. Similarly, in Yellowstone Park, Ray Paunovich reported seeing a red-tailed hawk with a ground squirrel. Two ravens approached. One distracted the hawk from the front while the other handily snatched the squirrel from behind. Carsten Hinnerichs saw the same maneuver repeated three times in a row in a field near Brück, Germany, where a fox was catching field mice. Terry McEneaney, Yellowstone Park ornithologist, observed two ravens circling an osprey nest where the female osprey was incubating. One raven landed on the nest rim and took a fish, then while the osprey was distracted by this thief, the other raven swooped down and stole an osprey egg.

Vermont naturalist Ted Levin saw a similar type of opportunism in raven pairs raiding the nests of whimbrels on the tundra near Churchill, Manitoba. One raven would flush the bird from the nest while the other snatched eggs. Wisconsin ornithologist and Arctic explorer Ludwig Kumlien saw raven pairs succeed in killing young seals that lay basking near their ice holes. One raven would at first leisurely circle over the seal, then drop down beside the seal's escape hole in the ice. The raven's partner then drove the seal to the hole, where the first raven killed it by pecking it on the head.

David Hatch reported in the *Winnipeg Free Press* (December 19, 1992) watching four common eider ducks on the Hudson Bay near Churchill while he was eating lunch with a group of tourists in a tundra buggy. "Two ravens landed within a few meters of the eiders—the eiders were alarmed but after five minutes settled down, burying their heads in the feathers on their back. Suddenly one raven made four or five large hops toward them—and drove its bill into the eider's eye—

both ravens then continued the relentless attacks on the eider's head and within a minute the eider was dead."

Was planning involved in the teamwork? An anecdote reported to me by Professor Dieter Wallenschläger, a biologist at the University of Potsdam, seems to suggest that it is sometimes possible. Wallenschläger saw a pair of ravens on the Island of Werda in the Baltic Sea attacking a mute swan incubating its eggs. The swan hissed and lunged, but stayed on its eggs. If the swan did not budge from its eggs, the raven's attempt would fail. One of the ravens did something unprecedented. It feigned injury, dragging a wing as plovers do routinely when leading a predator off its nest. The swan lunged after the seemingly crippled raven, and its mate rushed in and took an egg. Feigning injury as such is not remarkable. A great many, if not all, shorebirds and innumerable other ground-nesting birds do it. But in them it is likely an at least partially or largely programmed response. That argument can't be made for the same behavior in ravens, especially when it is used in a totally different context and where it serves a totally different purpose.

In all of the above instances, there is no proof that the ravens anticipated the consequences of their actions and behaved in a conscious way in mutually agreed-upon plans. For the most part, there is no necessity to invoke such a scenario, but neither can it be excluded. The simplest

A raven pair, with the the male (at right) in self-aggrandizing display.

hypothesis is not necessarily the right one. Indeed, there can be considerable debate on what the simplest hypothesis may be. My own bias is towards the idea that the cooperation displayed by these anecdotes does not usually involve planning and forethought, for the simple reason that in most cases planning may not be needed, although that logic is somewhat forced in the examples of the ravens and the swan.

According to some views, the birds are not cooperating if they each seek only to satisfy their own chances of success without a conscious regard for the other. From an ecological-evolutionary perspective, however, the effect is critical and intent is irrelevant, the latter being another question altogether. Cooperation might occur whether it is intended or not. What matters for practical cooperation is payoff. The question boils down to whether just one bird of a partnership feeds on the spoils that the second bird, by its chasing, blocking, or distracting, has helped to get.

To that question we *do* have an answer, and that answer is derived from observations in the aviary. First, I found that the raven partnerships can indeed last years, and as previously supposed mated pairing appears to be almost irreversible. Furthermore, mutual tolerances among partners ensure that food secured or held by one is also available to the other; food is not defended against partners. Therefore, from the standpoint of evolutionary ecology pairs cooperate, while from that of psychology they may not, but could.

I examined the power of pairs in detail in thousands of interactions with my group of six birds, when all were about two years old. In this group, Blue was a male and the undisputed dominant of the group. The birds probably wouldn't nest for another year, but Blue had already paired with a female, Red, a year earlier. All six birds normally got along amicably, but if they were hungry and I gave them a choice hunk of food like a calf haunch or a woodchuck carcass, Blue immediately perched on top of it and began to feed, excluding all others except for his partner. A year earlier, he still had excluded her mildly, but for only a short while. Now, he immediately let her feed beside him, even at the choicest meat. The "rule" was: Blue feeds first and he chases others off. The exception was if the food was feared, in which case he waited until subordinates "tested" the food and started to feed; then he chased them off. His mate, Red, was highly subordinate to almost all the other birds, but she joined him at all times and fed at his side, regardless of the presence of others. He chased off all others that dared to come near him, and since she fed right next to him, she effectively fed unmolested under the umbrella of his protection. When on one occasion I removed him, she received attacks from the others above her in the hierarchy, and she had to contend with them herself. At that point in their relationship, he did not bring her food. By the time they begin to nest, however, he would feed her as he would his nestlings. What one would catch both would eat, and together they would become even better providers for their young and for each other than if they were alone.

ELEVEN

Hunting and Foraging

THE PASSERINE, OR PERCHING BIRDS, the group that ravens belong to, are characterized by species-specific, largely innate, or "hardwired" foraging specializations. Ravens are unique passerines, not only because they are large meat-eaters that lack the physical toolkit of raptors, but also because they are opportunistic generalists that can feed on almost anything from fresh carcasses and the insects feeding on rotten carcasses, to tomatoes, Cheetos, and dog droppings. Even more impressive than their flexibility are the many ingenious techniques they have been observed using to get their food. These methods are not the ways of all ravens, because many raven individuals are unique and any one person observing them can hope to see only a very small portion of the species' amazing repertoire of behaviors.

All of the ravens I have raised from nestlings started out catching crawling insects and other invertebrates (noticeably exclusive of angle

Raven feeding on beef suet.

worms, which they disdain). Then they went on to larger prey, the only apparent requirement being that they could catch it. There are many published anecdotes of the raven's predatory exploits. They include attacks on reindeer (Ostbye, 1969), lambs (Hewson, 1984), and seal pups (Lydersen and Smith). But most of the reported predations are on other birds, including grouse (Allen and Allen, 1986), seabirds, and common pigeons (see Notes and References). The prey is often struck in midair (Elkins, 1964; Schmidt-Koenig and Prinzinger, 1992; Jensen, 1991).

I have never personally observed a raven catch a bird in the wild, but I once surprised Goliath in the woods just as he was plucking a blue jay. There was fresh blood on the jay's feathers. The bird had bled, so Goliath could not have found it already dead. I could not imagine how he might alone have outmaneuvered a jay in the woods. I also found a raven pellet consisting entirely of the fur and bones of a flying squirrel. One single pellet contained skulls of three moles! On another occasion, I found a gray squirrel skin lying on the ground under Goliath's favorite perch. The fresh skin had leg bones still attached and unbroken, and on the right hind leg the skin had a bright red spot, a blood bruise. There was another hematoma on the skin from between the eyes. This squirrel apparently had been pecked on the leg and the head with a sharp object before it had died. It was not a roadkill.

Adam Farrington, a raven-watcher from Poland, Maine, witnessed a raven's attack on a red squirrel. Walking in a pine grove to examine a raven's nest, Adam suddenly saw a raven diving and crashing through the branches in the manner of a goshawk, to pounce on a red squirrel. It knocked the squirrel out of the tree, and pursued. By then, the raven was close to Adam on the ground, and veered off and aborted its attack. Near the same area, Adam also saw a raven apparently clutching a thick tree trunk. The bird had its wings stretched out to the sides around the tree, blocking a tree hole where a squirrel had sought refuge. As long as the pursued animal remained inside the tree hole, it would be safe, because ravens cannot chip solid wood with their bills like woodpeckers. They can, however, use their bills to excavate soil. I received one report of a raven digging up a ground squirrel in Colorado, numerous reports of ravens on Maine islands digging petrels from their nesting burrows, and one report of ravens digging bank swallow nests out of their tunnels in sandbanks along Alaska's Kenai River.

The open country of Yellowstone Park affords good visibility for witnessing ravens' foraging activities. On April 17, 1987, Terry McEneaney saw ravens trout fishing. A pair of ravens caught, killed, and cached a total of twelve cutthroat trout. The two ravens fished by standing on a sandbar of a tributary of the Yellowstone River in the Hayden Valley. When a trout ascended a riffle of the creek, one raven would wade out into the water, grab the fish by its dorsal fin, and pull it onto the sandbar. There it would kill the fish by pounding it with its bill, and fly off with it into the sagebrush to cache it out of sight.

Fish are sometimes caught through the use of intermediaries. Bob Landis, also in Yellowstone Park, filmed a raven catching a cutthroat trout in a somewhat circuitous manner. An otter had caught the trout in the river, and as it took the fish out of the water to eat it, a bald eagle stole it away and perched to start to eat it. A raven, seeing this as its opportunity, came and harassed the eagle by yanking its tail feathers. The startled eagle turned toward the raven, momentarily letting go of the fish, and the raven rushed in and took the trout.

The previous example could show the vagaries and sheer luck involved in trout fishing. However, as is usually the case, the raven made the best of it. Luck is seldom as haphazard as it may seem. It means being at the right place at the right time, and most of all, it means being prepared to take advantage of opportunities.

Sometimes ravens force their luck, as the following observations by Terry Goodhue and John Drury illustrate. The two biologists had much opportunity to watch ravens on Seal Island off the coast of Maine during their years of implementing and studying the resettling of puffins there. The island is barren of trees, about a mile long, and twenty-five miles off shore. It is the home of one pair of ravens, who nest there in the spring on a ledge by a small pond. In early October, migrating flickers regularly come to the island, as well as peregrine falcons who hunt them. John and Terry noted that as soon as a falcon caught a flicker, the ravens approached to try to relieve it of its catch and/or to pick on the remains after the falcon was done feeding. The biologists further saw a refinement of the raven's use of peregrines: The ravens regularly flushed flickers out of hiding on the ground and the peregrines then caught them. Similarly, John Marzluff reports that in Idaho, ravens immediately appear at half of all rabbit kills that golden eagles make, although whether they also flush hiding prey for the eagles is not known.

If ravens quickly find prey killed by a raptor, might they also find roadkills as quickly? My data are inconclusive. I did take notes in June 1995 on a road trip through raven country between Baker, California, through Las Vegas, Nevada, and on to Salt Lake City, Utah. In all of the 960 miles of road-watching, I never once saw a raven picking at a roadkill. I counted twenty-three roadkills in all. Although most of these were smudges of unidentifiable (at 70 to 80 mph) fur that had been caked onto the highway for days or weeks, I did see four fresh snakes, three jackrabbits, one dog, two skunks, one raccoon, one small bird, and several probable ground squirrels. Not one was attended by ravens. The overall density of roadkills was only one per forty-two miles, and considerably less than that before coming out of the desert and into the green fertile Mormon Valley. Overall, it seemed that the Mojave Desert does not support its many ravens on roadkills.

On a trip in late September of the same year to New York City from Burlington, Vermont, a distance of three hundred miles, I also ticked off every roadkill I saw. I counted the following: three cats, fourteen raccoons, one porcupine, eleven skunks, one bird, one red squirrel, two woodchucks, one hundred and seventy-eight gray squirrels, and ninety-six unidentifiable smears. I had never before seen so many road-killed gray squirrels, and even *Newsweek* magazine commented on their large numbers that year in the September 5 issue. I saw one hundred and eighty-five big black birds on the highways and in the air. All of those along the highways were American crows, and twenty of those in the air were turkey vultures. As on most trips to Maine, I did not see a single raven on a roadkill, but most roadkills were well attended by crows.

The last week of August 1996, while traveling to Ann Arbor, Michigan, from Vermont through Quebec and Ontario, and back through Ohio, Pennsylvania, and New York, I saw not one run-over squirrel. I did see two road-killed dogs, a skunk, and about two dozen raccoons. I saw two live ravens, one in Ontario and the other in Adirondack Park in New York State, but neither was on a roadkill.

The ravens and roadkill study has a long way to go, but my initial probe could disprove the commonly held hypothesis that the distribution and abundance of ravens is determined by roadkills. It also suggests (but does not prove) that if most raptor-kills immediately attract a raven or two, then it is because they cue in on the hunter itself, not just the dead animal.

Roadkills are undoubtedly an important food supplement for some individual ravens. There are probably some ravens that go on a roadkill cruise daily, as did one or both ravens of the Hills Pond pair by my cabin in Maine. I used to encounter them regularly on my daily jog.

If ravens cue in on the predations of raptors, they must be keen observers indeed. The dozens of other types of large birds that are not raptors would be a waste of time for a raven to watch, and I'm inclined to believe that the ravens' skill in identifying flying birds of concern to them rivals that of many competent ornithologists. In Baja and Mexico, some ravens even distinguish the zone-tailed hawk, *Buteo albonotatus,* from the turkey vulture, *Cathartes arura.* The zone-tail is a black buteo that appears to mimic the turkey vulture in both plumage and habit of soaring. Soaring zone-tails tilt their wings in a V, and rock from side to side, as turkey vultures do; and their tails when in flight also mimic vultures'. Gary Clowers, a ravenphile from Oregon who migrates to Baja every winter to watch birds, found these hawks flying frequently with groups of vultures. Most potential prey animals are wary of hawks, but they still ignore the vultures. The hawk appears to use the cover of flying with the vultures to hunt. At first, Gary did not pick out the vulture imposters, but he learned to recognize them only after watching ravens. Near their nests with vulnerable young, ravens rose into the air and attacked the zone-tails, ignoring the vultures.

One pair of ravens impressed me with their ornithological skills while I was on a trip to Umiat, a former center for oil exploration on the Alaskan tundra along the Colville River. One of this pair always perched, guarding their young, on a piece of abandoned machinery. On two different days, I saw the bird suddenly come alive and fly off into the distance making the sharp, short *kek-kek* predator alarm calls. Straining my eyes, I saw a small black dot in the sky toward which the raven hurried. With my binoculars, I identified the dot as a golden eagle. The raven presumably sought to drive the eagle away, repeatedly diving down onto it. The chase continued, and soon both were out of sight of my ten-power binoculars. I was surprised that the raven not only saw the eagle at least two miles away, but also could distinguish it from such other large birds as Canada and white-fronted geese, sandhill cranes, glaucous gulls, and various ducks and sandpipers, all of which flew by and prompted no reaction.

Of the various ways in which ravens get food, none has received more attention than their presumed attacks on incapacitated large mammals. These sometimes involve grisly scenes. For example, on February 10, 1985, McEneaney in Yellowstone saw a pair of ravens feeding from a bison, taking the only meat they could get, the eyes. The bison was stuck in the mud on the Madison River near Mount Holmes. Its head was lying on the ground, and its only sign of life were clouds of steam coming from its nostrils. No park official would commit the humane act of shooting the suffering beast, because park policy forbids intervention into the affairs of nature. Due to the park's previous eradication of wolves, thousands of bison and elk had overgrazed the range drastically, altering the whole ecosystem. Now they were starving by the thousands, wandering in search of food and being shot by human hunters. Taking the no-killing policy literally is well-meaning, mindless inflexibility, neither of which ravens can be accused of.

Ravens do perhaps kill dying cattle outside the parks, and that possibility has raised outcries. In the 1980s, two ranchers in northern Arizona reported to the Animal Damage Control Department that ravens were killing their cows, and for several years the Department carried

out an ambitious raven eradication program, paying ranchers cash for
the cattle they claimed the ravens had killed. Similarly, in Germany
there was from 1994 to 1996 a flurry of newspaper articles and television news reports about a gang of fifty or so "killer ravens" that had
invaded the idyllic Schwäbische Alb near the town of Balingen.
"Nature had turned to horror," said the press. A shepherd there, Walter
Rehm, had described how, "disciplined like soldiers," the local raven
gang would "wait for the signal of the raven boss," and when he gave
"his hoarse signal" they would descend like a regiment and fearlessly
attack a victim and "bore into its skull with bills sharp as scalpels" to
kill it and pick out its eyes. Of course, all of this was nonsense. There
are no disciplined raven gangs, no raven bosses that give descend signals, and certainly no stiletto-sharp raven bills. Raven bills cannot pen-
etrate even the skin of a gray squirrel, much less the skin and skull of a
sheep or a calf. Some of my ravens eat roadkill gray squirrel by reaching
in through the mouth. By crushing bones and pulling out meat as they
feed backward, they eventually turn the squirrel inside out, leaving the
skin like an unrolled sock, with not a single hole or tear in it. But only
some of my birds could do this neat trick.

Gruesome newspaper tales of bold killer ravens were illustrated
with clever pictures of flying ravens splashed across the front pages.
The ravens were said to be in "attack flight" with bills wide open—
which is a sign of overheating or fright. The television accounts of the
raven menace were straight out of a Hitchcock movie. I was sent one
gory headline after another. The German Jäger (hunters) soon took up
the cry to defend the poor farmers and the poor threatened animals
from the new menace. In a letter to the editor in local newspapers, the
Jäger offered their invaluable services to maintain the balance of
nature, and to end this awful scourge that was, they claimed, also deci-
mating the songbirds, the rabbits, the partridges, and so on.

Preliminary studies revealed that ravens were indeed near or sometimes even in fields where sheep or cattle were giving birth, which was
proof enough for those wishing for a cash reimbursement for dead cattle. In northern Germany farmers were, unlike in former times, allowing calving to occur out of doors in the winter, and farmers got cash

for dead calves or lambs. Ever more livestock was found in the pastures with their eyes and tongues pecked out, and ravens were as before in attendance. It was an opportunity with something for everyone—the news media, the farmers, the Jäger, the politicians, and even the biologists. When asked to give comments and advice, I said that at least one documentation of a raven killing a cow might be handy. Ravens would feed on afterbirths and on dead or dying lambs. Luckily, our cattle and sheep in New England give birth inside barns, so that ravens cannot be blamed for the innumerable calves that die on their own during birth.

Ravens and other corvid birds have long been persecuted in Germany by the Jäger, a well-organized group that proudly think of themselves as keepers of the natural order. In a disturbed ecosystem that no longer has its natural predators, they do indeed justifiably control deer and boar populations that would otherwise inhibit regeneration of trees, remove underbrush, disturb soil, devastate crops, and affect breeding bird populations. They traditionally also persecute the corvid birds because they are nest predators on other birds. With this latest uproar, the Jäger were clamoring to have the raven taken off the endangered list, so that they could again be shot as in former times, when they were almost totally obliterated from the whole country.

When I visited Germany in 1996, ravens were still getting bad press, and Dieter Wallenschläger, professor of animal ecology and nature conservation at the University of Potsdam, undertook an investigation. We corresponded about the public menace. I personally was dubious of the killer raven stories, because the ravens I was familiar with are cautious even of unattended carcasses. Finding one, they jump up and down in front of it, as if to provoke a reaction. Although I suspect veterinarians have even better means of detecting life, the raven's method neatly discriminates healthy from just barely alive or dead animals. They would not normally touch a calf that moved, although it is possible that if ravens were to feed often enough on dead or nearly dead calves, they might eventually learn to feed on half-dead ones with less hesitation. Seeing a bird in the act of picking an eye from a dying calf, one might well believe the raven was the cause of the calf's death.

Dieter heard I was visiting Berlin, and invited me to come to Potsdam. Not only that, he organized a mini-symposium with the title, roughly translated: "Killer Ravens in Brandenburg—Legend or Occupation of a New Ecological Niche?" I was glad to attend. After the introductions, a farmer, Frau Breme, talked of the results of her recent 1996 questionnaire of raven damages that she had sent out to 141 farmers. She got seventy-one returns. Of those, fifty farmers saw ravens, and 41 percent of those saw *"Veränderungen"* (changes) of animals; i.e., ravens feeding on an animal. In the ensuing discussion of this report, I mentioned the high mortality of calves in New England and the high percentage of *Veränderungen* of these calves by my ravens, despite absolute proof that these ravens had not killed the calves. German nature conservation people agreed that there was so far not one proven case of a raven causing the death of a healthy calf.

The government had by now eventually insisted on proof of cause of death before making cash reimbursements. Veterinarians were consulted to make autopsies. Certain examinations of lungs and eyes of newborn calves could determine if the animal had been able to stand when born. Given the examinations, almost all of the calves *verändert* by ravens had something wrong with them to begin with. They were mostly found to belong to careless farmers who did not take adequate care of their animals. Good farmers didn't have a raven problem. After the government finally insisted on proof, and after none was forthcoming, the raven damage subsidy became unavailable. Money to study ravens evaporated as well.

The ravens' dependence on large animals is central to their biology. When we were hunters, ravens were revered companions who inspired poets and engendered creation myths. The presence of ravens meant large animals were near. They meant meat and merriment. All that changed when we became settled herders. Ravens soon became a suspected destroyer of lambs, and prophets of doom and gloom. They were relentlessly persecuted because they were *associated* with death, although not, as it now seems since scientific study, because they caused it. Ravens' physical power to kill had been overestimated; and the subtleties of their responses, where their real power lies, underestimated.

TWELVE

Adoption

IN MANY SEABIRDS, BATS, AND OTHER colonial species in which parents leave their young among others, there is much opportunity for accidentally feeding others' young instead of one's own. These animals have evolved to be capable of identifying their own offspring in a crowd, and they reserve their precious food offerings strictly for them. I wondered how ravens would react to missing or extra young. Could they count? Otto Koehler and his colleagues at Freiburg University in Germany had concluded that they could count to seven, having trained a raven to retrieve food under one of several covered vessels, after training it to expect the food in only one container that had the correct number of spots on the lid.

An unusual clutch of seven nearly indentically aged young.
The different young have small differences in the amount of white
on bill. Later the bill will turn totally black.

I played a numbers game with Houdi in Vermont, using her four young when they were already partly feathered-out toddlers at near-adult weight. While she was out foraging from a calf carcass I'd left nearby in the woods, I would rush out my bedroom window to her nest and bring two of them back inside. The little birds were still too young to be alarmed. They were at an age when they remained calm as long as they did not hear the parents' alarm calls. Since Fuzz, her mate, was missing (see p. 122), I could be confident no parent had seen me handling the young. When she returned to the nest with half her brood missing, she fed her young in several seconds as always, then hopped to the water tray for a drink, perched nearby to preen and shake, and then went out of the aviary to the calf carcass for another load of meat. As soon as she was out of sight, I rushed out and took the other two young, replacing the first two. When she came back, she again made no vocalizations and showed no emotion. I offered her a chicken egg. She took it eagerly, cracked it, and fed the two young three times. After she left the nest, I replaced the second pair of young to complete the clutch to four. Houdi again showed no reaction to the sudden change in the number of her brood. For all appearances, Houdi didn't count, didn't care, and/or didn't recognize the difference in appearance of a nest with two versus four young.

As I explained earlier, Houdi's four young later were orphaned. The responsibility of tending to the ravenous quadruplets that required hourly attention was not on my agenda at that time. I sought to pawn off my responsibility and hoped to do an experiment at the same time. Goliath and Whitefeather in Maine had only two off-spring, though much younger and still naked. Would they accept Houdi's four orphans in spite of the age difference?

I drove to Maine on May 6, and delivered Houdi's four feathered young into Goliath and Whitefeather's nest in the middle of the night so that they would not see what was happening, temporarily replacing the two much younger offspring of Goliath and Whitefeather. I took their own young out because I feared the parents might compare the newcomers against their own, and then reject and harm them. I went into the nearby observation hut at dawn to watch what would happen.

By 5:10 A.M., it had been light for at least ten minutes, and the pair perched quietly in the shed by the nest. They could not help seeing the four new strange young in their nest, nor miss that their own two were missing, but they showed no visible reaction.

At 5:29, one young began to beg. Whitefeather stretched. Another young began to beg. Ten minutes later, when the sun first peeked over the ridge and shone directly into the nest shed, the adults suddenly stood immobilized like statues. The young were silent, and their blue eyes blinked as their heads rested on the nest rim. The adults stood as if dumbstruck, intently examining the young. Both moved their heads quizzically from side to side, looking at the young in their nest first with one eye, then with the other from a different angle. Several young soon stood up and begged. Whitefeather's head then went fuzzy. She was agitated. At first tentatively, then vigorously, she started making *kek-kek-kek* alarm calls. The young, responding to her noise and movement, begged even more. She then made pecking motions at their heads, vigorously gesticulating her anger without hitting them. With her head still fuzzy, she partially opened her bill, as birds do when frightened or alarmed. This was not going as I had hoped. She was not happy. Goliath, in contrast, showed no reaction.

Five minutes later, both birds finally flew out of the nest shed and made long and rapidly repeated rasping alarm calls, such as when an intruder comes near. Whitefeather started hammering at branches in anger, and she also hung upside down by her feet from the wire screening. I had never seen her do either of these behaviors before. I then came out of hiding from my observation shed and sought to blunt their anger by bringing a roadkill. Both birds greeted me (and the dead beaver) without alarm, and both immediately fed. They then began to carry loads of meat up to the nest. The young begged unabashedly, and Whitefeather now readily fed them as if she had completely forgotten her alarm of a few minutes earlier. I was flabbergasted. The four strange young were apparently going to be adopted after all.

It then seemed safe and appropriate to return Goliath and Whitefeather's own young back to the nest, making it a clutch of six. Would the ravens now preferentially feed their own? To the contrary, in the first thirteen feeding trips to the nest, twelve by Whitefeather and one by Goliath, all of the food went to Houdi's young, not their own babies. Whitefeather probably simply fed the loudest and most insistent beggars, apparently finding a big pink open mouth irresistible. After the young stopped begging, she stood at the nest edge looking at them, repeatedly making soft low *krr-krr* sounds that induce hungry young to gape. She was making sure they had enough. Goliath was soon also feeding all the young. The question, and my dilemma, had been resolved: Whitefeather and Goliath had totally accepted the newcomers. Throughout the next few months, I brought them other treats besides the beaver carcass, but the pair were now on their own, and they successfully reared "their" six young to independence.

The next year, in spring 1997, Goliath and Whitefeather had started to rebuild their nest in the aviary, but they later abandoned the breeding attempt. I brought other young ravens I was then rearing for later observations and experiments to test insight, curious to see if these would also be openly received. At first, I kept these youngsters inside the cabin. Goliath responded quickly to their loud begging when I fed them. He came and perched on the birch tree outside the cabin, making long, high-pitched, trilling, upward-inflected alarm calls. He came several times right next to the cabin, as if wanting to come in, exhibiting the male dominance display with fuzzy head, bill-snapping, and bowing simultaneous with flaring his tail and wings.

On this same day, I had seen him chase a red-tailed hawk, giving the rapid staccato calls normally given to aerial predators. When three turkey vultures had come by, he immediately flew to the nest by the aviary, making sharp alarm calls and fluffing himself out. Other ravens had come, and he had vigorously chased them off as well.

I finally brought two of the young, nearly grown ravens out of the cabin and let them hop around on the picnic table. As the young begged from me, Goliath, perched right next to me in the top of the dead birch, showed anger. He stared at the youngsters and made single long calls, one after another, with a high pitch and an upward-inflection normally given as a territorial advertisement. He flew over the youngsters, who ducked and momentarily stopped their begging. Next, he perched forty yards away in a spruce tree, puffed himself out, and angrily attacked cones and branches, hacking them off. Curiously, he made no attempt to hop onto the table to harm the young.

Ravens and most other birds have not evolved behavior to reject young from their own nests, even when these young may look odd to them, because foreign young would normally almost never be deposited there. It is sometimes a different story with eggs that could become stranger's young. An egg could be laid into a neighbor's nest accidentally, and some bird species specialize in parasitizing other parents by dumping their eggs off to surrogate parents. They no longer build any nest of their own. The best-known examples of this are cuckoos in the Old World and cowbirds in the New.

As is usual in evolution, almost every strategy has a counter-strategy aimed at neutralizing it, and so on in a continuing tit-for-tat game that continues until a mutual balance is reached in a messy "real world," or continuing until one of the contestants becomes extinct. In birds, the arms race between bird nest parasites and their hosts is played out largely with egg color. There is the strategy of color matching to deceive by the parasites, pitted against that of detection of color differences to recognize deception, in what is a classic example of coevolution, where one species evolves in response to another or others. The variety and the beauty of the coloring of birds' eggs is one of the marvels of nature, and it is likely the outcome of the conflict between egg-dumping parasites and their hosts of those birds with open nests where the eggs are visible (most hole nesters have uncolored eggs).

At the population level, color variety between *species* creates difficulties for the local nest parasites, and such differences could easily evolve. For example, if one parasite species specializes to dump its eggs into victims with blue eggs, say species A, B, C, and D, then an individual bird of one of these victim species C would be protected if it has a mutation that results in the production of eggs with purple speckles. That mutation would then quickly spread, and individuals of that species would eventually all have purple-speckled eggs.

At the present time in evolution, the arms race in egg coloring in songbirds has resulted in a great variety of egg colors. Most familiar to us in North America, for example, are robins with pale blue eggs, phoebes with pure white eggs, and kingbirds with white eggs spotted and blotched with purple, black, and lavender. Similar color variety among species exists in the eggs of European birds. In Europe, the common cuckoo is the main egg parasite. The cuckoo's eggs closely match their hosts' eggs. They need to. If cuckoos laid white eggs in Europe, they could not parasitize birds laying blue eggs, because those species, after a long history of parasitism, would instantly recognize a white egg among a clutch of their own that are all blue. The cuckoo would lose that unmatched egg, and that line of cuckoos who persist in laying white eggs into nests with blue eggs would go extinct. In Europe, the outcome has been that each female cuckoo lays eggs of only one color, and lays her eggs in only the "correct" species—the one with those colors that her eggs will match. However, the population of cuckoos has different individuals who lay different-colored eggs, and who dump their eggs in species providing an appropriate match. Of course, as expected, the eggs don't always match perfectly—and some of the hosts, some of the time, are able to detect an off-color egg and toss it out, because in Europe the chances of an off-color egg being a cuckoo's is great. The host's chances of making a correct choice—tossing out the cuckoos' and not their own egg—increase enormously if it does two things: memorize what its first-laid egg looks like, and have all eggs of its own clutch look nearly alike. That is the case.

Ravens have greenish-blue eggs variously mottled with grays and blacks. They are often, but not always, quite variable within any one clutch (see p. 120), so given the rationale explained above, it seems unlikely that raven egg coloration has been standardized to provide a uniform background against which a strangers' eggs would stand out. In North America, there are no raven nest parasites, so there is no reason to suppose that common ravens have evolved egg-rejection behavior. They might still recognize a strange egg, but they face no risk of raising a stranger's young and the cost of making a mistake and ejecting one of their own is great. Evolutionary logic therefore dictates that while they should defend their nest from possible egg-dumping females, a strange-looking egg should *not* be ejected, because it is most likely one of their own.

I decided to experiment with egg recognition. I began by trying to climb to the raven nest at the Melcher farm, where I knew the birds should be incubating, and to give them a chicken egg. At Melcher's barn, I stopped to talk with Paul about his raven nest, then crossed his large rolling field, still covered with snow, and went on down to the brook at the other side. The brook was gurgling loudly underneath bridges of ice. Crossing one of these ice bridges, I ascended the steep hill on the other side through woods shaded by tall hemlocks. A pair of ravens had built their nest for tens of years in a grove of tall white pines near the top of the hill. Having seen the pair circle over the hill earlier in the spring, I expected to find the nest again.

The birds' staccato *kek-kek-kek* alarm calls commenced as I came close to the pines. To my surprise, the huge stick structure was on the very same spot it had been several years ago.

When I finally made it up to the solid live branches just beneath the nest, I was exhausted, but my spirits soared. The nest contained six eggs. These, like most raven eggs, were in a deep nest cup of shredded cedar bark and tufts of deer and cattle fur, all mixed together with the shredded inner bark of dead ash trees. The eggs were greenish-blue with a variety of irregularly shaped gray and black spots and blotches. Some of the dark spots had an olive tint, others were faintly purple. The spots and blotches varied from a smoky haze of gray to black, and ranged in size from much smaller than a pinhead to larger than a housefly. Like Houdi's, the six eggs were individually distinct. Some of them showed a background color of light blue-green, and in others most of this background was obscured by dense blotching of dark olive green. I could hardly imagine anything more beautiful, nor could I suppose any more tangible evidence that these birds likely did not have a long history of egg parasitism. These eggs were too variable in color.

After taking pictures of the nest and eggs and of the view, I hung my knapsack on a branch and with blessedly solid footing under me, got out a yogurt container, unwrapped the white chicken egg it held, and placed the egg into the raven nest. I knew that ravens treat round objects as if they might be food, which they usually are. I also knew that chicken and other bird eggs are one of the raven's favorite food. Would the raven eat this chicken egg, incubate it, throw it out, or desert the nest?

There is nothing like anticipation to flavor a new day, and it was with high hopes that I returned the next day to check on the chicken egg. I can invent reasons to explain almost anything a raven might do with a big white chicken egg miraculously dropped into its nest in the top of a pine tree. In truth, I had not the slightest idea what this pair would actually do. The raven made alarm calls as I neared the nest. When I got to the top of the tree, the eggs were warm. All seven of them.

I had come prepared for the possibility that the chicken egg might be gone. If that had happened, I'd have ended up wondering if the birds might have accepted an egg that looked less strange. Regarding that as a possibility, I had brought up with me a second chicken egg, this one colored greenish-blue and spotted like a raven egg. Since the ravens had already accepted a pure white egg, it seemed remote that they'd now reject a raven-like egg. I assumed they'd accept it. Nevertheless, it had been a hard climb, and I had nothing to lose by exchanging the white chicken egg with the imitation raven egg.

The preceeding day, after I had climbed down and left the pine grove, I had heard a raven make the musical double *glug-glug* call as the bird returned to the nest containing the white chicken egg. This time after I left the tree and the female returned to her nest, I heard instead the loud, long, deep rasping calls that are associated with anger. Was the raven indifferent the first time or merely puzzled? Was she angry this time because she recognized the new egg as being more like a raven's, and thought a raven had been up to mischief? There were also plaintive calls, made by both birds. I had never heard either of these calls near the nest before. Strange. The raven's much different reaction to this egg meant I'd have to climb the tree once more on another day to see if the new egg would be accepted. But where was the limit? Might they even accept a red egg?

On Easter Sunday I painted a chicken egg all red. I had found in previous trials that even robins will throw out red eggs while keeping green- or blue-colored control eggs. I now brought the red egg to the ravens' nest. I did not expect that it would remain unnoticed for long, and it wasn't. I hid in the foggy woods within earshot of the anticipated fracas.

When the female flew back to the nest to resume incubation, she erupted in long series of rapid *kek-kek-kek* calls. These are high-alert alarm calls, normally given in the presence of nest predators. She flew away from the nest, and the alarm calls somewhat subsided. The same calls resumed full vigor on each of her several approaches to the nest. It was only after fifty minutes that the alarm calls eventually stopped. I then heard the soft, conversational *gro* calls.

The red egg, too, was accepted, as I learned on my climb to the nest the next day. My final nest offering was a black film canister, filled with water to mimic an egg's weight. This was ejected, even though it was close to the size of a raven's egg. The canister was not just thrown out; I could not find it on the ground near the nest tree.

There are limits to what is allowed to remain in the nest. At this raven nest and at two others that I have examined subsequently, if it is an egg, it is accepted, even if it is a very oddly colored egg. In subsequent experiments with captive birds nesting for the first time, I saw *no* reactions to chicken eggs or the facsimiles in their nest, perhaps because these first breeders had not yet learned to recognize their own eggs. Egg-rejection behavior apparently has not evolved in ravens, even though at least one pair of many years appeared to recognize and be alarmed by strange eggs. The ravens' behavior might sometimes look dumb to us, but it is not always a reflection of their intelligence. Even little girls accept and cuddle rag dolls. We presume they know the difference between their doll and a real baby. Or do they?

THIRTEEN

Sensory Discrimination

Birds live in a sensory world that overlaps ours. Their vision and hearing are acute. They should be able to see and hear roughly what we see and hear, although their attention to specific detail undoubtedly is different, as is the neural processing of that information. They probably pick out details that we are not aware of and miss others that seem obvious to us. In addition, many can detect the earth's magnetic fields, polarized light, and ultrasound.

From day one out of the nest, Goliath and Whitefeather's young avidly pursued, pounced on, grabbed, crunched, and swallowed all manner of arthropods. With little hesitation, they ate beetles and maggots, dun-colored moths, brightly colored butterflies, dragonflies, grasshoppers, flies, caddisflies, and all larvae. They even ate carrion beetles, but showed caution in approaching their first mouse. After

LEFT: *Chicken egg in Melcher farm raven nest, for egg-rejection experiment.*
RIGHT: *One- or two-day-old raven chick. The "egg tooth" on front of bill is used to cut open eggshell. Eyes are still closed. Note ear and nose openings, and sparse fluffy fuzz on head.*

dismembering and eating the first one, though, they approached and ate others with gusto.

Despite the sometimes disgusting things they eat, ravens do subject potential food to careful taste tests. They delicately crush and palpate new food in their bills for many seconds before either swallowing or rejecting it. Some potential food cannot be ingested, not because it tastes or smells bad, but because it causes such unpleasant reactions as stinging in the mouth. A series of photographs in biology textbooks chronicles a toad's behavior with a bumblebee: The toad swallows the bee, then spits the bee up. Presumably the bee has stung the toad's tongue. The toad ducks down in fright of the next bee that comes close. Simple. Ravens should do no less.

My first group of four, Goliath and his nestmates, were avid insect-feeders. Yet I found it strange that they showed little interest in the first bumblebees they ever encountered. It was unusual for them to ignore anything at all. I was shocked to see them only limply pick up a few of the dead bees I had scattered in front of them to test their reactions. They quickly dropped them after shaking their heads violently and puffing out their head feathers, indicating disgust. They could not have been reacting to stings; dead bees don't sting. They were not afraid. Since there was still a remote possibility that all four had somehow already gained experience with my presumably dead bumblebees' stingers. I needed to make another test. I did in 1997, with my group of six new ravens.

Luckily for my test, bumblebee drones, which are male bees, don't have stingers (modified ovipositors), while worker bees are females, and sting. I had learned through several painful lessons that drones and workers look alike, when I tried to show off by popping what I thought were drones into my own mouth. Having as a consequence become alert to the differences between male and female, I collected a handful of drones. All the bees crawled and buzzed. The six ravens gathered around and didn't hesitate to grab them. Live bees were more to their liking than dead ones, so they obviously had not been conditioned against bumblebees by prior experiences. After grabbing the bees, however, the ravens instantly fluffed out and shook their heads, as the previous group of four had done with the dead bees. Bumble-bees apparently tasted disgusting to the ravens! To me, they taste okay.

As with bumblebees, the ravens readily picked up dead honeybees, then dropped them uneaten. My hive had large numbers of drones, and I gave my birds handfuls of live ones that I had rendered flightless by damaging one wing. As with the bumblebees, the ravens showed no fear. They picked up the bees, crushed them, then spit them out. I offered them the hover flies, *Eristalis,* which mimic honeybees closely enough that few people can distinguish them. The ravens took them, carefully macerated them, drooled saliva, and swallowed. So—flies that looked just like bees were palatable, suitable food.

Wasps? I tried yellow jackets (*Vespula*) and white-faced hornets (*Paravespula*). The ravens were not afraid of either. They thoroughly crushed one or two, sometimes ate a little bit, then spit out the rest. As with bee-mimicking *Eristalis* flies, syrphid flies that mimic these wasps almost perfectly were eaten whole. With each test, I ran a control. I offered an insect other than the test objects. The other insects were always eaten eagerly.

Some other insects are well known to be protected by tasting bad, at least to birds, their potential predators. Their noxiousness is often derived from chemicals they absorb by feeding on chemically defended plants. Usually they advertise their noxiousness so the bird won't kill the insect before rejecting it because of its bad taste.

The summer of 1997 was a very good one for milkweed in Vermont, and also for the bright, brick-red cerambycid beetles and the flashy, black-white-yellow striped monarch butterfly caterpillars that feed on milkweed. The milkweed-eating beetles and monarch caterpillars are both sufficiently flashy to warn predators against eating them; they should taste noxious from the toxic milkweed, their food.

I raised a dozen monarch caterpillars to adulthood, then released the butterflies in the aviary. The ravens eagerly chased after the big butterflies, caught them, then pulled off their wings. They were not eager to eat them, though. They kept tasting and crushing them, pulling off the abdomen, shaking their heads, then dropping the butterflies piece by piece without eating any part. Other butterflies were eaten enthusiastically, wings and all. These results were classic— "according to the book"—just what I had expected. But these were not the only results. I also offered them the brightly colored monarch *cater-pillars* and the red milkweed-eating beetles. Did they reject them? Not at all! They ate both avidly. I had expected them to relish the deactivated stinging insects, and to reject the insects that advertise themselves with bright colors because they are chemically defended. Clearly, the ravens hadn't read the texts. They had it mostly backwards.

Some things attract the ravens by their appearance alone. Young ravens that had never before eaten eggs were instantly attracted to them, as though hardwired to recognize this nutritious food. On the other hand, they were also attracted to potato chips, which they vastly preferred over raw beef liver (in agreement with me on that!). The young ravens I raised were, through their periods of rapid growth, fed almost exclusively on chopped whole animals of all kinds. Meat is the mainstay of their diet. It was therefore surprising that although they would gobble down chopped chicken guts, they would turn their beaks up at fresh liver, while treating potato chips like a great delicacy.

What did they "see" in an egg? When I first had offered them a sparrow egg, all six of my young ravens had rushed over to try to be the first to get it. A goose egg was treated similarly. All gathered round and pecked at it. I wondered if a more oblong, smooth object that was new to them would generate the same response. A ripe banana! All looked at it from a distance and ignored it for several minutes. How about a *huge* egg? I put down an ostrich egg. The dominant male immediately approached it, and the others followed. Then one of them flew up and they all followed suit, making alarm calls. They looked at the ostrich egg for a few more minutes from their perches above, then ignored it.

How about something else roundish but not quite so smooth? I tried hickory nuts. All six immediately approached them, pecked them, and quickly left them. A handful of pistachios? For more than ten minutes they played with them. I say "played," because as they tried to crush them, the nuts popped out of their bills and shot some five or six feet off to the side. The birds would chase after them, only to have the same thing happen. It was a comical scene. You'd think the nuts were jumping around on their own. Only one bird tried to hold a nut fast with her feet, trying to peck into the crack of the shell. She got a few tiny morsels of the nut inside, but none of the birds succeeded in opening a nut on the first try. I continued to give them the pistachios, and a week later some of them opened the nuts as efficiently as I could, and ate them with the same gusto.

The smooth, round eggs and the moths and butterflies that they all pounced on are all conspicuous items to my eye. I suspected the ravens were attracted to bright, conspicuous objects. Would they pass up bright, colorful flowers? As I walked through the aviary, I dropped as unobtrusively as I could behind me five or six each of red roses, purple phlox, and yellow pea flowers. A couple of the birds hopped down and picked weakly at them, but all lost interest within one minute. An hour later, I repeated the experiment with blue irises, daisies, red clover, and yellow hawkweed. The result was similar, although one bird picked up four individual iris flowers and another took three clover flowers and deposited them into the water pan, then picked them out again. Two more pulled the petals off one daisy each. Three birds took no interest in the flowers whatsoever. Were they just generally not responsive today? A quick test with a handful of pistachio nuts told me no—*all* sprang into action. I tried a batch of *different* flowers, wondering if the birds would generalize now and treat all flowers as unworthy of attention. African violets and red cyclamens? They were all ignored. Just to be sure, I tried another batch of flowers, pink and blue lupines, and white clover. One bird hopped down, tore apart a lupine inflorescence, then cached another one, covering it with a leaf. A minute later, he and the rest of the birds were back to playing with rocks, bones, nuts, sticks, and bark. There were still a few flowers left in the garden. I decided to try again. The next batch were red jewelweed and blue petunias. Two birds picked up a couple and flung them aside. Their interest in flowers was nothing like that shown to such toys as a Ping-Pong ball, a lightbulb, two film canisters, and one red and white fishing bobber that I gave them next. Then I tried a skunk.

It was late winter, when skunks emerge for their first walkabouts, and there was plenty of roadkill reeking with typical skunk effluvium. My ravens had never before seen or smelled a skunk, but they didn't hesitate. They were into a skunk in seconds, yanking fur out and poking into all orifices. They gorged. Skunk is fat, and fat is good. Smell be damned, if they smelled it at all.

My aviary was divided into two sections by an opaque partition, so I could shut them up in one side without them seeing what I did in the other. There was a foot of snow on the ground. Whenever I had buried a squirrel in their view, they invariably had found it and dug it back out. On the other hand, squirrels, peanuts, bread, and chunks of calf meat that they had *not* seen me bury always remained buried. I concluded that ravens don't use scent to discover food. But then, I hadn't yet tested skunk.

Four times I buried the by now partially eaten skunk out of their sight, and when I let them into the approximately 1,500-square-foot aviary section where the skunk was located, they always dug in the right spot within minutes after I let them in. They'd quickly exhume the skunk from the snow and resume their feast. I thought they smelled it, because they had not seen me bury it. But I'm an experimental biologist. I'm slightly obsessive. I wondered if they could possibly be so highly motivated to eat skunk that they'd be observant enough to find it by some other cue.

In the fifth trial, I made four fake burials, plus one with the skunk. As before, when I let the birds back into the arena, they immediately charged around as if looking for the skunk. And as before, they found it. Not only did they dig where it was buried, they also dug holes in the snow at all of the fake burial places. So they had learned a second, even more subtle cue than scent: my disturbance of the snow that marked the spot.

I didn't immediately follow up on these experiments with the exhausting detail of hundreds of trials, as is proper for scientific protocol. I was satisfied, and the reek of skunk around the aviary and our house after just one week was also by now sufficient. Deciding I'd better curtail the experiment, I hauled the skunk far off into the woods.

A snowstorm came a few days later. When it started to snow, I knew right off that it was too good an opportunity to pass up. The snow would hide my tracks. I immediately reclaimed the discarded skunk, to give the ravens another crack at it, because now my disturbances in the snow could be camouflaged.

This time, after I led the birds out, I made six fake burial sites in the snow in the experimental aviary and one hole where I buried the skunk. I left the ravens in the side aviary until two inches of snow had accumulated, obliterating all tracks. I presumed that when I let the birds in, they would immediately expect food and start looking for it, as they had all the times before, but if I was correct in my hypothesis, they should now not find it.

When I let the birds into the experimental aviary, they did act like a bunch of kids on an Easter egg hunt. They looked everywhere, as though expecting I had hidden something. They didn't all converge on the spot where the skunk was, as I would have expected if they used scent as a beacon. Nevertheless, they dug into all seven places, despite the fact that the sites were hidden under the fresh snow. The spot where the skunk was buried was the fourth one they went to. Each of the six birds dug in at least three different fake burial places. As one bird started to dig, others would join in, as though thinking that the other, because he or she dug there, must know something. I had no idea what any had known.

After the skunk was partially exhumed, I took it away to try the same experiment again later. As it continued to snow, I buried the skunk in a different place and made six more fake burials, then waited overnight, when nearly *eight* more inches of snow fell. There was no hint of any mounds in the morning, except a four-foot-high snow pile I had made for them. This time, the birds did not find the skunk in even two days. Instead, they dug only into the huge snow pile. Apparently, they cued into not only the disturbance of snow, but perhaps also to bumps. They had now not noted any bumps in the snow except the ultra-large one. Did they think I'd buried a giant carcass there?

Contrary to what I had presumed from the first tests, the ravens did not provide one shred of evidence that they use scent to locate food. Nor did they appear to smell skunk's effluvium, much less mind it. What I did learn, though, is that ravens are very alert to subtle visual cues. Reacting to subtle cues is admittedly not intelligence, but it is a prerequisite for many kinds of intelligent behavior.

Mostly I learned that one never really knows what might be relevant to them. One can't predict what they perceive, what cues they use, or how they will react. There are always surprises. I suspected ever more that what they *do* has less to do with what they perceive than with how they process information in their minds.

FOURTEEN

Individual Recognition

I<small>F A</small> <small>NUMBER OF RAVENS WERE LINED</small> up in a row, most of us could not distinguish one from another. I know I can't, and I've had lots of opportunity. I could no more distinguish them as individuals than identify individual peas out of the same pod. I may see a smudge on a wing feather here, a blemish on a tail feather there, but all of these markings are temporary. I see differences in behavior, but I can't assume behavior is constant, especially if that is what I'm interested in studying.

The birds appear to treat many of their own kind as individuals. One hint of individual recognition can be seen in the evening at communal roosts. Some of the birds' most vehemently vociferous and sustained squabbles relate to who sleeps with whom. Pairs and preening partners always sit close to each other, and they repeatedly chase off specific individuals while allowing others to perch nearby. Each bird responds to every other bird differently, as if it knows each bird individually. I have

Raven bills tend to be more individually distinct
(to our eyes) than other of their facial features.

observed, for instance, that while the most dominant birds chase any and all other birds to try to induce them to drop the food they are carrying, those lower in the dominance hierarchy chase only those still lower until they drop their food. I have seen many hundreds of such chases, and I have never seen a low-ranking bird take off to try to chase a high-ranking bird. Obviously, it would not be successful in taking its food, but the important point is that it never takes a mistake for them to find out.

Unfortunately, I had no proof of individual recognition, and I was anxious to devise a study to see if my hunch was right. But what might I use for criteria that could be statistically evaluated? What kind of data would a hypothetical raven scientist gather in order to prove whether or not even humans, who we know recognize each other as individuals, actually do? As Paul Sherman, Hudson Reeve, and David Pfennis have pointed out in a recent review, the only objective measure of individual recognition is a demonstration of differential treatment. If we discriminate individuals one from another, we must first recognize them. That is probably a conservative approach, because even though we may recognize each other, we might still treat each other democratically.

In January 1998, when my group of six ravens were eight months out of the nest, I weighed them and also determined their dominance hierarchy. The latter is easier to do than the first. When the birds feed together at a carcass, there is a constant shuffling as the birds try to feed next to specific individuals and make aggressive jabs and jumps at other specific individuals. I can score who feeds next to whom and who yields to whom, and in a couple of hours know who is top and who is bottom raven, and who likes to be next to whom. In that case, after only 155 interactions the hierarchy was clear. Blue made seventy-eight challenges, and never backed down from any bird. He was top raven. White, his sister, made no challenges but was challenged a total of fifty-three times by all birds, and she backed off to all of them. She was bottom raven. The other numbers also fell into place to reveal a dominance hierarchy from top to bottom, of: Blue, Orange, Green, Yellow, Red, White. Of this group, only Blue and Orange were males, and they were the largest birds.

In August, eight months later, I reexamined the dominance hierarchy of the same six birds, who had been kept together during the intervening time. One thing had changed; Blue and Red had become friends (they would probably attempt to nest in two years) and regularly fed together, played together, and preened each other. For a week and a half, I brought the birds a frozen calf haunch each day and tabulated 678 dominance interactions. Blue was still the undisputed top bird, dishing out 366 aggressive interactions but never once being challenged. White was still the under-raven, being at the receiving end of 382 of the 678 interactions and never once challenging any other bird. It was with the four ravens in between that relationships got interesting. Orange, the next dominant bird, was still second. He was challenged only by Blue, but he now *rarely* challenged any bird, although before he had challenged Red. Blue's friend, Red, had risen one rung in the hierarchy. She registered fifty-six aggressions against Yellow and seventy-five against White, but received none from them in return. Furthermore, Orange and Green, the other two birds above her, hardly touched her—I scored a total of only ten aggressions against her. That is, Red was now dishing it out to two underlings, Yellow and White, and she had become practically immune from the attacks of Orange, Green, and Blue. The reason was easily apparent: She fed under the protective umbrella of her dominant mate, Blue. I couldn't wait to see what would happen when I removed Blue from the group.

I led Blue into the adjoining aviary, separating him from the group by a wall of wire screening. When I fed the group, he did not come to the wire as if trying to get through. Instead, he kept his back to the feeding group and showed his anger by making long rasping calls. After I had tabulated 863 aggressive interactions within the group without Blue, I saw significant changes. First, Orange went from his previous 4 percent of the aggressive interactions to 33 percent. A quarter of his attacks were now directed against Red, whom previously he had not touched. Red had sunk in status, and the dominance hierarchy was back to what it had been in January 1998. Red was being "hit" much more, not only by Orange but also by Green. Curiously, even White, the most subordinate bird, who had previously backed away from all in thousands of interactions, scored five "hits" against Red. That doesn't seem like many, especially since Red hit White 184 times, but White had not once before hit any bird.

After these experiments, I opened the door to let Blue back in, having first put down a hunk of meat in one spot and another equally large piece about three yards away. Blue rushed to one meat pile and started to feed. His partner, Red, rushed up to feed alongside him, and he tolerated her. All four of the other birds went to the other feeding spot to avoid him. But Blue, greedy as always, wanted both meat piles. He had trouble having both at the same time. In the one hour that I watched, he switched back and forth seventeen times between the two. Red accompanied him, tagging along behind him. Usually, the instant he left a meat pile the others rushed to it, even with Red still there, to grab bits of meat. As soon as he came back, they rushed off to return to the second piece. Red was still an underbird at the meat pile where Blue was *not* feeding. She went near only those individuals that would tolerate her, pointedly distancing herself from Orange and Green.

The birds left their piece of meat when Blue came, not because they had just been hit by him, but before they might be hit. They left when he came within about a yard or two of them, and none went to the other piece if Blue was already there. That is, the birds appeared to *recognize* each other from at least two yards away, and this ability allowed them to stay clear of aggressive individuals as well as to seek out and stay with tolerant ones.

The experiments also showed that the female, Red, benefited from her partnership with Blue. He tolerated her, and she followed him. Under the conditions of the aviary, it seemed like a one-sided relationship, because Red provided nothing to Blue. However, you will recall that in the field (see Chapter 10), teamwork is advantageous in capturing food and/or taking it from powerful adversaries, so the partnerships would be reciprocal.

If the birds recognized each other, as my observations suggested, the next question was: What is the basis for that recognition? We use faces, voice, gestures, gait, and clothes. We can distinguish one individual from any other, even though our facial expressions can vary tremendously, and even though those expressions communicate a great range of other information besides identity. Ravens also vary their "facial" (head) expressions enormously. Their skin is hidden by feathers, but the head feathers can be rearranged by muscles in the skin to convey a wide range of moods, emotions, and intents. Body language does the same. I can see at a glance if a raven feels afraid, self-assertive, attentive, angry, or contented. Behavior changes as well. When Goliath, in the presence of Fuzz, fell from a dominant to a submissive status, his head feather patterns, body postures, and vocalizations all changed dramatically. But the mates still recognized each other, showing that specific status, as such, is not the one critical feature that serves as an identity tag. Had these birds not been banded, I would, without a shadow of doubt, have thought Fuzz was Goliath and I would have missed an important behavioral event.

What accounts for the ravens' individuals recognition? Positive individual identification of ravens for us humans must necessarily rely on markers. The best markers are, of course, those already on the bird. I have seen two birds that had white wing feathers, and one with a partially white wing feather. One, Stumpy, had a missing foot. One had a missing right eye. Another had useless curled toes on one foot, and several had unusual bill shapes. For the most part, I have had to mark the ravens in order to ID them. At first, I attached commercially available colored plastic leg rings, but these wore off after about two years. I tried freeze-branding. Freezing kills pigment-producing melanocytes; in mammals, hair that regrows at a once-frozen site is white. Marking domestic animals with a very cold branding instrument to cause regrowth of white hair is a well-known marking technique (Farrell and Johnson, 1973). Lacking a branding instrument, I tried dry ice, which maintains a temperature of minus 57 degrees Celsius. I tried to brand different feather tracts of four young ravens by holding a chunk of dry ice to them for five to ten seconds. I had tried the same technique on myself, pressing the chunk of dry ice to my scalp for thirty seconds. I felt no pain. Unfortunately, when the young ravens feathered out, they stayed totally black, and my hair remained as brown as before. (I later learned that in order to get regrowth of white hair, one first needs extensive damage to the skin.) As for the white wing feathers on individual ravens, they had all grown where I had previously attached a wing marker with a metal rivet. All the birds molted and regrew black feathers by the third year.

The method of choice for making ravens individually recognizable seemed to be to attach onto the wings plastic tags of different colors with numbers on them. These lasted up to ten years, after which the birds melted back into the anonymous crowd. I also attached numbered aluminum Fish and Wildlife rings onto birds' legs, but these can generally be read only when the bird is in the hand.

I had been able to recognize some individuals among twenty-two wild-caught ravens in my aviary complex in the winter of 1992, after I associated with them every day, even when I didn't see their markers. I had at first presumed they would all be anonymous, were it not for their colored rings and wing tags. Thanks to these markers, I could eventually recognize some individuals even without seeing their identity tags, and I started forming attachments and having favorites among those "characters" that stood out.

For a while, my favorite was C48, a big, dominant bird who had always been the first to investigate anything new, as well as being first at the food I'd bring. Gradually, Yellow O stood out, too. She was uncommonly tame for a wild-caught bird, and she appeared to make a point of perching in front of me, as if to place herself between me and the rest of the crowd. The others tried to keep their distance instead, except when I brought food, but Yellow O came to perch near me even if I didn't bring food. She sat, calm and composed, with partially fluffed out feathers, indicating her relaxed state of mind. She watched me. I wondered why.

Eventually, Red Dot became my favorite bird of this group. I nicknamed him "Hook" because his bill, which was conspicuously longer than any of the others' and had a prominent hawklike curve at the tip. Unlike Yellow O, Hook never perched near me. Whenever food was near me, he showed no hesitation in walking up to it, keeping his head tilted so that his bill was pointed slightly up. He walked slowly, deliberately planting each foot one at a time, and his pant feathers and breast feathers seemed unusually long and flowing because they were not pulled up tight against his belly. His head feathers were often slightly raised, giving his head a fuzzy appearance. In contrast, the other dominant males usually maintained sleek head feathers. Unless displaying, they showed no pants and walked with a quicker stride. Although feather posture often varied radically between individuals, it could hardly contain the key information for individual identification by the ravens themselves, because it often changed from moment to moment.

Individually identified ravens have always been special to people, and other methods of color-coding them have been possibly even stranger than mine. The Athabascan Indians have a myth about why the raven is black and the world imperfect. They say that in the time before man, when the world was young, Raven (an individual) was white as snow. He was the creator of mountains and a lover of life whose soul was filled with light and beauty. All of this goodness made his evil black twin brother jealous, and the evil twin killed the white one with an ax. Ever since, the world has been imperfect and the raven black. In other early American cultures, Raven also was originally white, later turning black and becoming numerous and anonymous. The Tlingit say he was transformed by smoke.

Color is the most superficial of possible characteristics used to identified ravens, if only mythical ones. We tend to see all ravens as stereotypically black, although from up close they have a greenish, blue, and purple sheen. We do not distinguish one individual from another; and until we do, we will see their behavior as stereotyped and programmed.

There are rare ravens in the wild that are identifiable by their color. I received a photo of a young white raven from Josina Davis of the Queen Charlotte Islands in the late summer of 1997. The bird was white as snow except for its pink eyes, feet, and bill. An albino. Davis wrote, "He is hanging around in town [Port Clements] and I have never seen him harassed by others of his own kind. I saw him by the post office and rushed home for my bags of bread and dog food. He didn't call to other ravens. He picked up the largest pieces of bread—

he chased away two crows—he cached bread and dog food in an empty lot nearby. When he came back, four crows were eating the food I'd put down, and he made no attempt to chase them. The 'caution white raven' signs are up."

Others have written to me about seeing a white raven flying in a flock with others. One young emaciated white raven was found in New Brunswick, Canada, and ended up in an animal rehabilitation center. Mary Majka, the raven's keeper, wrote, "Albie is tremendously shy and scared of his own shadow. We've had regular ravens here before, and they were very intelligent, inquisitive, bold individuals. But this bird, although he's been with us for two years, still displays the same shy behavior." I wasn't surprised that Albie had idiosyncratic behavior. Most do, once you get to identify them and to know them.

I was sent a photograph of a chocolate brown raven that an Alaska newspaper article described as "the first off-color raven in Anchorage," although a dark silvery metallic one had been seen there in the 1970s. The bird is "very noticeable. When the sun hits him he's almost like a golden raven," Alaska raven biologist Rick Sinnett wrote. "This bird seems to be a loner that other ravens seem to pick on." The Anchorage newspaper headline read, "Off-Color Bird Faces Raven Racism."

The rare off-color ones notwithstanding, ravens must normally distinguish one another by cues other than color. Movement might be one relevant cue. But that is a difficult cue to interpret, except that you know when it is absent.

Given the ravens' predilection for poultry of all sizes, I provided my birds a recently killed raven that had been shot by a crow hunter. They reacted to this raven with loud, deep, rasping alarm calls. After some hours, they still ignored it. It was not eaten within minutes, unlike other birds I had given them that were pounced on in seconds. It was not eaten at all.

They would easily recognize a live crow. But a dead one? Would they eat that? My curiosity aroused, I had to observe their reaction to a dead crow. I presented a young frozen crow from my stock of roadkill in the freezer to Whitefeather and Goliath. Both birds erupted in

harsh, deep, rasping alarm calls, and just as with the dead raven, neither bird fed from the carcass. It eventually rotted in place. They did
not even dig for the maggots.

Two months later, I presented another crow, a headless one, to the
same two ravens plus Fuzz and Houdi. The reaction was the same.
Two days later, on September 18, I again presented the headless frozen
crow to observe their reactions more closely. That time, the reactions
were weird, to put it mildly. As I entered the aviary with the crow in
my hand, Fuzz erupted in the deep, long, rasping alarm calls normally
given in response to ground predators. Within seconds, all four birds
had joined in. At first, all made only the deep rasping caws. After a
few minutes, they also made the rapid *rap-rap-rap* calls they usually
give when a raven comes by. Fuzz then initiated several series of high-
pitched, fluting *quork*s that I had never heard him give before. Whitefeather, the dominant female, suddenly went into an extreme
crouching position (often referred to as a precopulatory display) in
which she flared her wings widely and rapidly vibrated her tail. She
crouched in place like that for minutes, as if glued to her perch. I had
seen the same display used in entreaty, as when a subordinate bird was
denied access to food or a bath. Then she erupted in knocking calls,
the female power display. Houdi, the subordinate female, also did a
mild version of the crouching display, but she didn't knock. Fuzz did
a lot of bill-wiping on the perch, a displacement activity common
during times of conflicting emotions. I had never seen these behaviors
toward other bird carcasses—those that they ate, and they usually ate
at once.

After several minutes of intense interest, both Fuzz and Goliath
went down to investigate the crow carcass. Only Fuzz, the most dominant male, touched it. He approached it in his macho, erect, male-
dominance stance with smooth head feathers, gently touched the dead
crow with his bill, and shouted directly at it. He stared at the carcass
while doing several series of rapid *rap-rap-rap* calls normally given
when a raven flies by in the distance, a call that I believe says, "Notice,
I'm here." He looked at the carcass for a time, then casually walked
off. After that, none of the other birds paid the crow carcass any more

visible attention. I next gave them a dead cottontail rabbit, and they began feeding on it in seconds.

Seven days later, I presented the same frozen, headless crow still again. The response was similar to the previous experiment, except that Whitefeather knocked but did not crouch. I left the crow. Two days later, they still had not eaten it. After I skinned it and removed the wings and legs, Fuzz approached it in less than ten seconds and ate it.

Do ravens have inhibitions against cannibalism, and did they mistake the dead crow for a raven? Or are they upset by the black? I gave them a piece of black rubber they had never seen before. They made deep rasping alarm calls, but no *rap-rap-rap* calls. Then I tried red flagging tape. Same response. When I presented them with an almost pure white bird—a ring-billed gull—they again reacted with deep, rasping alarm calls. They did not eat it right then, either. But when I gave them a second ring-billed gull two months later, Fuzz approached it within several minutes, plucked it, and ate it. When I presented them with the weirdest bird facsimile I could think of, a plastic pink flamingo purchased specifically for them, they made no alarm calls. They ignored it. I then brought them a big, black, battery-operated raven facsimile that gave the speech, "The end is near, the end is near—ha, ha, ha, ha, ha," when you touched a button. Long before I pushed this button, all four birds flew wildly about with their bills partially open in fright. Even though this fake raven was smaller than any of them, all acted afraid of it. None went near it or displayed to it. The "speech" it made later had little additional effect on them. Goliath, the big male, made deep rasping alarm calls. They apparently saw it as something strange and hence somewhat frightening. After I removed the toy raven from the aviary, both males went to their respective females and made throaty, hiccup-like calls with a downward inflection followed by a musical upward inflection. They showed their ears during these displays, which apparently were to reassure their mates of their strength and power. The ravens showed no alarm when I gave them the carcass of a black chicken. The black feathers didn't fool them one minute; they went right to it to pluck and eat. I also gave them a dead adult opossum. They had never before seen one, but in less than a minute, Fuzz approached it; then all came and fed. There is no easy explanation for these observations. That is why ravens are so interesting. Nevertheless, I think the experiments do show that body language is important to them for recognition, even to the extent of allowing them to distinguish between a raven and a crow, an otherwise easy task for them in live objects.

I am confident the birds do not identify each other by the markings I put on them, such as leg rings. No bird ever looked down to examine another's legs before cozying up to it. Goliath never looked over Whitefeather's shoulder when he approached on her left side, to check for the white wing feather on her right wing. When she molted and lost the white feather, he did not change his behavior towards her. Mutual preening (allopreening) is often accompanied by soft comfort sounds that

Whitefeather's white feather
was mostly black.

may have individual characteristics that the birds recognize, but they did not "call" to identify themselves before beginning to allopreen.

The relevant cues they use become more specific with age. Young birds in the first weeks or so of life are blind, and respond to any sudden vibration of the nest by stretching their necks up, begging loudly, and opening their mouths for food. The vibrations are normally caused by the adults' landing on the nest edge. The young nestlings do not yet identify the parent as such, because their necks are directed straight up, regardless of the parent's direction. After the youngs' eyes open, they gape toward almost any moving object. With time, they learn increasingly specific details of their parents. By the time they leave the nest, they have learned to respond to numerous cues and to reliably identify whoever has been feeding them, be it raven parents or human foster-parents. Soon, they follow specific *individuals,* flying away in fright from strangers, both human and raven.

Vision is important in recognition, but I have no clue what they look for or see in another raven. They recognize me—I am the only individual who can regularly walk up to them to within one yard while they are feeding. If any other person comes within fifteen yards, they fly up in fright. I have experimented with them by giving or withholding clues about my identity, to see what cues were important to them. When my young ravens were less than a year old, they showed fright if I wore different clothes. They were startled and flew away from me when I wore a hideous Halloween face mask. Perhaps they used clothes and faces to identify different humans.

Tests with my four tame birds after they were two years old showed that more was by then involved. In one experiment, I came into the cage after having been absent for a week, wearing my blue jacket, snow pants, and snow boots, but I had pulled a knitted green stocking cap they knew well completely over my head, with tiny peepholes for my eyes. As I entered the aviary "faceless" on this and other occasions of the test, I was careful not to say a word to them, because I didn't want acoustic cues to override all others. It was clear that my face was not the sole criterion they used to identify me; the birds were quite at ease, although they appeared to look me over more carefully than usual. In my next trial, I showed my face but wore new clothes that I had never worn near them before. The birds were still quite at ease. On the other hand, when I came dressed up in a bear suit they were quite alarmed, especially when I did the "bear walk" on all fours.

Ann, a neighbor, obliged me by taking part in the next experiment. As I expected, the same birds were wild with fright when she walked into the aviary. So far so good. We left, exchanged clothes, and came back. I wore Ann's blue shirt, and she wore my boots, jacket, and snowsuit, and pulled the green stocking hat with the one peephole in it over her head. Then she walked into the aviary again. She was masked and dressed exactly as I had been when the birds were quite nonchalant. The birds flew about in their cage in fright with open bills, although not quite as wildly as when she had come in dressed in her own clothes. Next, I walked in wearing Ann's blue shirt and the ski mask. The birds were not fully at ease, as they had been when I wore my own clothes and a mask, but neither were they wildly afraid. They were just moderately uneasy. When I pulled off my mask, they became totally at ease.

I next tested the responses of another group of birds that were wild-caught but had become tame after a year. I had on four previous occasions chased and caught these birds using a long-handled smelt net. I had tried to conceal my identity by wearing a black mask and wig during the chases so that they would later allow me to come close to them for my behavioral observations. When I came near the birds at other times, I had worn my "good" hat, the orange one that left my face bare. They always flew away from afar when they saw me coming either with the smelt net or wearing the black mask with wig. Did they know just the smelt net, or me, or my mask? Wearing my "good" orange wool hat, I entered the aviary and threw down a handful of oatmeal. They gathered around me like chickens. Okay, they were not afraid of me and the orange hat. As soon as I pulled the black mask out of my pocket, they all flew off. So they feared the *mask.* I asked Chelsea, one of my Winter Ecology students, to enter the aviary while wearing my orange hat. Results were clear: As soon as she entered the aviary, the birds left the main aviary, flying off into the side aviaries. She spread the oatmeal, and they all stayed away. When she came out of the aviary, the birds streamed back in to eat the oatmeal. I reentered and they stayed fully relaxed, feeding all around me. It was clear that the birds identified individual humans, regardless of what they wore or carried. However, one's trappings can themselves be frightening to them, especially if they were associated with fearful situations.

The hiding of my face had not frightened one group of my tame birds, but I wondered if perhaps they had identified me by my clothes. What about a *strange* face? I put on a grotesque face by crossing my eyes and rolling them up. It made no difference. I could still walk to within a yard of the feeding crowd (a later group). With dark sunglasses, I got to three yards. Apparently, my eyes didn't contain the important clue of my identity of them. I tried limping in to them. That also got me almost equally close, but when I hopped on one leg they flew up at seven yards already. So they at least noticed my gait, but my walk was either not the one deciding characteristic that held my "signature" for them, or else they knew very well it was me, and they were uneasy merely because of my weird behavior.

Some things, though, scared them at the same distance as strangers did: carrying a gun, wearing a long dress, and carrying a broom. The birds were then one and a half years old and no longer afraid of my regular clothes. Perhaps they recognized my consistent style or fashion. But a kimono? When I wore that, they flew up at fifteen yards, but it didn't fool them for long. After my thirteenth approach in the kimono, they again allowed me to get next to them. They never allowed me to get close if I was carrying a broom, even though I had never chased or hit them with one.

I had not pinpointed any precise cues they used to identify me, or that they might use to identify each other. A big complication was that whenever I took away one cue I was really adding another. I could not just remove my face. I could only substitute. I concluded that the ravens recognize me without seeing my face and without identifying my clothes, although both cues are used when they are available and relevant. Presumably, the cues they use to identify each other are also multiple.

Vocal signatures are surely important in individual recognition, such as at a distance or when pairs make barely audible whispering-wimpering calls to each other when they are being intimate with each other. Males and females and dominants and subordinates have displays and vocalizations signifying sex and status, but none of that explains how they distinguish one another while they are engaged in the business of feeding, when, as the previous experiments indicated, they discriminate each other at a glance.

Can they recognize themselves in a mirror? We think of self-recognition as a higher mental faculty. It is not a faculty demonstrated in any bird. To the contrary, birds routinely and convincingly demonstrate their incapacity for self-recognition, by battering themselves aggressively against a mirroring surface, quite often until they bleed, in apparent attempts to dominate a rival. In North America, almost everyone with a window to a dark cellar has had the experience of seeing a male robin or song sparrow bash himself against such a mirroring window every morning, sometimes for hours and many days in succession, during the breeding season. There is no evidence whatsoever that the unfortunate birds ever catch on to their misguided behavior, despite the repeated daily brutal punishment they give themselves.

Not even crows are immune to attacking themselves in the mirror. The May 12, 1996, Western Australia *Sunday Times* shows a picture of an Australian raven presumably "caught in the act" of vandalizing a car, although to me the bird appears to be attacking its reflection in the car window. Wondering how a dominant northern raven, *Corvus corax,* who always vigorously challenged every other dominant raven that he met would react to his image in a mirror, I planned to videotape Fuzz, the super-male, when he met his unyielding and identical match.

It was June 25, 1995, and for three days I had been involved in the very frustrating process of trying to videotape the raven's reaction to itself in a 17" x 36" frameless mirror that a glass vendor had kindly donated expressly for this project. I had set up the mirror, but hid it behind a plywood panel of equal size. A dead chipmunk was tied to a stick in front of it. It took a full day before Fuzz, the boldest raven, ventured to yank at the chipmunk in front of the plywood hiding the mirror.

I presumed I could then pull the board away to reveal the mirror and stand back to film the show. Fuzz would want not only to impress any rival in the aviary, but also to repel any competitor daring to try to take his choicest food. What happened? As soon as I removed the plywood panel and the reflecting surface became visible, all the birds went bonkers. They retreated into their loft. After a few hours, they started begging piteously. They were hungry for chipmunk, yet they did not dare to come down to feed.

I tried again on July 5, and the results were almost an exact repeat of the previous week. I left the mirror set up with the plywood covering. The birds fed every day in front of the plywood, and were thoroughly habituated to feeding at that spot.

On September 20 at 7:02 A.M., I provided Fuzz, Goliath, White-feather, and Houdi with delicacies—chopped squirrel—at the usual place, but this time I exposed the mirror surface. The meat was within inches of the polished and unscratched mirror, and it was situated so that any bird walking up to it could see the "other" raven approaching from the opposite direction.

The birds looked down from their perches. They hopped back and forth nervously. At last, after fourteen minutes, Goliath descended to the ground, walked cautiously to the mirror, and took a piece of meat. Fuzz instantly chased him until he dropped the meat, but it happened to fall back in front of the mirror.

After four more minutes, Fuzz went to the front of the mirror and grabbed a piece of meat. Goliath then followed and got one, too. Next, the two females begged and received their food from their respective male partners. Curiously, neither male acted as though he saw what was *in* the mirror when he approached it.

On September 23, I again revealed the mirror, and they went to the food placed two inches from it within two minutes. They acted only slightly nervous. Even though all four birds stayed feeding directly in front of the mirror, they at no time overtly acknowledged anything they might have seen in the mirror. I was somewhat puzzled by these results. I could conclude that they didn't attack the mirror reflections as they would if they saw strangers, but I didn't think I could conclude they recognized themselves because they didn't attack.

I tested another group of ravens at the same mirror on October 25. As before, when I first brought the mirror into the aviary and set it in front of plywood, the birds were afraid of it. I turned the reflective surface away to let them first get used to the mirror as a strange object. When I finally reversed the mirror after a week, to expose its reflective surface, they were again afraid, staying away from food placed in front of it. The next dawn, they came up to the mirror and took the food, appearing to ignore the images of themselves, as the other group had done; but then two of the six birds ambled back to mildly interact with the mirror. These two each peered into the mirror intently, bill to

reflected bill, then both repeatedly reached up with their feet as if trying to grab their reflected images. They were silent and they didn't
seem aggressive. These birds had been born that spring, whereas the
first birds in the experiment were over two years old. I'm not claiming
that age is relevant. I suspect it isn't. It is just the only difference that
seems tangible enough to mention.

Soon after I had made these observations, I received a call from a
man in northern Maine named Matt Libby, who told me about a peculiar raven problem. For about a hundred years, his family had rented
out wilderness camps, and there had never been any raven problem
"until three or four years ago." The problem was that one or two
ravens perched on his camp porch railings, leaving considerable white
deposits, tearing up his cedar wood chairs, hacking window framing
to bits, and dirtying up the windows under the shaded porch. Ravens
had always been around, he said, but they had never been a problem
before. I asked if there was anything different now. There wasn't, he
said. Then he thought a bit and came up with one tiny detail: "Only
that I've now got thermopane windows rather than single glazed
panes." I checked with glass merchants and learned that most thermopane sold these days is "very low reflective," called "Low E." It has
a coating on the outside that allows heat waves in and then reflects
them inside, so that less heat leaks out as radiation. Did that have
something to do with the ravens under the dark porch attacking or
being attracted to their reflected images in the camp windows? My
data do not answer the questions with which I started, but they point
to interesting studies that would be fun to do, not only with mirrors
but with television screens. For the time being, I had to be—and
was—content to observe ravens in the field, where they routinely
make other, perhaps even more vital, identifications.

FIFTEEN

Dangerous Neighbors

RAVENS ARE FLEXIBLE. THEY NEST on pine trees in Maine's inland wilderness, on tall beech trees in northern Germany, on power lines, on radar towers, on buildings above busy streets, under highway overpasses, in active railroad trestles and abandoned buildings. They've nested in the trunk of an abandoned car in the Mojave desert, and on a baseball stadium in Elmira, New York. There is even now an active raven nest under the MJ section of Penn State's Beaver Stadium.

Whenever possible, ravens prefer to nest on rock shelves tucked under overhangs on cliffs. These are the same sites also preferred by some of their mortal enemies: golden eagles, gyrfalcons, and great horned owls. Strange as it may seem, raptors and ravens are often coinhabitants of the same cliff, sometimes nesting within yards of each other. Very often,

Houdi's nemesis. The female raven at her cliff nest, about a mile from my house in Vermont. Photograph from a blind in a nearby maple tree.

falcons and owls take over an old raven nest because the raptors cannot build their own nests. The ravens, when evicted from their nest, will often build another one nearby. Thus, ravens provide the raptor a nest.

In the city of Bern, Switzerland, ravens started nesting above the busy public square in the middle of the city on a large government building, the Bundeshaus, in 1988. By 1995, the raven pair had built five nests in different cornices of the building, and at least five pairs of falcons of two different species (*Falco tinnunculus* and *Falco peregrinus*) had moved in to use each of the nest places the ravens had constructed. In 1998, they still nested there, bringing off three young, and entertaining the townspeople.

The associations between nesting ravens and raptors, and between neighboring territorial ravens, are complex and "personal." Relationships probably change with time as disputes are resolved, antagonisms subside, and truces develop. As one example of a truce among traditional enemies, I once had a tame great horned owl and two tame crows, all of whom lived free in the woods close around my cabin. Crows are one of the owl's favorite and common prey. Their first encounters were tense, yet owl and crows eventually ignored each other altogether and peace was restored.

The raven-raptor association is not always one-sided. There is also advantage for ravens to associate with raptors, who often provide kills

Raven nest on the Bundeshaus in downtown Bern, Switzerland.

from which to scavenge. Perhaps even more important, if one fierce species becomes tolerated near the nest, it becomes a "watchdog" for predators of the other, seeing and repelling strangers.

Most raven nests do not have a "dear enemy" guard, and at least one member of the raven pair stands guard at the nest at all times. In Denali Park in Alaska, I had the privilege of watching a raven nest with several half-grown young for a week at a cliff where I occasionally saw peregrines, gyrfalcons, and golden eagles fly by in the distance. No raptor nested on the cliff itself. One of the raven pair was always on guard to greet me loudly whenever I came near the nest. The second bird, summoned by the first's commotion, would come immediately and join in the clamor. Then one day, one of the adults disappeared. Since the remaining bird had to leave the nest to forage, the young were sometimes left unprotected. After just one day, the young disappeared from that nest. Only a large raptor could have taken those young, since the nest was inaccessible from the ground.

Ravens also come in contact with numerous other potentially dangerous birds at food. What follows is a set of observations of one raven, several crows, a turkey vulture, and two broad-winged hawks, *Buteo platypterus,* feeding at the same carcass near my Vermont house in 1994.

On April 2, I did my part like all the other people who feed birds in the winter. I laid out bird food. I put a dead, cut-open calf within sight of my bedroom window. The first birds came at dawn: three crows and

Young ravens taking the measure of a turkey in my aviary.

one raven. The raven was soon feeding and flying off with chunks of meat. The crows alternately perched in the trees and fed when the raven had left. The raven, the male of the pair that nests nearby, had already come to feed on my offerings here for a number of years. The female would now be incubating, and he daily brought her meat from this calf.

On April 3, one raven along with one crow and one turkey vulture came in the early morning. The vulture fed almost continuously for at least two hours in the presence of either the crow or the raven. As before, crow and raven fed alternately, but both fed alongside the vulture, acting as if it were invisible. The vulture in turn ignored both corvids.

Eight crows arrived in the early forenoon. As many as five of them fed simultaneously alongside the vulture, but all the crows left the bait whenever the lone raven came back to feed. Crows vigorously chase ravens one-on-one, especially in spring and summer, but sometimes also in the fall and winter. But here, none of the eight ventured onto the ground when the one raven was near. The raven paid no attention to the crows, nor have I ever seen ravens aggressive to crows or try to chase them away from bait.

The next day, eight crows and the raven were back, behaving like the day before.

On April 11, after I had been away for a few days the calf carcass hosted not only five crows, a turkey vulture, and two broad-winged hawks, but also one coyote. The latter left just as it was getting light, before the birds arrived.

Again, the crows flew up and did not feed when the ravens fed, but the crows fed amicably and without sign of alarm or caution directly alongside both hawk and vulture. The second hawk waited its turn perched in a tree, and fed only after the first had left. As always before, the raven took precedence over all. It unhesitatingly came down to feed regardless of whether a hawk was on the meat or on the neighboring tree; but when it fed, no crow, singly or in a crowd, ventured near this male raven. The raven never bothered to nip the tails of the hawks or the vulture next to it.

Ravens are well known to pull the tails of raptors near baits. Their reason for doing so is obscure, but their effects on the raptors are clearer, though varied. In northern Arizona, recently released captive-reared California condors, *Gymnogyps californianus,* are provided with animal carcasses to feed on. Quite often, the condors are flushed from this food when golden eagles appear, and they may not come back for days. However, Amy Nichols, working for the Peregrine Fund on the condor project, reports that although ravens may "continuously torment the condors by pulling relentlessly at their feathers," they also viciously mob any golden eagle that comes near. The eagles retreat from the ravens, giving the condors time to feed, while the condors are not displaced from the carcass by the ravens.

Ravens do not normally interact with chickens, but naive ravens may have no way of distinguishing them from eagles. What might they do with them? Our neighbor had a dozen Rhode Island Reds, each weighing almost twice as much as a raven, and I had my group of six ravens, six months in age. The chickens were past laying age, which is why my neighbor donated them in the name of science. She didn't want to kill, pluck, clean, cook, and eat them herself. After I had tried one, I knew why. My ravens had hardier palates. They liked the first broiled chicken meat just fine. Before I planned to give them the rest raw, I decided first to find out how they would respond to them live.

When the first two chickens were let out of the burlap bag I'd brought them in, they cackled excitedly, walked slowly and deliberately, and seemed unconcerned about a mere six ravens that happened to gather round. The ravens started edging closer, walking sideways and crouching, enabling them to make an instant retreat, which they did whenever a chicken as much as took one step toward them. Within an hour, some of the ravens got bolder, managing to sneak up occasionally to yank on a tail feather. If a chicken ran, the raven hopped comically behind the cackling hen. Soon the ravens were interrupting their usual play with sticks and other objects to "count coup" with chickens. This activity continued almost ceaselessly for the whole day. Except for losing a few tail feathers, no harm was done to the chickens. The ravens became ever more disinterested and the hens became ever more nonchalant. Before two days had passed, they were feeding alongside the ravens on a calf carcass. The ravens ignored them, making no attempt to chase them off "their" carcass, although they were often intolerant of each other.

To find out if the ravens became habituated to chickens in general, or just to these particular individuals, I replaced the first pair with two others. A new Rhode Island Red I released into the aviary at night seemed to be quite excited when it woke up the next morning to find ravens flying all around. This bird panicked, and it was briefly "tested" by the ravens as the other two had been initially. All the other hens I later provided were totally ignored.

The rooster was next—a robust black and white Plymouth with a brilliant red comb and wattles and a thick neck. He stood so tall, you'd think he would have toppled when I let him out of the bag. He spread his shoulders, flapped his wings, crowed six times, then ran to the hens. The ravens were unimpressed. They continued to play with sticks. The rooster acted as if the ravens weren't there. Whenever he happened to walk near a raven, it scuttled or flew away. No raven "tested" him as they had the first two hens. Nor did they approach another confident chicken, a buff Orkinton. Had they generalized, realizing that despite their superficial differences, they were all just chickens?

I left one chicken in the aviary with the ravens for a month. The ravens always yielded to her, and never even tweaked her tail. Whenever food was provided that she and the ravens wanted, she walked up to it and they briefly scattered. This is not exactly the behavior one might expect of ravens, who boldly pull eagles' and wolves' tails, chase golden eagles, and feed among wolves at the kill.

Several months later, I provided them with two new Rhode Island Reds who were the same size and appearance as the others, but who were still young. One acted unsure of herself. The ravens drew closer. She panicked, and then all the ravens attacked her relentlessly. After two minutes and an apparent imminent slaughter, I felt obliged to remove her. Like an ostrich sticking its head in the sand, she had sought refuge by sticking her head behind a loose board. The other chicken, who was a highly confident individual, fed alongside them and was ignored.

The next trial was with an adult turkey of wild-type coloration. The huge bird, seeing the ravens, fluffed out to make herself look even bigger, fanned her huge tail wide, and while holding her head horizontal, hissed and walked slowly and deliberately. The six ravens couldn't resist. In the first thirty minutes, they tweaked her tail a total of sixty-two times. I sampled four more half-hour periods over the same and the next day that I allowed the turkey to stay with them. The total tail-pulling contacts per half hour steadily dropped from sixty-one to forty-six, twenty-eight, sixteen, and then to two. Two ravens contacted the turkey forty times each, while one bird did so only once. Together, the six ravens snagged only about a dozen tail feathers.

On May 16, I presented them with a fat, brown-backed, calm, and confident gander. The six ravens immediately came off their perches to surround the goose. All were silent. They tried to sneak up on his rear, but the gander turned. All jumped back. One raven finally nipped at his tail. As the gander made a feint at her, her mate rushed in and also yanked the goose's tail, but that was all that happened. In fifteen minutes, all the ravens had lost interest and wandered off. When I later let a tame Canada goose they knew well into their aviary—it had for two months wandered to the edge of their aviary—they attacked it vigorously at once. As with the unsure hen, I feared for its life and quickly removed it from them.

Their next meeting was with Murphy, my daughter's lively, wolf-colored German shepherd. The dog was treated with considerably more respect than the fowl, but with at least as much interest. At first, the ravens made high-pitched, upward-inflected alarm calls. But Murphy paid no attention. She sniffed all around and within seconds settled down to chew on a frozen calf as all six ravens swirled around her. The birds were excited, squeezing out one fecal dropping after another, which kept getting smaller with each one. Within a minute or two, all of the ravens except White had flown out through the open door and into the side aviary to perch safely and silently up on the high perches in their shed. White, the normally totally silent bird who was at the very bottom of the dominance hierarchy of this group of ravens, behaved entirely differently. Not only did she stay to watch Murphy intently, but she also kept following the dog, and she swooped over the dog repeatedly, making rasping calls. Blue, the undisputed large dominant male, continued to cower silently up under the protective shed in the adjoining aviary, along with the others. Within minutes, White's vociferations grew louder. Soon they were deafening and she became ever more animated! She not only chose to stay in the same aviary with the large dog, but she began an extraordinary display I had never seen her do before, which was out of character with her otherwise meek postures and demeanor. She stood tall, held her bill high, erected her ear feathers (a display of power and dominance), fluffed out her throat and spread her shoulders (also parts of the dominance display); then she backed up these postures and feather displays with an amazing range of vocalizations in an endless spirited, loud monologue, an unbroken series of long trills, loud yells, high, rapid series of *yip*s that switched now and then to low, rumbling, rasping growls. The sounds ranged back and forth from squeaks, so high in pitch that her voice broke, to low, deep, rumbling sounds. All of this was accompanied by wildly gesticulating head and wings. Every few minutes, she stopped for a few seconds to swoop once more over the dog's back, making uniformly deep, long, rasping calls, the same kind the birds make when confronting any strange animals, and sometimes also strange ravens. Her animated behavior startled me, because she had been almost silent for months. She also had never acted "macho" before. She had changed from a wallflower to an engaging, happy, confident, and powerful personality. I doubt that the display was for the dog, though; perhaps she realized that she was alone, and that the others would keep their distance (see p. 201).

After four separate ten-minute sessions of Murphy in the aviary, none of the ravens had come close enough to yank tail, as a previous group of much younger ravens did within a minute or two to an old, lame husky I had introduced to them.

My experiments with barnyard fowl and household pets may seem inconsequential, but they were informative. They showed that ravens

go to some lengths to try to get acquainted with strange large creatures they don't know. After they get to know them, they either
ignore them or try to kill them. Knowing this, however, is no substitute for seeing what actually happens in the field, where the real test
comes as they face big, fierce carnivores.

I had made plans years earlier to travel to Yellowstone Park and to
Oregon, where colleagues had kindly volunteered to set up carcasses
for me to watch wild ravens interacting with naturally occurring competitors and carnivores. Wolves had not yet been reintroduced to Yellowstone, but coyotes there (unlike in Maine) were diurnal, and I was
curious how they interacted with ravens. Unfortunately, when it came
time to use my airline ticket, I had reached a critical point in a radio-
tracking experiment and was unable to leave without putting this
ongoing work into jeopardy. However, I found an eager volunteer,
Delia Kaye, who gladly accepted the challenge to be my emissary,
with the understanding that she would take extensive notes.

Delia went to Yellowstone Park in mid-February 1992, where she
was hosted by John Williams, whose team of researchers was studying
the pack behavior of coyotes. The park ravens were not food-stressed.
As is usual near the end of winter when ravens begin breeding, there
were winter-kill elk carcasses all around. These were so abundant that
many had no ravens feeding from them at all. The birds apparently
preferred only the freshest, just-opened carcasses.

Delia was on hand to see one pack of coyotes begin to tear into a
bull elk that had just died. Within several minutes after the coyotes
had torn a hole into the neck, two ravens arrived and started feeding as
well. The coyotes occasionally lunged at the ravens, but the birds
merely jumped aside and came right back. Most of the time, the
ravens, whose numbers quickly increased to about a dozen, fed unmolested with the coyotes nearby. The birds seemed surprisingly relaxed
in the company of the coyotes.

In Eastern Oregon, the situation was entirely different. Gary
Clowers, our host there, had set out a deer carcass near Grandview on
the eastern base of the Cascade Mountains. From a blind made near

this carcass, Gary and Delia kept watch for four days. No coyotes came, but as many as twenty-six ravens were present at a time. Unlike at Yellowstone but much like in Maine, these birds did not quickly descend to the opened carcass. Instead, they loitered about in the vicinity, acting fearful of the unattended carcass.

Typically, all the ravens assembled on the ground about ten to fifteen yards from the carcass, then started to approach it as a group. Coming closer, two or three birds might haltingly edge toward the carcass, drawing others behind them. Then they would all jump back again and fly off. As in Maine, whenever they drew near the carcass, they walked hesitatingly, opening their bills in fright and performing what looked like numerous jumping jacks. Even on the fourth day, with twenty ravens routinely near the carcass, only two individuals approached it close enough to feed. These results were almost precisely as those I was used to seeing in Maine. But there was a big difference. Instead of all of the crowd eventually ending up at the carcass, most of the ravens got their meat without ever needing to go near the carcass they feared. They used intermediaries. The dozen or so magpies at the site showed no hesitation about feeding from the carcass. Neither did four eagles (two immature and one adult bald, and one immature golden). The eagles freely walked up to the carcass to feed, showing no hesitation at all. As a magpie, raven, or eagle left with meat, members of the timid raven crowd would take up a chase. Only birds leaving the carcass with food were chased. Up to four ravens might chase a single eagle or magpie. Magpies quickly dropped their prize when pursued by ravens, but whether all the eagles dropped their food could not be determined since the eagle-chases went far out over the countryside.

In New England, there are no magpies, and eagles are rare. Blue jays and crows, unlike magpies, don't come near a carcass with ravens. The strategy of stealing food from intermediaries rather than approaching a feared carcass is less effective in Maine, but I have on numerous occasions seen ravens that have secured a piece of meat, when chased by other ravens, drop their meat routinely to terminate the chase. At dumps gulls are similarly chased by ravens until they give up their food morsels.

Along the Maine coast, bald eagles were fed at supplemental feeding stations for several years to prevent them from migrating south and becoming contaminated with pesticides. These feeding stations became prime *raven* magnets, attracting huge crowds of them. The ravens did not ignore the rare eagles. They treated them as my young aviary birds had treated the chickens and turkey when they first met them—they edged up behind them to yank their tail feathers. Were they trying to tell them something, discourage competition, find out something, or impress potential mates with their daring? There were hints here that the ravens' lives are interdependent with other animals at and around carcasses, although I did not yet suspect the true relevance of these interactions.

SIXTEEN

Vocal Communication

COMMUNICATION CHANNELS IN THE animal world include those of touch, sound, sight, and scent. Electric eels even use electric pulses. Insects communicate their sexual readiness and their location by scents, sounds, and movements. In our social interactions, we communicate all sorts of information unconsciously, using our eyes, gestures, tone of voice, and facial expressions. Ravens also are very expressive. By a combination of voice, patterns of feather erection, and body posture, ravens communicate so clearly that an experienced observer can identify anger, affection, hunger, curiosity, playfulness, fright, boldness, and (rarely) depression. The ravens' calls have one basic message, which is to draw attention to themselves. Beyond that, they

A still-young Goliath (note light-colored mouth) play-vocalizing
(i.e., "singing") to himself. His feather and body postures reflect
a confident, self-assertive mood. He had, at this time, already
lost most of his tongue in a fight with Fuzz.

also indicate functions: feed me, stay away, come here, recognition. Increasing specificity comes from context. The specific calls are not used as language. Ravens don't have calls symbolizing carcass, eat, come, meat, et cetera. Ravens can't say, "Come with me to that carcass to eat some meat." Such communication implies complex thoughts. Ravens don't think with words. If they think, it is with images, as we do when we don't use words; but the basic logic of communication remains.

As an example of the logic of communication, let us reconsider a simple case: the loud begging cries of young ravens. The begging is understood by the parent to mean that the young need to be fed. Of course, the young can "lie"—in their competition with one another, they can try to outshout the others to get more than their fair share. I mean this in an unconscious sense, because I'm referring to evolutionary logic. A high cost of that loud begging could be the attraction of predators who would eat *them*. That would be *information transfer* to predators, but not *communication*. Information transfer occurs between young and parents, and in that case there is communication, because both signaler and receiver benefit. But the costs and benefits to the participants may vary, and evolution within both participants acts to minimize costs. Why raven young are especially noisy relative to most other young birds can be seen from the "experiments" that evolution has conducted over millions of years. We find, for example, that young woodpeckers safely ensconced inside fortresses of solid wood are even noisier for their size than ravens. Young woodpeckers make a din that scarcely ever stops, whereas the young of all ground-nesting birds, who are extremely vulnerable to predators, are almost totally silent except for the few peeps they make at the precise moment that a parent visits the nest with food. We can conclude, therefore, that young ravens, like young woodpeckers, are proximally noisy because they are hungry, and ultimately very noisy because of their relative safety in their hard-to-reach nests. There is little cost to their being loudmouthed; they can "lie" and act as if they are starving when they merely have an appetite. No mental awareness is implied in any step of this process.

The communication described above conforms to the theory called the Handicap Principle, as promoted by Israeli biologist Amotz Zahavi.

Much of animal signaling makes sense according to this theory. It says that in order to be effective, signals must be reliable. And in order for the receiver to know they are reliable—for the receiver to treat them seriously—they must be costly to make, for example, by attracting predators. If they were not costly to make, then signalers could "cheat" or give false information. This logic, however, is not always easy to apply to ravens' vocalizations, as the following examples suggest.

At dusk on September 7, 1997, a cougar crept up on Ginny Hannum as she was working at the back of her cabin at the head of Boulder Canyon in Colorado. The cougar crouched low among the rocks, facing her from about twenty feet, and it was ready to pounce. Hannum, at ninety-eight pounds and four-feet-eleven-inches tall, was a well-chosen target.

Although Mrs. Hannum was unaware of the cougar's presence, she had become "somewhat annoyed" by a raven "putting on a fuss like crazy." "I never paid much attention to ravens," she told me, but "this one was so noisy that it was downright irritating." The noisy raven kept coming closer, having started its commotion twenty minutes earlier from about three hundred yards away. Hannum had never before noticed ravens "cackling like crazy." Was this raven trying to say something? She started to listen more closely.

The cougar was ready to make its kill, but the raven was close, and it made a pass over the woman, calling raucously, then flying up above her to some rocks, where she finally saw the crouching cougar. As the cougar glared down with yellow eyes locked onto hers, Hannum quickly backed off and called her three-hundred-pound husband. The surprise attack had been averted. She had been saved. She recounted, "The lion moved his head just a little bit as the raven flew over it. That's when I saw him. I never would have seen him otherwise. He was going to jump me. That raven saved my life." The event was declared a miracle in the news.

A miracle is any event the natural cause of which we do not understand. That provides an adequate number of miracles to some of us—certainly to me. Why did the raven call? To the religious Hannums, it seemed a miracle that a raven would go out of its way to deliberately save a human life. To me, raven behavior is still a miracle, although I have faith that this raven's behavior was within the realm of what ravens normally do. They are alert to predators that could potentially provide them with food, as well as to anything strange in their environment. Perhaps the raven had been luring the lion to make a kill, alerting it to a suitable target. If the lion had feasted, so would the raven. That is, both would have benefited, as expected in communication.

David P. Barash, a professor of psychology and zoology at the University of Washington in Seattle, wrote to me about two observations he made that suggest, but do not prove, that some ravens may follow cougars to feed on their kills. Barash was working on his dissertation research on the sociobiology of marmots in Olympic National Park when he saw a cougar stalk and kill an adult female marmot. He recorded in his notebook that "within seconds," the cougar carrying the dead marmot was followed by two ravens. Who followed whom? Had the cougar perhaps at first followed the ravens?

A similar account with a possibly similar scenario appeared in the *Anchorage Daily News* (December 29, 1998). In this incident George Dalton, Jr. came face to face with a grizzly bear on a hunting trip near his village of False Bay. George had wounded a deer and he followed its blood trail into the brush where the deer went to die. The bear found the deer and also wanted to lay claim to it. After some tough negotiating with the bear, who was stomping angrily on the ground, George told him (in Tlingit) to please leave him alone. The bear came closer nevertheless. Soon George could smell the bear's breath, and fearing for his life, then said to it: "OK, you can have him. He's yours," while backing away and retreating into the brushy muskeg. George recounts that the bear made a charge: "Ravens were following me and squealing. I thought they were guiding me and telling me that the bear was still following me."

My interpretation here is also precisely the opposite. I suspect the ravens were not warning the man, but informing the bear of a potential victim instead. The ravens have a lot to gain if a bear makes a kill. They were probably guiding it to an intended, perhaps prechosen victim. Everything I know about ravens, as well as folklore (see Afterword), is congruent with the idea that ravens communicate not only with each other, but also with hunters, to get in on their spoils.

Whatever else these two incidents illustrate, they show the difficulties of interpreting communication, and how much interpretation can depend on the mind-set of the receiver. The Hannums and George Dalton thought the ravens were communicating with them. Instead, the ravens were probably informing the predators. To make sense of communication, the first relevant questions to ask are: What are the costs and the payoff to the givers and the potential receivers of the signals given?

Since prehistoric times, ravens have been thought to have divining powers. Ravens were first kept in the Tower of London because of their vocalizations, which were thought to warn of approaching danger, much as the Hannums and George Dalton believed they were warned of the predator. No bird calls have generated more excitement throughout history than those of the raven. Even now, the calls of the raven seldom fail to excite those that hear them in field, forest, and mountaintops. In a 587-page book on communication in birds, titled *Ecology and Evolution of Acoustic Communication in Birds* by Donald E. Kroodsman and Edward H. Miller, published in 1996, however, there is not one peep about the raven, although it lists the genus *Corvus* as "particularly worth investigating." We know infinitely less about vocal communication in ravens than we know about the call of a frog, a cricket, or the zebra finch. That disparity reflects not so much lack of interest as our inability to get replicable data. The more complex and specific a communication system becomes, the more random-sounding and arbitrary it will appear. We will have trouble distinguishing it from noise. Any meaning that we can find should be welcomed.

In one study completed in 1988, Ulrich Pfister of the University of Bern in Switzerland spent one winter recording all the vocalizations of raven pairs living within about 1,000 square kilometers south of Bern. Of the thirty-four different call types that he recorded, fifteen were individual-specific, eleven sex-specific, and eight specific to the ravens of that area. His associate Peter Enggist-Düblin since then has made 64,000 additional recordings of raven calls near Bern. New calls kept turning up with each raven pair examined until, after analyzing seventy-four individual ravens from thirty-seven pairs, Peter brought the total to eighty-one calls. Even more calls would presumably have been found if more ravens had been sampled, even in just that one study area. On an informal basis, I still recognize new raven calls almost every year in my Maine study area, even after fifteen years. I hear distinctly different calls in every area outside New England where I've been. There are tremendous variations of intonation and dialect, and I'm not at all sure that what I perceive as one call type is not really many, or vice versa. But what are the meanings? Peter concludes from his work that ravens' calls do not all have the same meaning. Rather, some calls' meaning are context-dependent and established by convention. They are then culturally transmitted.

In a letter Peter acknowledged to me that: "We have difficulty publishing our ideas, which, in our own opinion, go beyond, and are therefore also a critique of the current understanding of communication in animal behavior science, which seems for some referees hard to handle." I would soon enough find out the truth of his observations.

I made only an informal short glossary of some seventeen common raven calls from my two tame pairs, Fuzz and Houdi and Goliath and Whitefeather (see Table 16.1). Seven were restricted to one sex only (four to male and three to female), and of these seven, four were given only by one of the four individuals, and then only after the birds were more than a year old. That is, the trend was for sex-specific calls to appear when the birds were older, and some of these calls were also individual-specific. I was unable to decipher innumerable other nuances that occurred routinely, and was forced to be a "lumper" of calls rather than a "splitter." Mine was only a rough personal probe without systematic recordings. I wanted to distinguish the calls by ear, so that I could then routinely "watch" these birds in the field with my ears when I could not see them in order to possibly do a more systematic study later if I should detect an interesting hypothesis to test. Despite the crudeness of my method, I did find something significant: Regardless of the type of call, dominants of each sex may effectively silence all or almost all calls of others of their respective sex in their presence (see Table 16.1). For example, while confined exclusively as a pair with her mate Fuzz, Houdi made 804 of the 2,309 calls I tabulated in May 1995. Four months later, the Fuzz-Houdi pair was combined with the Goliath-Whitefeather pair in the same aviary. Houdi (no calls) and Goliath (13 calls) went nearly silent as Fuzz and Whitefeather then made most (467 and 338, respectively) of the calls. I again separated the two pairs, Fuzz-Houdi and Goliath-Whitefeather, and during December–January Houdi and Goliath became vocal again, making 380 and 902 of the total of 6,570 calls I tabulated.

By and large, my efforts were a confirmation of what Pfister and Enggist-Düblin also found. Some of the raven calls they recorded were common to all the birds, but the majority could be learned and culturally passed on. Even within the 1,000-square-kilometer Swiss study area, there was an east-west geographical separation in distribution of call types. The greater the distance between nests, the fewer call types were shared between them. Some individuals at the dialect boundary were "bilingual" for certain call types. Since some of the calls were strictly specific to males and others strictly specific to females, it was concluded that there is a tendency for males to learn the calls of other males, and females to learn those of other females. There was also a tendency for mated pairs to share calls.

Ravens are well known for their capacity to mimic, especially if they are isolated from others of their kind. Mukat, a lone resident in a cage at the Living Desert Museum in Arizona, makes a perfect rendition of portable radio static. A raven used for physiological research outside the biology building at Duke University perfectly mimics a motorcycle being revved up. A couple of the raven's perhaps more interesting and unusual vocalizations were related to me by David P. Barash. While David was studying his marmots at a colony in early June in Olympic National Park, he distinctly heard, "Three, two, one, *bccccchhh,*" (the last a guttural sound of about four seconds' duration, serving as an excellent imitation of an explosion). The sequence was repeated at least three times. He wrote me, "It sounded so realistic that I looked around for the speaker, even calling, 'Who's there?' out loud, despite the fact that this risked disturbing the marmots I was supposed to be watching." It turned out that the "speaker" was a raven, perched on a nearby snag. Park rangers had conducted avalanche control the previous week, and apparently the raven had heard, and been sufficiently impressed. David continued, "Later in the summer, I would commonly hear the rushing, gurgling sound of urinals flushing. Again, the culprits were ravens—at least two different ones this time. There was a picnic area about a half kilometer away, outfitted with toilets whose urinals automatically flushed every thirty seconds or so. Ravens often perched atop these structures." And they apparently were at least as impressed with *those* sounds as they were with those from the avalanche control crew.

An intriguing question is whether or not ravens can learn arbitrary sounds and then associate meaning to them. For example, babies may at first babble and make sounds like "mama" or "dada" that they only later associate with the appropriate subjects. Darwin, the raven that Duane Callahan is currently training for wilderness rescue work in California, gives hints of recognizing the meaning of sounds he makes, perhaps because he gets rewarded with certain coincidental associations. For example, he has learned what the words "Want to go outside?" mean, because Duane always uses them before he takes him out of the house for his free flight. He may also have learned "Duane, Duane" from Susan and Duane's brother Charles. Now Darwin says, "Duane, Duane, want to go outside?" when he wants to go out. He is, in effect, perhaps asking to go for a walk. Darwin also perfectly mimics Charles's raucous laugh. Sometimes when the telephone rings and Duane answers, "No, Charles isn't here right now," Darwin will erupt in the background with "a perfect rendition of Charles's laugh." Is it sheer coincidence? Does the raven hear "Charles," and knowing who that is, then think of his laugh?

The ravens' vocalizations invite comparison with our language, and with that comparison in mind I was especially interested in the development of language in my son. Eliot's first sounds were cries of emotion, signifying discomfort, surprise, contentment, or anger. Parents are able to decipher at least some meaning from context, much as

I can often learn meaning from a raven's calls from context. Eliot's other early vocalizations were sounds of recognition. On seeing a cat, a dog, a car, or a turtle, or almost anything else that surprised him, he said, "Da, da, da—" and the number of times he repeated it was variable, depending on his surprise. I suspect that like the raven's *rap-rap-rap* calls, his vocalizations meant, "I see, and I'm interested and surprised." The next stage in vocal development concerns specificity. For example, ravens' various levels of surprise and alarm are expressed at different predators or potential predators, although there are no raven words for them. If specific calls have meanings, they can be correctly interpreted only by those other ravens who regularly associate with that particular individual and thus know what sound is associated with what object. It was similar with Eliot. At age twelve months, for example, wetness or anything liquid was "juice." He divided the animal world into "dog" (all furry animals), "turtle" (reptiles and beetles), and "fish" (pisces of all the various orders as well as dolphins). "Dada" was dad, and curiously, also some men and any ape. In time, he would distinguish ever finer details, and the sounds he made would have specific meaning to an ever-greater circle of others beyond parents, relatives, and associates.

After Goliath and the other three ravens of his group settled into their roost to sleep at night (when less than a half year old), I often opened my bedroom window into their shed and talked to them as one might to a baby. They always answered with soft, low murmurs—*km, mm.* When the murmurs were very low, soft and long and almost whispered, I learned the birds were at ease, as could also be seen from their relaxed postures. The calls seemed to be contact calls meaning, "I hear you. Everything is fine." The young gave the same calls with a slightly upward inflection when we explored together in the woods and lost visual contact with each other. I presumed in that context they meant, "Where are you?" because when I answered the birds, they responded without that inflection, and I knew they had heard me and were still in contact. If they were being attacked by a predator, I'm certain I would have known it from their calls as well. I'm also certain nobody else would who did not know the birds.

Their intimate calls at the evening roost were usually almost whispers when we were very close together. I was reassured by their whispers and I reassured them. After our chat was finished, I sometimes heard a muffled shake of feathers, little zipping sounds of pinions drawn through bills, rapid dull scratching of toenails on the back of a head, and hollow-sounding footsteps as one shifted along its wooden perch. After I closed the window, I occasionally still heard a soft cough or a stirring on a branch. These sounds had meaning to me, because they said something about the behavior of the birds on the sleeping roost.

As my ravens got older and more independent, they chatted less with me. Instead, they woke me at the first sign of dawn with raucous calling. Perching on the windowsill, they pointedly peered in and made mostly bouts of *rap-rap-rap* calls and also deep, penetrating, long rasping caws. Both calls were otherwise given when raven intruders came near. Here in a different context, they had an entirely different meaning. My ravens wanted my attention and food. They got it, and thus I reinforced their specific vocal behavior. When they were hungry, they also gave long, drawn-out, high-pitched calls, which I call their "beg" or "yell," and they stopped after I fed them.

Still other calls draw attention and probably say, "Here I am," but from a very long distance. These are probably territorial calls because they can be heard for miles, and are usually answered by neighbors but never attract them. There are several of these loud, long, penetrating calls that are used equally by both sexes. I suspect they are less to get attention, as such, as to say, "This area is claimed." When these calls are directed at ravens within visible contact, the callers erect their "ears," flare their shoulders, and puff out their throat hackles. The macho display is omitted when the same calls are directed to me. I presume that their intent is to get my attention, because my tame birds stop giving them after I open the window and greet them.

Numerous other raven calls are given in what appear to be specific circumstances, when one emotion should predominate and be expressed as casually and without conscious intention as eagerly jumping off a perch at dawn. For example, one might expect alarm when a human or other predator approaches the nest. Instead, many different kinds of calls may be given, and the mix of calls varies from one pair to another. A seemingly much more alarming situation, such as capture, never evokes a sound from them. I suspect that calling when they are vulnerable and helpless could attract predators, not helpers, and so silence is then golden. There are numerous nuances of comfort sounds, yet recently captured wild birds give no "shouts of joy" on release from one's hand. In that case, calling to reveal emotion would serve no purpose because there is no potential listener that could benefit them for giving the call. However, that's an oversimplification—ravens often vocalize without any apparent listener near.

Raven Number 34 was an example. He was one of twenty-two wild-caught birds I had released after I had kept them in the aviary for over a year. After these birds left, Number 34 remained, perching alternately on top of the aviary, on a beech to the left of it, or on a big red maple to the right of it. For hours, he sang. His song was so uplifting and exuberant to my ears that I got out my tape recorder and started recording. I sat down less than fifteen feet from him and he paid me no attention. He was gurgling, chortling, yelling, trilling, bill-snapping, *quork*ing, and making sounds like water rattling pebbles. The bird made no female knocking sounds, so I speculated that it was probably a male, but I could not be sure.

As he sang, he raised his head high, often turning and gazing in all directions, alternately preening, stretching, picking at twigs, and gulping bills full of snow. I talked to him, telling him how beautiful his raven song was—that it was the most beautiful thing I'd ever heard. He didn't know precisely what I said, but I venture he got my message; he would have responded differently if I had made sharp, rapid yells toward him. As I had intended, he showed me not the slightest visible attention. He continued to sing, rising to another loud, rasping crescendo, then fading into a series of soft chuckles and gurgling sounds. A second recent releasee was engaged in a similar monologue far down the hill in the nearby woods, where it was also all by itself. I had never once heard these raven songs where there was a large audience, at a carcass, for example.

An anthropomorphic biased view could be that the birds sang because they were overjoyed to be released from the cage. If it had been freedom, then they all should have started to sing when first let loose. They knew they were free, because they perched in the woods all around, then returned into the cage, onto it, and all around it before dispersing. None of the twenty-two had, to my knowledge, sung at all during the whole previous year while they were in the crowd. On another occasion, I released three birds out of four, and the bird *remaining* alone in the cage was the one that erupted in song. I concluded that they started to sing because the others were gone. As previously mentioned, dominant birds may totally silence others of their own sex.

Their expressions of anger are usually directed toward another raven and hence they are a potentially far more immediately useful signal than expressions of joy for self. The raven's anger is expressed most palpably in those individuals that know their enemies and are brave and knowledgeable enough to defend the nest. When a human or presumably other predators approaches their nest, these birds violently hammer branches, tear off and toss twigs and cones, and give deep, long rasping caws that convey that the caller is powerful and serious. A raccoon trying to raid the nest would read this display correctly, and a human may do so as well.

One out of a myriad of other examples of ravens expressing anger occurred in the aviary complex where the wild-caught birds had become thoroughly used to my bringing them food. When I set food down, they always flew right over and started feeding. One day, I happened to be blocking the door to the side aviary with C48 inside, who wanted to get out by me to get to the food. He came up to me, looked me in the eye, and erupted in the same long, deep rasping alarm or anger calls that greet me when I intrude at nests with young. This demonstration would have been unthinkable earlier in the year, when he was still fearful of me and always yielded. Fright always won out over all other considerations. This time, he did not fear me and he dared to show his emotions. Given the circumstances, his message was clear to me: "Get out of the way—I want to get past you to the food." Without the context and my personal experiences with ravens, his vocalizations would have been meaningless. They would have contained no message.

He would not have given his message if he did not feel he had a chance of making me yield. He had in effect talked with me, because his message was directed at me only. I heard, understood, and stepped aside. He immediately slipped by.

Body language is also extremely important to ravens. Obviously, actions speak louder than words with them. Contrary to numerous accounts in the literature, I have never heard a raven give an alarm call when I have come

Alert and confident.

near a feeding crowd. If only one bird of the group sees me and flies up in alarm, the others also fly up almost instantaneously. I have on several occasions been hidden from the feeding crowd when a bird that is flying overhead sees me and changes the rhythm of its wing-beats, usually in rapid backpedaling of its wings. Without one vocalization being given, the crowd instantly flies up and scatters, even though they may have been feeding out of my view over the rise of a hill.

Since ravens are much smarter than insects, they don't need a long song and dance, as do honeybees, to alert them when one of their fellows has found food. Information is contained in simple action. Suppose a group of four to five ravens who know where there is food eagerly leave the roost thirty minutes before sunrise. Since all the birds who know where food is go first of all to feed in the early morning, these birds may realize that something is up when others leave so eagerly and so early. Not knowing where food is, the hungry birds follow those who demonstrate strong motivation.

My speculation is not without data. I spent one winter getting up every day hours before dawn to climb tall spruce trees near baits I had put out. Aside from the pure enjoyment of climbing snowy trees in the dark at subzero temperatures, I did it to count birds. I found that the *first* big crowds at a carcass always arrived before light. On succeeding days the birds came increasingly later and in smaller groups, even as individuals and pairs. Once feeding had begun, birds knew where the food was. From then on, they did not have to follow any other bird, and they did not have to leave the roost long before dawn when the first birds left. They were free to come on their own at any time. Some would say the ravens were not conveying information, and that this was information-parasitism instead. That implies that one benefited at the expense of the other. In the case of the ravens described above, both followed and follower birds benefited; they both got to feed by overcoming territorial defenders and/or reducing their fear of the food. So it was a communication, even if nonvocal.

Seventeen of the Most Common Vocalizations of Fuzz, Goliath, Houdi, and Whitefeather

1 = loud, hollow-sounding *cark*s that were given singly or widely spaced in time
2 = soft *mm*-sounding intimate calls
3 = low honks
4 = nasal honks
5 = rapidly repeated *caulk*-sounding calls
6 = very rapidly repeated *rap-rap-rap* calls
7 = deep, short rasping calls
8 = long, undulating territorial advertisement calls
9 = long, deep, loud rasping calls
10 = loud begging yells
11 = knocking calls, like stick in a bicycle wheel
12 = *oo-oo*-sounding calls
13 = dog whine-like calls
14 = *cheow*-sounding calls
15 = soft upward inflected calls given singly
16 = whine-thunks
17 = continuous sing-song monologues

Keys to Calls (see legend)

	1	2	3	4	5	6	7	8	9	10	11	12	13	14	15	16	17	Total
When Fuzz and Houdi were together alone, May 1995.																		
Fuzz	0	0	0	0	0	170	199	87	65	495	0	138	0	71	0	10	10	1,245
Houdi	0	0	0	0	0	220	23	33	16	25	351	0	112	0	24	0	9	812
Goliath and Whitefeather were in the same aviary with Fuzz and Houdi, September 1995.																		
Fuzz	0	0	0	0	0	20	15	17	15	380	0	9	0	5	6	0	0	467
Houdi	0	0	0	0	0	0	0	0	0	0	0	0	0	0	0	0	0	0
Goliath	0	0	0	0	0	0	0	0	1	12	0	0	0	0	0	0	0	13
Whitefeather	0	0	0	0	0	39	27	120	0	0	124	0	0	0	28	0	0	338
Fuzz and Houdi were again alone together, December 1995.																		
Fuzz	1	3	0	30	0	35	15	17	80	10	0	33	0	65	0	0	0	289
Houdi	0	0	0	10	0	37	10	90	50	0	130	0	0	0	15	0	0	342
Goliath and Whitefeather were again alone together, January 1996.																		
Goliath	87	68	0	0	14	523	2	98	105	5	0	0	0	0	0	0	0	902
Whitefeather	0	127	380	575	918	652	19	161	6	487	827	0	118	0	300	0	0	4,570
Fuzz and Houdi were alone together, January 1996.																		
Fuzz	0	0	40	0	13	357	606	96	254	0	0	10	0	0	0	0	0	1,426
Houdi	0	8	43	164	0	508	84	212	0	0	289	0	22	0	0	0	0	1,330

SEVENTEEN

Prestige Among Ravens

W̲HEN ANY GROUP OF JUVENILE ravens is put together in an aviary for the first time, they immediately challenge each other, and they soon sort themselves out into a dominance hierarchy. Notwithstanding the dominance switch between Fuzz and Goliath, and later Red and Yellow, changes of status are rare. In general, a low-status bird may improve its status only by leaving its associates and joining another group. The reverse, resulting in a loss of status, can happen as well, as the following incident indicates.

A raven crowd had slept in the thick fir grove a hundred yards from calf carcasses, but only ten birds came to feed at them on March 1, 1993. One was a new bird I had never seen before, with a long, curved upper bill that projected about a half inch in front and then curled down. "Hawkbill" contrasted with others that had appeared on occasion, including "No Bill" (half of upper mandible missing), "Peg Leg," "One

Typical male macho display, showing "ears," elevated head, spread shoulders, and puffed-out throat feathers.

Eye," and "Crump Leg," and still others who may have been victims of gunshots or traps. Hawkbill, being new among this crowd, was confident, launching into a knocking call duel with another female.

Suddenly, a violent chase ensued through the forest surrounding the carcass. The chaser had singled out a specific bird with whom it stayed relentlessly, weaving in and out among the trees and past all the other birds. I suspect the chased one was Hawkbill, because she was then missing from the feeding crowd. When she returned only twenty minutes later, she stayed at the periphery of the crowd, where she was ignored. I watched the feeding crowd for another half hour. Throughout this time, she stayed well out of the others' way, maintaining a submissive posture. She had changed from an apparently

Female raven in power display, making knocking calls.

dominant bird when she first arrived to one who always yielded to the others. Since the chase, she had not made a single knock.

In the male peacock and most other male birds, the showy and self-aggrandizing behaviors are under the control of hormones, principally testosterone. To maintain high output of testosterone, the testes increase some thirty times in size. In ravens, high-status birds suppress the sexual development of others of their sex, not just their behavior. I wondered if the dominants' brains might be testosterone-soaked at an early age. To find out, Michael Romero, an endocrinologist now at Tufts University, and I collaborated in making a probe into raven's blood hormone profiles.

After only one winter and innumerable imaginable and unimaginable mishaps later, I managed to get sufficient blood samples from ravens of known status. I sent the raven blood, and for fun and curiosity a sample I had taken from myself, to Romero, then at the University of Seattle, for the hormone assays. The results were exciting, because I had not anticipated them. The blood samples had been taken in late winter, the breeding season for birds (although ours were not breeding). I had sampled adult (black-mouthed) and juvenile (pink-mouthed) birds of both sexes and of both high and low status. Testosterone levels were low, and we detected no statistically significant difference in testosterone levels in the blood between the birds' status and sex. The big dominant black-mouthed males had only slightly greater testosterone level than the pink-mouthed immature females. Testosterone is therefore not the hormone involved in *maintaining* status in ravens. Nor can it be involved in achieving status, because ravens sort out relative status within minutes after they meet and size each other up. Unfortunately, reviewers of our manuscript felt our results would "confuse" the current ideas rather than support them, so it was not accepted. I, on the other hand, felt that if they *had* supported them, they would not have been newsworthy.

We had also assayed corticosterone levels. Corticosterone is a hormone that is released into the bloodstream with increased stress. Unlike testosterone, this hormone can be released quickly, in minutes. We sampled within a minute of capture and later, and found, as expected, a quick rise in blood corticosterone levels after the birds were captured, then this hormone again as quickly declined to basal levels. There were no differences either in basal or rise in stress hormone levels between the birds at the bottom or the top of the pecking order. I was pleased to see the birds' physiological stress response decline so quickly, and also to learn that low-status birds did not seem to be more stressed than high-status birds.

Dominance in ravens is not maintained by hormones, but by body size. In birds, real size is not easily apparent, because they are enclosed in a thick layer of feathers. The feathers may be fluffed out or depressed, making a bird seem any of a number of sizes. During the initial greeting ceremonies of males, the newcomer to a group often flaunts his size, walking slowly and deliberately, strutting with his head held in a grand, self-assertive manner accentuated by gestures. The head is held high with the bill angled up, which adds considerably to the birds' stature. The elevated "ears," partially puffed-out head feathers, and greatly broadened neck with extended throat hackles enhance his apparent bulk. The throat hackles glisten from reflected light and vibrate from swallowing motions, thus drawing attention to the neck, which seems inflated to nearly the thickness of the body. Bulk at the lower end of the body is accentuated by covering up the skinny stick-like legs with long, hanging belly feathers that cover the legs like trousers. Approaching a possible rival, the male may forcefully snap his bill and also flash his strikingly white nictitating eye membranes like headlights turning on and off. He walks, as one raven-watcher quipped, "with an air of cockiness that makes him resemble a street tough hogging a sidewalk."

If the bird that is approached is smaller and is suitably impressed by all this show, it backs down by pulling its head tight into its shoulders and pointing its bill down. That usually settles the encounter. In contrast, if a more confident bird is approached, it may rise to the challenge with a similar self-aggrandizing display. If both are suitably puffed up, a physical contest may ensue where the two antagonists grapple and fight. Full use of the bill, a potentially lethal and powerful weapon, is seldom instigated. The contest is lost by the bird who is first to show subordinate gestures.

*A subordinate bird—
here in process of blinking,
as the white nictitating
membrane is sliding across the
eyeball, forward to back.*

Do smaller birds "cheat," and try to bluff larger birds? The answer is, not much. Any individual who puts on a show is sure to be challenged by others who consider themselves a notch or two above. A prestige display (as could potentially be stimulated by testosterone) in a weaker bird can be costly, because it will be challenged. A status display in ravens is thus generally an "honest" advertisement of size and power. It is not a testosterone puff, and cannot be one.

One of the elegances of raven behavior, as I've mentioned, is that the birds reveal something else entirely besides their confidence and rank and prestige when they first meet and size each other up. They simultaneously reveal their sex. That is, there are two prestige displays, one for males and one for females. As a result, since male and female ravens have to our eyes identical physical appearances, only those ravens who have high rank are allowed to reveal their sexual identity. As shown in the following example, the chronically low-ranking birds remain effectively genderless.

In April 1992, I had a group of wild-caught birds in the aviary that included White Slash, a female (so name for a wing mark designation), and Blue Diamond, a male. They perched together, preened each other, and the female always begged from him. They were the only solidly mated pair of the group, and they never offered their preening favors to any other bird. Blank White, another female, was loosely "going with" No Tag, who had previously preened with several others. Yellow 0, as named earlier, a big juvenile male, preened Green once, and she preened him back. Otherwise, he made no displays to any birds, nor did any females sidle up to him. He seemed genderless and anonymous. That changed in minutes after a little mix-up.

I tried to chase some birds out of the side aviary and into the main aviary for an experiment on caching behavior (see Chapter 22). During this transfer, Yellow 0, the silent unassuming male, was left behind alone with a female, One Dot. It wasn't planned. They simply didn't make it out in the rush. To avoid causing more disturbance than necessary, I left them there. Once they were alone, they put on quite a show. I had never seen anything like it from those two before, when they were with the others. Both were juveniles with bright pink mouths who had not yet molted into their glossy feathers, yet from their behavior now it might have been easy to presume they were fully adult. The two perched side by side. He stood tall and erect with his bill up in the air, his throat hackles puffed out, his flank feathers and wings spread broadly to the side, making himself look big. He erected the feathers on his head. At times, he also erected his "ears" in the typical fashion of a dominant male. He flashed the white nictitating

membranes of his eyes at her, and went through the male vocal repertoire of choke sounds, gurgles, bill-snaps, grunts, honks, and *quork*s of
high and low pitch. He gave inflected, deep, and nasal *quork*s, deep
rasping *quork*s, and hollow gong sounds. Apparently impressed, One
Dot bowed with fuzzy head and made the typical female knocking
sounds. Since the sexual displays look virtually indistinguishable from
the social display of dominance, I now understand why they did not
perform them in the company of the crowd; Blue Diamond and others
would have instantly attacked to squelch them. Here, both could strut
their stuff without interference.

If the largest birds can more easily display to defend food, secure mates,
rear offspring, and keep warm at very low temperatures (because large
body size aids in conserving body heat), why isn't there runaway selection for ever larger body size? In ravens, as in all others, everything has
a cost. Like a male peacock's tail, a male raven's dominance is at times
highly beneficial, but most of the time it probably has negative value.
The energy cost of achieving and maintaining a large body mass may
not always be obvious, because at large carcasses where ravens feed,
food is unlimited, even to large dominant birds.

Perhaps when the food resources that support large size and dominance are removed, then the potential payoff is gone and the smaller subordinates "win." That there is a cost to the raven's dominance as achieved
through size was suggested during an unplanned experiment. I had left a
group of fifteen captive ravens and one crow to be cared for by a responsible neighbor-friend when I had to leave for three weeks. To make it
easier for Ron, my helper, I procured five calf carcasses, dragged them
up the hill, and put them in cold storage in a pit in the ground. I covered the pit with an insulating layer of last fall's maple leaves to make
sure the meat would last. I also left several bags of potatoes that could be
boiled and fed to the birds. I had explained the necessity of cutting
the calves open to expose the meat, and I left a sharp knife for the task.
The welfare of the birds assured, I left feeling that all was in order.

I saw my friend when I returned, and he felt that he had faithfully
carried out his duties as raven feeder, eagerly telling me how "tame"

some of the ravens had become. When I entered the aviary, I was shocked. One of the ravens had just died. Others, who seemed "tame" all right, were merely weak from hunger. Then I noticed the calves. Although they had been cut along the belly, most of the meat still remained inaccessible. I hastily cut the calves' hide to expose more meat, as a predator would, and within two days the rest of the birds had fully recovered their strength.

The surprising thing, besides how unobservant this caretaker had been, was that the starved raven was "NT," the undisputed, most dominant bird of the whole group of fifteen. The crow, a very much smaller bird totally subordinate to all the ravens, had remained in full health and vigor. Similarly, the smaller, very low-status ravens were also little affected. The ravens could easily have caught and killed the crow, since they have caught and eaten all sorts of birds—including blue jays, grouse, robins, and a saw-whet-owl—that had entered the aviary through the large meshed wire. They also could have eaten the dead NT.

Rank has its privileges in ravens. High-status birds can monopolize a carcass if they put in the effort to do so. For example, on April 23, 1995, eight vagrants suddenly showed up at 7:30 A.M. at the calf carcass in back of my house in Vermont. Within minutes, the male of the resident nesting pair arrived. (The female was then brooding on her young and could not come.) He chased one or the other of the eight vagrants without pause. The vagrants flew through the trees but always came back near the calf. A nonstop battle lasted for sixty-two minutes. Finally at 8:32 A.M., the male was victorious—the eight vagrants left without having fed once, but all of his costly efforts likely benefited the pairs' offspring.

High-status birds can also bide their time before starting to feed at feared carcasses, thereby reducing risk. In late April 1992, I worked on the ground, watching ravens in the aviary complex, after I had provided them with a carcass. It was not until a full hour had passed that one of the birds approached the carcass and started feeding. Within the next minute, most of the crowd had also fallen upon the calf. All, that is, except White Slash, a very high-status female, and her mate, Blue Diamond, the most dominant bird of the whole group. She con tinued to nudge up to him and beg in her distinctive, yell-like beg call. Her yells were low, more throaty than the others'. She was the only bird in the whole crowd I saw giving anything resembling the food-indicating yell. All her yelling was directed at him. As she nudged up to him, she often preened him, and was preened in return. It looked as if she were entreating him to go to the calf. He soon did, and they both fed.

I was surprised at that time that the highest-status birds were *not* the first to feed, and started to note status interactions more closely. The first pattern I saw during the next three days was that indeed the bird that fed first was always a *low*-status bird. After feeding started, White Slash and Blue Diamond always rushed in and fed wherever they wanted. They had simply let the others take the risk of approaching the feared carcass, while they took the rewards it had to offer.

A male peacock advertises its sex and its vigor by carrying around an extravagant tail, the embodiment of Zahavi's "honest advertisement" in communication. Most of a raven's interactions with others of its kind take place while feeding at carcasses. How can it advertise its sex and vigor? I've already noted various ways. Another occurred to me as I watched my captive group feed in May 1992.

At a carcass, there would often be one bird perched on top, in what almost seemed a precarious position. I had always assumed the dominant bird held this perch because it was the best feeding spot. When I reviewed in my mind what I had observed of birds I knew as individuals, I realized that I may have had it backwards. There are constant squabbles near a bait. I had interpreted the behavior as fighting over food, but had I really seen any bird excluded? No! Not even wimpy X, whom I suspected had a shotgun pellet lodged in his right breast muscle, because his right wing drooped. There was not a single fight at the carcass in three days of feeding. Usually, ten to fifteen birds fed amicably side by side. Almost all of the numerous jabs were by dominants at near-status birds, apparently causing them to make their fuzzy-headed submissive display. After the submissive birds' displays, they then fed along with the dominants. There was no evidence that the subordinates got less food than dominants, as I found out in a series of experiments where I captured and weighed the birds immediately after feeding bouts. If anything, the dominants took smaller bites and had to feed longer to get the same amount of food. There was an order at the dining room table. With birds wanting to be at the "head of the table," not only to eat, but possibly also "to be seen." The squabbles were less over whether certain birds were going to eat than where they would perch. To perch on top of the carcass was equivalent to sitting at the head of the table, a high-prestige place. It was not necessarily the place where food could be grabbed most easily and quickly.

One could watch the birds for days on end, even in the confines of the aviary, and have no idea of the status of one bird relative to another; but as soon as they begin feeding, anyone can pick out the alpha bird in thirty seconds. The alpha bird is positioned on top of the bait. While feeding there, its shoulders are spread, making them seem broader, which results in the wing-tips being *crossed* over the tail rather than lying parallel to it. The tail also tends to be slightly elevated, rather than straight out or slightly drooping.

The top-bird position is evident in other contexts as well. In my aviary complex, there is one large dead beech tree with a top stub that seemed to be a favorite perch for Green 67, the most dominant bird in a later group of wild-caught birds I held. This bird, unlike all the others of his group, was restless in mid-April as though wanting to leave. He flew laps round the aviary, always returning to the same stub. When he flew at the stub while another happened to be perched there already, he simply knocked the other bird off. It would flutter off or fall down into the brambles. Since Green 67 always faced in the same westerly direction from that stub, I presumed he had a destination in mind, and the stub was a convenient place for him to face in that direction. (I had captured him four miles from the aviary in that direction, and when later I released him there, he circled and indeed flew off continuing in that direction.) Meanwhile, after Green 67 was removed from the aviary, Blue 110 became dominant and took the same perch. There was no shortage of perches to choose from. When he wasn't on it then NV, the next in line, perched there. Finally, when I let Goliath in with these birds, he assumed the new dominant position, and the perch became his.

I concluded that status is central to many aspects of raven behavior. High status is costly to show and to maintain. In ravens, it is based on large size, and large body size requires much food. High status is shown by displays, and displays invite challenge. Nevertheless, the benefits of high status are many. Only high-status birds can consistently reveal their sex, and thus only they can be sexy to potential mates. They become good providers because they can defend food bonanzas when it is critical to do so. Status determines who may feed first at prized food, and also who can afford to allow others to act as a shield by "testing" dangerous food.

EIGHTEEN

Ravens' Fears

THE RAVENS' FEARS WERE MOST PE-
culiar to me. Ravens, I learned, not only were reticent to go into traps
baited with carcasses, they also acted impressively shy of carcasses
alone. They would fly over repeatedly, not landing at all. When they
eventually did land hours or maybe even days later, it was usually more
than ten yards away, and then they approached cautiously, stopping
often and looking all around, and preferably in a crowd. Finally, when
within a yard of the carcass, they would jump up and back, flapping
their wings as if startled. They would approach again, repeating the
jumping jack maneuver, eventually giving the carcass a jab with their
bills before retreating, and then approaching a little quicker the next
time. It always took a long time before feeding began. I did not know
why they were so cautious. For ravens to fear carcasses seemed almost as

---

*Young ravens have a strong innate tendency to "challenge" dogs, testing
their reactions, and then they quickly become used to the dogs, and vice versa.*

bizarre as for rabbits to be afraid of carrots. A friend and colleague, Paul Sherman at Cornell University, suggested to me that ravens were perhaps not afraid of carcasses as such, but of ground predators that might be nearby waiting to catch them. This idea seemed plausible, so I dragged a calf up into a tree. To my surprise, the ravens appeared even more afraid to go near it than before. I have friends who place the remains of deer and moose carcasses on trees to feed woodpeckers, jays, and chickadees, and they told me that ravens regularly come at dawn and call, but never touch this food. I was back to square one.

Another hypothesis might have accounted for the birds' fear. Instead of innately fearing carcasses, they might have learned individually to associate carcasses with traps and then passed the fear on to others. If fear was learned, then ravens who were never subjected to danger near a carcass should not fear carcasses as adults. On the other hand, if fear of carcasses is innate, they would. I wanted to raise nestlings by hand and examine their reactions to carcasses to distinguish these hypotheses.

Ravens specialize in feeding on animal carcasses in winter, yet to my amazement, the young ravens that I raised on dog food and chopped roadkill showed fright near any roadkill I presented them. They would hide in their shed and venture out only after a day or so. When they finally did approach, they jumped away from the dead animals. Just to be sure that it wasn't a fluke, I repeated the experiments with three groups of ravens on all sorts of carcasses and variously shaped objects. One group of birds was raised by John Marzluff, another by Bill Adams, a neighbor and friend since my childhood, and the third by myself. The results were the same. We determined that while feathered and furry things were feared, round, smooth objects were highly attractive. Long, thin food objects were treated with indifference. We were pleased with the consistency of our results, which suggested that fear of carcasses was innate, a counterintuitive and puzzling conclusion.

When I later got Goliath and the three other young ravens, I saw something else entirely. I raised them by feeding them with the same food in the same way the previous groups had been raised. However, I led them through the woods from the day that they hopped out of the nest. Unlike the other young ravens, these four were not afraid at all.

They went out of their way to contact any carcass or anything else I pointed out to them. They rushed up to peck a huge blue water bucket, brown paper bags, or a dead woodchuck. I saw wild ravens behaving the same way, but only toward edible items.

Those things my ravens had not experienced under my tutelage, they remained afraid of as they got older. Turtles, for example. At just over two years of age, Fuzz and Houdi were ready, I felt, to face a turtle. They had by then contacted most kinds of roadkill, and showed little hesitation to begin feeding on them, although a dead crow was greeted with rasping alarm calls, then never touched at all. On June 6, 1995, just as the painted turtles were leaving ponds to lay their eggs on shore, I secured a mature, 6.5-inch-long specimen, to observe both the turtle's and the ravens' responses. When I picked the turtle off the road, it extended its legs and head as if trying to escape from my hand. It did not withdraw into its shell.

When I put the turtle deep into the aviary, Fuzz and Houdi were curious as well as mildly alarmed. Both made long, deep rasping calls, hers slightly higher in pitch than his. The turtle walked forward tentatively and stopped at a log as both birds flew down and looked closely. Although Fuzz never attacked Houdi at food, he attacked her now. She ceded and became passive and watched as he approached closer.

The turtle, meanwhile, withdrew into its shell. Twenty-one minutes later, Fuzz still had not touched the immobile turtle, but he did yank away sticks and leaves and other debris from it, all the while continuing to edge closer to it, then jumping back nervously. The turtle did not budge. After thirty-three more minutes, both birds finally lost interest in it. But twenty minutes later, the turtle slowly and cautiously moved six inches closer to some cover of nearby grass. Both birds resumed their rasping caws, and Fuzz again approached it. The turtle pulled back into its shell. This time Fuzz left it alone after only seven minutes. After he left, Houdi again became active, hopping back and forth above the turtle and making rasping caws. After thirty-two minutes, at 7:10 A.M., the turtle again poked its head out, but by this time Fuzz was bored. Finally, when at 7:33 the turtle crawled off, the birds, perched on their tree above it, showed no more interest. I then released a turtle hatchling. In less than a minute, Fuzz flew down, grabbed it, and crushed it in his bill. Knocking, Houdi sidled up to him. He responded with a macho choking display, then cached the hatchling, which she immediately retrieved and ate.

Nineteen days later, on June 26, I found another adult female painted turtle that had just laid her eggs, and put her into the aviary. Fuzz walked to her and nonchalantly contacted her in less than one minute. Within two minutes, he reached under the turtle with his bill and deliberately lifted her up and heaved her over onto her back in one smooth motion, much the way the ravens of Yellowstone Park flip buffalo patties to find insects. He danced around her, lightly picking at her shell. Then he left her and showed no more interest.

Normally, a painted turtle rights itself in seconds. In the presence of the ravens, this one played dead. The turtle stayed upside down and immobile for ninety minutes before I removed her and put in a dried cecropia moth instead. The dead moth evoked deep, rasping alarm calls. Fuzz danced around it for at least ten minutes, lunging and retreating, jabbing and pecking like a matador. He clearly was afraid—of a moth! Finally, after a number of tentative pecks at its wings, he grabbed it and ate it. As always, Houdi stayed back and watched intently from a safe distance. Was she learning about this object, the moth, by watching Fuzz?

There is no doubt that the young originally learn from the parents' example. On June 7 in a later year, 1996, when they were parents, I spread out a gray squirrel, a woodchuck, a calf, and a porcupine at different locations in my clearing by the cabin. Goliath and Whitefeather fed unhesitatingly from one carcass after another, and their six young followed close behind them. The young had never seen any of these animal carcasses. The six of them, unlike all of my hand-reared ravens of this age, did not hesitate to hop right up to the carcass with their parents to be fed at the squirrel, the calf, the woodchuck, and the porcupine, and to peck them themselves. They showed not a flicker of concern. When the parents left, they continued to feed at the different carcasses. Just like that, the young had overcome all fear that would have kept them from feeding without the parents. No wonder that when Goliath had been just out

of the nest, he had followed me closely and examined and pecked everything that I touched, quickly feeding on it if it was food. His following
behavior ensured he'd learn what *not* to fear and what was food, from an
older individual with experience, a parent, who in his case was me.

Ten days later, I spread more roadkills around in the field in front
of the cabin—two woodchucks, one painted turtle, and one large
snapping turtle. Within minutes, the whole crowd of eight made raucous cries and came boiling out of the woods. Goliath was first down
to the woodchucks. The young followed and started to pick meat on
their own. Goliath then left them and went first to one turtle, then to
the other. The young followed and fed unhesitatingly at the turtles
also, quickly and effortlessly having learned from him.

Eight days later, I dropped off another carcass the young had never
seen before: a roadkilled cat. At 7:45 A.M., one of the young circled
over it and flew on. Goliath and Whitefeather were away that morning. At 9:00 A.M., five of their young were near the cat, perching in a
spruce, looking and yelling forlornly. They wanted to feed but were
afraid to go near it. Goliath returned at 10:44 A.M., saw the cat immediately, walked to it, pulled at it, rolled it over, and started to feed at
the cut-open belly. Within seconds, the whole crowd of five young
rushed down, crowded around him, and started to feed themselves. (I
presumed Whitefeather had left that morning with the sixth young.)

Wild ravens do not touch a carcass if they know it has been associated
with someone they don't trust, as the following example showed. At
the end of October 1992, I had fourteen newly caught wild ravens in
the main aviary with thirteen wild-caught veterans of the previous
year remaining in a side aviary. Even though the birds were left alone,
it took the newcomers three days before they started feeding on a calf
they had seen me drag in for them. I had also brought them two calf
lungs and a gray squirrel, items the tame ravens would have taken in
seconds. These remained untouched even after five days.

I placed one of the calf lungs directly next to the calf from which
the fourteen had started feeding. Would they feed from the lung at the
"safe spot," or avoid it because they had seen me handle it? I soon had

the answer. They almost immediately took the one lung I left, not touching the one I had laid next to the calf. Had they forgotten that I had handled it three days before, since they just saw me carry one and not the other? Two days later, nothing was left of the lung but a picked-clean windpipe. They still had totally avoided the other lung that they had seen me carry to the calf from which they continued to feed.

The most parsimonious explanation of my conflicting experimental and observational results is that ravens go through developmental stages. When they first leave the nest, they are, like human babies, highly curious of all objects. They are fearless and "get into" everything that crosses their paths. At that time, they are accompanied by their parents, who steer them away from dangerous things. Because of this parental protection, the young are free to explore and learn. During these early months, they also learn from their parents' reaction what is dangerous, what is harmless, and what is a source of food. The parents have experience that might span decades, and that has been passed down partly from their parents in turn, so that in effect the young learn from grandparents as well. The first group of juvenile ravens I examined were afraid of carcasses because they had skipped the tutorial stage during which the young follow their parents. Those young ravens had had to do all of their learning on their own, and so had little guidance in determining what was safe or not. It might take them two days instead of two minutes to begin to feed on an opened raccoon carcass.

As parental bonds are severed, the young are eventually on their own. They might encounter new things that are precisely what their parents had avoided. They enter a stage where they are neophobic, fearful of new things. They no longer have the backup of their elders' guidance to steer them away from potential harm. They must determine the nature of each new thing independently. They cannot just walk up to any immobile, possibly sleeping wolf and try to peck out its eye for a meal.

A bird that has evolved to live around wolves and people and all sorts of other carnivores has much to fear. Neophobia, or fear of the new, could have been especially strongly selected in ravens of the north-eastern United States in the last several hundred years because of humans. Starting about 250 years ago, a raven could not just walk up to any meat and ignore a contraption next to it. It might be a trap, or the person near the meat could have left poison. These new dangers would have been most easily and quickly avoided by being shy of *all* things new and "strange." As a result, the surviving birds would fear many irrelevant things. All of the seemingly irrelevant fears would be a necessary cost, or evolutionary baggage, for avoiding the potentially few but unpredictably dangerous. Following are a few experiments I did to explore some of the fears of ravens.

FEAR OF QUANTITY OR CONFIGURATION

Throughout their first summer, Goliath, Fuzz, Lefty, and Houdi ate sixty-six chopped-up red squirrels that they swallowed guts, bones, fur, and all. Then on August 2 and August 15, I offered them *whole* red squirrels. Thinking that a whole squirrel would be a much better prize than a piece of one, I expected an enthusiastic response. Instead, they refused the squirrels. They appeared frightened. They did not go near the squirrels either time.

I had fed them spaghetti before, and they eagerly had picked up and eaten all the spaghetti strands I flung to the ground. When I put a quart of spaghetti onto the ground, they jumped up in fright whenever they wandered near it. At times, they approached the little pile haltingly, feet braced for a hasty backward retreat. They advanced hesitatingly, but always jumped back in the last second just before making contact. It was not until a full day later, when hunger finally overcame fear, that they at last fed from the plate-sized spaghetti pile.

The four tame ravens eagerly fed on inch-long, yellow-orange cheese puffs. Whenever I brought this factory food and scattered it over the ground, there was a great rush among the four to be the first to gobble it. Those individual puffs that were not eaten right away were cached for later use. Every single cheese puff I ever left was either eaten immediately or cached. One day, I dumped a whole bagful. Did they all rush in? No way. They showed alarm. After a while, the bolder among them cautiously edged closer and picked up a piece that was several inches removed from the pile. The pile itself was left inviolate. I did not feed them anything else, and they begged piteously. Saliva was dripping from their bills—they could almost "taste" those cheese puffs. I listened to their constant begging cries for five long hours before I finally took pity on them (or on myself). As I spread the pile, they rushed in and took every one. The second time I offered piles of spaghetti or cheese puffs, there were no problems; they went right to them.

The piece-versus-pile experiment was easy to do, so I repeated it three more times with similar results, once with dog food, once with freshwater clams, and once with cornflakes. I had thought that after a while their irrational fear of a pile of food might wane, but it didn't. They remained cautious. Each *kind* of pile was evaluated separately.

On November 14, 1993, I dropped a raven wing primary feather into the aviary. That stirred up quite a commotion in the eight-month-old birds. All, especially Fuzz, made deep, long, rasping alarm calls. They flew all around the feather, both curious and fearful. They stared at it, flew down to the ground near it, then flew up again to a high perch. Although Lefty and Houdi soon lost interest, Fuzz and Goliath persisted in making alarm calls. After about ten minutes, Goliath finally grabbed the feather, manipulated it with his bill for thirty seconds, then dropped it. After that, neither he nor any of the others ever gave the black feather a second glance.

By mid-January 1994, the ravens had broken off all the twigs and peeled most of the bark from the branches and trunk of a six-inch-thick pine tree in the aviary. Thinking they might now be ready to debark and delimb a new tree, I brought in another, smaller pine tree. Instead of tackling it, they all hid in their shed and stayed holed up there for one and a half days, without once coming down to feed. Nevertheless, when they did finally come out of seclusion, they seemed hell-bent on the tree's destruction.

PHANTOM MOVEMENTS

Fuzz and the others had instantly pursued live prey. They had chased chipmunks and caught mice, shrews, and birds that entered their half-acre aviary through the wire mesh. They were hunters. Knowing this,

I brought a dead short-tailed shrew, *Blarina brevicauda* (a favorite and common snack of theirs, and an unusual mammal that has a poisonous bite), tied it onto a long white thread, and threw the shrew out onto the snow, while holding an end of the thread that I hoped would not be visible to them because of the snow. As I had expected, they all raced after the shrew. Goliath snapped it up first and flew off with it. I gave the thread a tug after he had gone ten to fifteen feet. Feeling the shrew tug in his bill, he dropped it instantly, landed, and watched it from a respectful distance as it lay on the snow. Something was not quite right here. The others seemed to know it, too. They all retreated from this shrew as though it were some frightful apparition.

Hoping to entice them to grab it again, I dragged it slowly over the snow, as one might drag a bass lure through the water with monofilament line. They still stayed away. Was I too close? Did they think I was the cause of the unexplained behavior of this weird shrew? To try to put them at ease, I walked fifteen yards away from the shrew, keeping the end of the spool of white thread in my pocket. Still no takers. As I walked near them holding the spool of thread in my hand, they even flew away from me. After about five minutes, Houdi hopped down to the snow, cautiously walked to the shrew, picked it up, and dropped it instantly and flew off. She did not come back. I rolled up the thread, untied the shrew, and tossed it back onto the snow. Houdi then immediately flew back to it, picked it up, tore it in half, and greedily swallowed the two chunks.

Wondering if they had been afraid of me with the spool of thread because the spool was new to them, I tried other new things. I dropped a nickel onto the snow. All came to me instantly as though nothing had happened, and Lefty was the first to grab the nickel. For added enticement, or as an added control for my experiment, I pulled out my keychain with four keys on it and jingled it in front of them. They had never seen keys or keychains before, and they all crowded close to me to try to get these toys. I did not relinquish my keys to them, but having lured them close, I next pulled the roll of white thread out of my pocket. Bedlam! They all flew off in alarm. When I unrolled the thread and laid it across the snow, they were not afraid in the least. Apparently they had associated the new thing, the spool, with the phantom movements of the shrew, which they did not understand and that frightened them.

To further examine fear of phantom movements, I tied a string to a stick that contacted a horizontal perch in the aviary being used by Fuzz, Goliath, Houdi, and Whitefeather. Similar string was already supporting the plastic netting of part of their aviary, so the ravens were used to it and were not alarmed by it. I threaded the string from the stick in their aviary to my desk in front of a window. From the house, I then caused the stick and perch to jiggle on a wind-still day. Their response was dramatic: They jumped up as if electroshocked, even when perched several yards from the jiggled and unused perch. After jumping up, rapidly twisting their heads and looking in all directions, all four birds retreated into their shed. I tried the stick-wiggle test four times, and each time they panicked. In contrast, when I went into the aviary and jiggled the string and the stick or the perch directly with my hand, they showed no visible reaction whatsoever. Similarly, I can make all sorts of noise in the aviary. When they can see me, they are unperturbed. If I do even a little banging around in the nearby woodshed where they can't see me, I can hear them flying around in apparent panic.

A raven is afraid of what it knows to be dangerous, but that seems to be the least of its fears. Most of all, it is afraid of events that violate its expectations. Perhaps, like us, they fear what they do not understand. If they can fear the unknown, then that implies that they know. Aside from drawing that inference, I was no closer to answering my original question of why ravens fear carcasses, yet can be so bold as to court danger by pulling eagles' and wolves' tails, riding on the backs of boars (Dathe, 1964; Steinbacher, 1964) and bison (D. Stahler, personal communication) or pulling fur for upholstering their nest from a donkey at pasture at Bill Chester's farm in Tunbridge, Vermont.

NINETEEN

Ravens and Wolves in Yellowstone

THE WELL-GROOMED YOUNG MAN sitting next to me on the plane to Bozeman, Montana, identified himself as a used car salesman from Memphis, Tennessee. He'd been reading a Bible before he introduced himself, and he told me he was traveling with a group of other Tennesseans to a convention in Bozeman "to learn how to talk to people to save them so that they can go to heaven."

He asked me what I did. "Study ravens," I told him. The conversation lagged.

"How does one get to heaven?" I asked.

"By believing in the Lord Jesus Christ." And he added, "Do *you* believe?"

"I don't believe, I *know*," I told him. "I'm going to heaven right now—to Yellowstone National Park to see ravens with wolves."

Ravens and eagles soaring about five miles outside Yellowstone Park, near Gardiner, Montana.

As soon as we landed in Bozeman, my wife Rachel and I rented a car and drove to Gardiner, at the north entrance to Yellowstone, to check into our motel. I'd never before checked into a motel with blood all over its parking lot. Next to our new white rented Nissan were several pickup trucks and campers carrying gutted elk carcasses. There were also four trucks from the Montana Department of Livestock, with men wearing wide hats and toting long rifles. They were the state's hired guns, here to kill bison from the park. Inside our motel room, next to the cellophane-wrapped plastic glasses, were a few bits of rags with a prominent sign saying: THESE RAGS ARE FOR YOUR USE. TO CLEAN YOUR GUNS. PLEASE LEAVE THEM WHEN YOUR (sic) DONE. PLEASE DO NOT TAKE THESE RAGS. I wouldn't think of it.

It was still dark when we got up the next morning, a Sunday. There were snow-covered hills all around, and the town seemed dead. We went to the Town Cafe for breakfast. A bold headline in the *Billings Gazette* for sale inside said: "Three Killed in a Bar Brawl." A big deal. In small print below the headline it said that 665 of the park's bison had been shot so far, because they had migrated out of the park seeking food. They had been shot by the men from the Montana Department of Wildlife returning nightly to our motel. About as many more bison would be butchered by spring. That's nearly a third of the entire Yellowstone herd.

Not one of the dozens of tables was occupied. "Where is everyone?" I asked the waitress. "Gone elk hunting," she said.

The walls of the cafe were decorated with row after row of heads: bighorn sheep, bison, mule deer, and elk. We left at about 7:30 A.M., as it was getting light, after eating what we could of the biggest pancakes I'd ever seen in my life. Stepping out of the cafe in the gray dawn, I saw a raven fly overhead, heading directly over a herd of bison grazing just in front of us, near the stone entranceway to Yellowstone Park—the gateway to heaven.

The wolves had been in Yellowstone only since the fall of 1994, when they were reintroduced by the very same federal agencies that had only fifty years earlier engaged in a very costly but successful campaign of exterminating them with poison, traps, and dynamite. The wolves were now being brought back at an expense of millions of dollars, even though they would surely have returned on their own. If they had been allowed to come on their own, they would have been protected under the Endangered Species Act. By *deliberately* introducing this endangered species, the government could legally shoot them when they left park boundaries. The wolves would not be inviolate, but under government jurisdiction. The introduced wolves had formed packs, and would soon have pups. Lots of pups. They would prevent the overabundant elk, deer, and bison from creating a hell of the ecosystem. Already, the ungulates had multiplied unchecked and were denuding the park. With the wolves cutting their numbers, poplars would again grow back. Beavers could again find food. Streams would be dammed, and more ducks, geese, and swans would breed, meadows would form, and rails, wrens . . . and so on. It would be a community, and the community I feel psychically a part of includes all of these, plus bears and moose and ravens.

My contact and friend, Doug Chadwick from *National Geographic* magazine, had told me that it was "guaranteed" we would see wolves in the Lamar Valley. They were all radio-tagged, and between aerial and ground-based surveys, everyone knew where they were and what they were doing practically all the time. Ironically, these few wolves—now fifty-two in the whole park—were celebrities, whereas before all wolves were "vermin." Now every tourist brochure about Yellowstone bragged about every detail of their lives. There was even a Web page on the Internet devoted to them. Every wolf was identified and its history known as though it were someone's pet.

On this first day in the park, we did not see wolves, not even the famous Druid pack of the Lamar Valley. We saw many bison and elk in the open valleys under snow-clad mountains that were crisscrossed with dense networks of elk and bison tracks. At Mammoth Hot Springs, bison were wandering down the street in front of the office buildings. Bison lying along the side of the road seemed oblivious to us in the car, staring at us through big, dark, bulging eyes. The animals seemed to be in a torpor. We passed occasional deer, coyotes, and bighorned sheep.

Finally, we saw four coyotes in the distance with several ravens near them. Within minutes, I heard four different raven calls I had never heard before. The knocking of these ravens was a series of only four rapid sounds that seemed more wooden than the more liquid and rapid calls of those with the Maine accent that I was used to. Although we did not see the wolves just yet, we found one recently wolf-killed elk calf. Only skin and bones were left, but a dozen ravens flew up. By the end of our ten-day stay, I had seen nine wolf-killed elk cows and calves. All had ravens feeding on them.

The next morning, we headed to the Lamar Valley as it was getting light, and we repeatedly saw ravens flying overhead. We saw singles, pairs, and small groups of up to a half dozen. Whenever there were two flying together, we'd occasionally see one dip its wing and then we'd hear a *glug-glug-glug* call. During the next week, we'd see most of the ravens traveling in pairs in the early morning, and the *glug-glug-glug* call and wing-dipping were characteristic. I had heard an Eskimo legend about ravens dipping their wing to indicate prey to human hunters, while at the same time making a certain call. I wondered if that was the display.

We again saw no wolves, but just beyond the Lamar Valley we found two just-killed elk by "following" ravens. The Soda Butte pack that had made the kill had left most of the meat. The two female elk carcasses lay almost side by side, up the slope from Soda Butte. One was ripped open at the neck only, and the other had its flank torn open. Dozens of ravens and four bald eagles and a golden eagle were all around them. The ravens ignored the golden and bald eagles, but at the cliffs near Mammoth Hot Springs we had just seen a pair of ravens attack a golden eagle and escort it away. Just a mile or two up the road, near Pebble Creek, we found three more fresh elk cow carcasses and many wolf and coyote tracks. One carcass had a large hole cut into the rear end, and ravens were using it to get deep into the elk. Two more elk carcasses were within a half mile of this one, and ravens were with them also. The wolves here kill often, eat the choicest part at each carcass, then move on to kill again. It *is* a raven's heaven.

The long weekend was over when we returned to Gardiner in the afternoon. It was also the end of the second or late annual elk hunt, which takes place only on four-day "weekends," Friday through Monday. The bison shooters had left, and the Best Western motel seemed almost deserted. Looking out the window beyond the parking lot and over the town, I saw about fifteen to twenty ravens gamboling in the wind rising off the high hill just northwest of town. Ravens circled high in the sky in all directions. I was quickly drawn out to take a drive to observe them more closely. Above the ridges around town I saw other crowds of them swirling, diving, playing, flying with bald eagles against the blue-black snow clouds. One eagle had a red tag on its left wing, and a raven kept flying at the circling eagle, trying to peck this tag.

The eagles were heading into a communal roost in a valley filled with large Douglas fir trees. The hillside behind the towering firs was dotted with elk slowly descending out of the park, and it was braided with a network of tracks, as more and more elk were leaving the park. They came in long files, following one another's tracks through the deep snow to conserve energy. Hungry animals.

The eagles came singly, diving with still and partially folded wings. They extended their wings just after passing the roost, then wheeled around to brake their descent, and came in for a gentle landing on a tree. As each bird landed, it made high chittering or peeping calls like those of gulls or sandpipers, which some think are contact calls for their mates. Scores of magpie nests dotted the willow thickets in washes between the sagebrush-covered hills. In this area, barely a five-minute drive from the motel, I saw many blood-smeared drag marks in the snow where hunters had pulled elk to the road. There would be many gut piles here, but no wolves. Would there be feeding ravens?

In the park, just a one-minute raven's flight to the south, the wolves of each pack kill an elk on average every 1.5 days. Ravens always arrive immediately and start feeding at each fresh carcass. Ravens often arrive even before the elk is down. That evening, in town, wildlife filmmaker Bob Landis showed us film footage of a pack of wolves trotting leisurely among a group of elk, testing them for several minutes. One of them suddenly shifted gears and picked out *one* elk to pursue relentlessly even as it rejoined the herd. Eventually, the wolf grabbed the elk by the throat, and another wolf helped pull it down. Ravens were flying over the wolves as they made the kill.

Were the ravens completely tied to the wolves or would they feed at the gut piles outside the park? To find out, I drove up the little dirt road toward Jardine early the next morning. As soon as the dawn came, I heard and saw ravens. Most of them were in pairs, flying side by side from one end of the horizon to the other. Where might they be traveling, and why?

I started to walk up into the rocky, sagebrush-covered hills on a well-worn bison trail. Then I soon found and followed the red lines through the snow where gutted elk carcasses had been dragged. In three and a half hours, I located a total of sixteen elk gut piles, most containing liver, spleen, lungs, intestines, stomach, and diaphragm. There were no ravens feeding on any of them, though some had been fed from. Why hadn't the birds descended on this food bonanza in large numbers and quickly obliterated it all? I was surprised, but what I saw was congruent with a set of experiments I had done in Nova Scotia.

I had gone in mid-winter to the Canadian Wolf Research Center in Shubernacadie, Nova Scotia, to observe ten semitame wolves in a ten-acre outdoor wooded enclosure. In experiments for testing raven's preferences, I had put down two meat piles simultaneously. The wolves had fed as a group, and only at one pile at a time. There, as here at Yellowstone, the ravens had a choice—to feed with the wolves, or to feed without them. They invariably had chosen to be *with* the wolves. They had fed where the wolves were currently feeding or had just recently fed. These observations, given the raven's fears that I had observed previously, were about as nonintuitive and surprising as anything I had ever seen or heard about ravens. It didn't make sense.

Then I thought of my Maine ravens, who were so shy near carcasses that I had thought they might be almost paralytically afraid of dangerous ground predators. Did I have it backward? Had they been afraid because there were *no* wolves at the Maine carcass? I became excited, wanting ever more data. Maybe ravens are "wolf-birds!" Maybe they had evolved with wolves in a mutualism that is millions of years old, so that they have innate behaviors that link them to wolves, making them uncomfortable without their presence.

During my present attempt to increase my sample size of gut piles, I kept going higher and higher. I eventually reached deeper snow, and open sage country, then stands of Douglas fir where I heard red squirrels, red-breasted nuthatches, and Clark's nutcrackers, in addition to ravens. In the Douglas firs, I came across blood flecks on the snow without carcass drag marks, indicating a wounded animal. I followed the spoor and soon found a dead elk cow with rifle bullet hole through her paunch. One or two ravens had left tracks on the dusting of fresh snow, and the birds had taken an exposed eye and a piece of the elk's tongue. Given that no wolf or coyote had been near this carcass, that is all they could have taken.

The carcass had not yet been opened. I lanced it with my jackknife and skinned it on one side to let the birds get at the hundreds of pounds of red, unspoiled meat. While I was working on the elk, three ravens came near and called loudly. One landed in a Douglas fir nearby, looked, then made both short and long rasping caws, and flew off. I went back to the motel, imagining seeing a huge crowd of ravens when I'd return in a few days.

In the meantime, I had seen thirty ravens at an elk fawn within one hour after it was killed by wolves inside the park. Additionally, Doug Smith, leader of the Yellowstone wolf recovery project, had taken us to see the wolf pen at the Lamar ranger station, and there we found a large congregation of ravens, even though there was not a scrap of meat left inside the pen. When I hiked up to the elk cow carcass the second and then the third and last times, I was surprised that there was still no crowd of ravens, despite the plentifully available meat.

Shortly after breakfast one morning, we drove east toward Cooke City. Near Soda Butte we came to two parked vans, one with a radio antenna on top, which indicated researchers and was a good sign of wolves nearby. We stopped and made introductions. Nathan Varley, Lisa Belmonte, and Dan McNulty were training three spotting scopes on the hillside, and they invited us to take a peek through their spotting scope.

I saw all five Druid pack members lounging in the snow like one big happy family on siesta. Two of the wolves were jet black, two were gray, and one was tan. That morning at about 9:30 A.M., the pack had killed an elk cow. This was wilderness at last.

The elk lay well exposed under an aspen tree. The wolves had eaten little meat so far, and the second act in this drama was beginning. Three or four ravens and at least as many magpies were already feeding on the carcass. The ravens made no attempt to chase the magpies. A golden eagle swooped down and landed on top of the carcass. The ravens and magpies flew up briefly, then walked round and round the carcass, moving closer, and gradually resumed their feeding. The golden eagle left in a half hour, and a bald eagle took its place. Ravens and magpies remained. Not once did the ravens chase another bird to steal food from it, as they had done in Oregon. Not once did the ravens hesitate to go in and feed, as they do in Maine. With wolves present, the ravens had no fear of the carcass. They could go in and get their own meat.

A few hours later, it was snowing hard. The wolves started to get up from their nap and stretch. One returned to the carcass and fed briefly, then they all walked off together in single file, fading like ghosts into the thickly falling snow as they rounded the crest of a low ridge. The next day, I walked up to examine this carcass, finding many fresh coyote and weasel tracks. Even if the wolves ate little of their kills, all of the carcasses would eventually be eaten. The carcass would feed weasels, grizzlies, coyotes, foxes, carrion beetles and corvids, eagles, maggots, and possibly a wolverine. The wolves provided for all, but I expected that here, precious little would be left for the maggots next spring.

I was told that the ravens sometimes followed the wolves, and that if they were not seen visibly following the pack, then they were "always" there "within minutes." Observers and students of wolves take ravens for granted, commonly remarking that ravens and wolves go together. So much is taken for granted that further comment, or data, have seemed superfluous. I tried to convince the wolf researchers otherwise, pointing out that nobody really knows the extent to which ravens may monitor or follow carnivores, or if they are simply opportunistic and sharp-eyed enough to see meat as it becomes available and then simply take it, as everyone assumes, but as my data disputed. My prod had an effect.

By the next winter, Doug Smith was in full support of appending a complimentary raven project onto the wolf project. Furthermore, Dan Stahler, who had already been following wolves and would continue to do so, agreed to take on the project by simultaneously gathering raven data. He would follow a protocol for getting systematic observations suitable for the first critical study on the raven-wolf association.

Dan reported to me after his first seven days of watching the fifteen wolves of the Rose Pack in late November 1997. He had then observed twenty-four activity bouts of wolves that involved traveling, resting, chasing, at kill, and near kill. In all but three of these, there were ravens with the wolves. In contrast, of nine activity bouts of coyotes, ravens were present at only two. By March, Dan had data on two dozen very recent kills, and ravens were feeding at all "in seconds or minutes." Raven numbers at wolf kills averaged thirty-two, ranging typically from fifteen to thirty, and were over eighty at one kill. The ravens routinely fed within feet of wolves, coyotes, and eagles.

Dan's description crackled with excitement: "Last Wednesday, I watched a large grizzly bear lay on top of a fresh wolf-killed elk for over four hours while nine wolves and about twelve to sixteen ravens tried to get in to feed. A couple times during the four hours the bear slept 'spread eagle' on the carcass, while the wolves bedded less than thirty meters away. The grizzly was fairly successful in keeping all away, but was extremely agitated by the ravens that kept grabbing food despite his lunging at them. This topped my list of best field observations. Doug [Smith] has started recording data from his flights [in light aircraft] whenever he observes wolves. The other day he saw three different packs involved in elk chases (two ending in kills). During two of the chases ravens were flying directly above the chase or perched nearby. During one chase involving eight wolves there were eight ravens and two bald eagles soaring directly above. Amazing!"

Yes, I'll second that. It is beginning to look as if ravens are dependent on wolves not only to kill for them and to open carcasses, but also to overcome their innate shyness of large food, whether in the form of a carcass or a pile (see Chapter 18). These facts hint at a relationship with an ancient evolutionary history.

Note: As this book was going to press, with the full cooperation and permission of park authorities, Dan had laid cut-open deer carcasses in established wolf territories where ravens, in over thirty previously observed cases, had always both shown up at wolf kills and fed with the wolves within minutes. In the twenty-five trials with the meat he provided that was unattended by wolves, no raven fed within the hour of observation period that his protocol called for. In the nine instances where one or two ravens discovered the unattended deer carcass, they circled the meat briefly and then left.

TWENTY

From Wolf-Birds to Human-Birds

DURWARD ALLEN, A PIONEER OF
wolf studies, remarked that the ravens of Isle Royale in Lake Superior
accompany wolves in their travels, feed at their kills, and sometimes
even eat their scats. L. David Mech from the University of Minnesota,
who has studied wolves for decades, has seen ravens chase wolves, flying just above their heads, and reported in his book *The Wolf* that
"once, a raven waddled to a resting wolf, pecked at its tail, and
jumped aside as the wolf snapped at it. When the wolf retaliated by
stalking the raven, the bird allowed it within a foot before arising.
Then it landed a few feet beyond the wolf, and repeated the prank."

Mech noted, as Allen did, that ravens appear to follow wolves, and
he speculated that both must possess the psychological mechanisms
necessary for forming social attachments, and that individuals of each
species include members of the other in their social group, forming
bonds with them.

Ravens and wolves almost ignore each other when they feed together.

Rolf O. Peterson, a former student of Mech's now at Michigan Technical University, also studied wolves on Isle Royale. He agreed: "There is more than playfulness between wolves and ravens. Ravens make their living by scavenging wolf-killed moose (the fresher the better) and as we start the day, flying along wolf tracks, we often overtake a raven doing the same thing. When wolves pause, the birds also stop, roosting in trees or swooping to the ice where they can watch and harass the wolves at close range. Once disturbed, wolves resume travel, which is what the ravens intended. Also, few wolf scats left on open ice escape the selective recycling provided by foraging ravens."

It is not surprising, as another former Mech student, Fred Harrington from the University of Nova Scotia, has shown, that ravens can be attracted to wolf howls. They also come to gunshots where there is much hunting of large animals. The wolves' howls before they go on the hunt are a signal that the birds learn to heed. Conversely, wolves may respond to certain raven vocalizations or behavior that indicate prey.

Dan Stahler, in his observations of wolves and their association with ravens in Yellowstone National Park, saw ravens not only following wolves on their hunts, but also hanging around at the wolves' dens. Whenever the wolves at a den got ready to hunt, they howled, and the ravens loitering there started to get lively and to vocalize as well. The pups start to come out of the den at about three weeks of age. They are smaller than ravens and could potentially be killed by them. Yet the ravens, who occasionally walk behind a young pup, only gently yank its tail.

There are innumerable anecdotes of dogs interacting with ravens as well. As one example I cite Graham W. Rowley (in *Cold Comfort: My Love Affair with the Arctic,* McGill Univ. Press, Montreal 1996). "One day I watched a raven flying low over the Inuit camp, heading into the wind. Chasing after it were about twenty dogs. The raven decided to rest and alighted on a large rock, ignoring the dogs dashing toward it. When the leading dog was only about three yards away, the raven turned its head toward them and gave a single baleful squawk. All the dogs stopped dead, turned around, and trotted back to camp like sheep. The raven rested a few more minutes before taking off again into the wind." Like the raven's fear of carcasses the mutual attraction between wolf and/or dogs is a reflection of ancient selective pressures.

The raven-wolf association may be close to a symbiosis that benefits the wolves and ravens alike. Wildlife photographer and writer Jim Brandenburg, in his book *Brother Wolf* (see Notes), described ravens coming to an unopened bear carcass. They could get nothing but the eyeballs since ravens can't open a carcass. The ravens then started yelling, and soon a wolf arrived and tore the carcass open. Brandenberg repeatedly saw wolves as well as coyotes come to a carcass that he provided shortly after it was discovered by ravens who were yelling.

Ravens may do even more than locate meat for the carnivores. Brandenberg says, "I can state unequivocally that, at a kill site, ravens are more suspicious and alert than wolves. In many instances, I have seen ravens become nervous at one of my small movements where the wolves seemed unaware. I believe that the birds serve the wolves as extra eyes and ears." Wildlife filmmaker Jeff Turner suspects the same: "I can sneak up on a wolf," he told me, "but *never* on a raven. They are *unbelievably* alert."

In the High Arctic, ravens follow polar bears and feed at their kills, and in Yellowstone Park they also feed not only near coyotes and wolves but also sometimes with brown bear. Doug Peacock in *Grizzly Years* described watching a large grizzly sow and her cub at the edge of Wild Goose Valley in Yellowstone in late May. The grizzly dug and tugged at the ground among sagebrush, with ravens all around. At times, she reared and swatted at the cloud of ravens, flailing away at the air with her paws. The ravens probably also wanted what the bear was searching for, most likely pocket gophers or their seed caches.

Seldom do those whose activity provides food to ravens try, like the above-mentioned grizzly, to chase them off, especially wolves and humans. The Vikings, who usually got the upper hand in battle, eagerly welcomed ravens. To them the birds were an omen of victory, not doom. Why else would they fly their raven banner as they went into battle? If the ravens followed people anticipating a glut of carrion, it was for the same reason they now follow wolves. As we shall see, the raven still follows people (see p. 241) and is said to follow deer hunters in the Scottish Highlands, where its presence is similarly regarded as a portent of a successful hunt (Ratcliffe, 1997).

From a raven's perspective, the closest thing to the aftermath of a Viking battle occurs every October in northern Maine at the annual moose hunt. I needed to see it, and drove to Greenville at the tip of Moosehead Lake to stay at Bob Lawrence's hunting lodge near Rockwood on the opening day of the hunt. The moose hunters streamed out at dawn, ranging up and down the dirt roads in their pickup trucks; nobody is keen to walk far into the woods to shoot a half-ton animal and then have to drag it out. I did not personally see a moose right after it was shot, but stopping to talk with hunters, I located six gut piles or their remains that were less than a day old. Ravens were feeding at all of them. One gut pile from a kill made the evening before had a stream of about fifty birds arriving when I checked it at dawn. One bird from this crowd pointedly came toward me, circled twice around my head at about ten yards, then flew back toward the others. This behavior made no sense to me. Did the bird mistake me for a hunter?

There are times when we perceive ravens to communicate with us. Craig Comstock, a raven-watcher from Starks, Maine, wrote of seeing a raven flying overhead. Craig called out, "Hey, how's it going?" The raven immediately pulled a U-turn, did a half-roll, then went back on course. Craig waited until it had gone a bit farther and called again. Immediately, the raven pulled another U-turn, executed two back-flips and a half-roll before again returning to steady flapping. Craig commented, "I can't prove the displays were for me. I understand the need for the scientific method, but . . . there are times when nature speaks just once, and it is a loss not to listen."

Although ravens in New England are conspicuously shy, I have on several occasions seen one inexplicably come close to me, checking me out minutely. A Canadian Arctic biologist, Don Pattie, told me of tending small mammal traps on the Canadian High Arctic tundra when a lone raven flew near and landed. Don held out his hand and talked to the raven, which walked up and "took a bite into my hand and then flew off." Don was perplexed, but he was not totally surprised. Ravens may do surprising things. A raven researcher from Austria said, "Ravens are incalculable—and uncanny magic emanates from them." Another biologist, Steven Wainwright, told me of taking a walk in the woods near Vancouver, British Columbia, where he heard three ravens overhead. One of them came down through the trees and "talked" to him, so he began to talk back, making "ravennoises. I sat down on a log and decided to talk in normal tones. I said, 'Hi raven,' that sort of thing. The raven came near, and when I got up to leave, that bird followed me while the others flew away . . . It was a mind-blowing experience."

As mentioned previously, my aviary ravens stopped pecking a chicken and a turkey within a day or two of being introduced to them, when they presumably "knew" these fowl. Yet the interest and interaction with some animals, including dogs and humans, lasts much longer. My six ravens "know" me. I am their "wolf." There is hardly a day that I do not enjoy visiting with them, and they in turn act as though they want to be with me. Of course, they come to me for food, but they still gather around me when they are fully satiated. When I bring them a pile of food and walk away, some of them almost invariably leave the food to follow me. Some of them act as if trying to engage me in play. There are several individuals (White, Yellow, and Green) who used to routinely sneak up behind me, nip me on the pant cuff, and then look up at me; or they skimmed closely over my head from behind. I then yelled at them and feigned aggressive moves toward them, and that seemed to induce them to come right back and try it again. When I walked away, they followed like puppies.

For another mind-blowing experience I now cite the experiences that University of Vermont graduate students Cindy Riegal wrote down for me: "On the morning of December 25, 1996 in the Langtang region of Nepal, a piece of bread was taken from my brother Jerry's side as he slept out under the rising sun. We know it was taken in the early morning because our guides from Kyangin Gompa who had led us (Jerry, myself, and Eric Busch) over Ganja La (pass) the previous day generously left it next to him before they headed home and woke him briefly to inform him of their gift. Our assumption was the one of the ravens we had seen soaring above us as we descended the pass had cleverly and quietly stolen the bread. As we started our trek to Tarkeghyand through remote alpine terrain that day, we noticed two ravens flying above us as we walked. Around noon, we stopped for some snacks near a stone structure used as a temporary yak herder camp in the summer. The ravens in pursuit flew down and landed on the stones about fifteen feet from where we rested. We subsequently lured them closer with the raisins from our gorp until they were virtually eating out of our hands. Eventually we started walking again but got off track for a few hours. I don't recall the ravens sticking with us through our harrowing adventures trying to find the trail down in a forested valley, but once we were back on the open alpine slopes, the pair seemed to reappear. They continued to follow us until we set up camp at dusk. We were up and walking early the next morning once again with a pair of ravens in tow. They followed us for most of the second day until we began to descend in elevation and entered more forested terrain. They received no food from us other than the "stolen" bread and a few raisins. We had traveled in a remote area; we saw no other people during the trek with the company of the ravens."

Do the ravens follow the moose hunters or their trucks, or associate either with food? At two moose kill sites I visited on that trip to Moosewood Lake, all the entrails other than stomach contents had already been removed, so unfortunately I was unable to tell if the birds arrived "immediately" after the kill. The next year, my friend Bill Valleau, who is also one of my former zoology professors from the University of Maine, drew a moose permit in the lottery. Bill hunted about a hundred miles father north, near Bridgewater, on the side of "Number 9" mountain. Two bull moose charged their truck as he and his sons rode around on the gravel roads. Bill wrote me, "The ravens were with us throughout the hunt. We felt that they were watching us, waiting for the kill. We shot a cow moose on the second day and the ravens were circling our kill while I was gutting it out!" Twice his son Dana, a lawyer for the Maine Department of Environmental Protection, shot a deer, and waited five minutes before he went in to make sure it was dead. When he got to the deer, a raven was perched nearby. "I had assumed the ravens were attracted by the shot," Dana said. Maine game wardens find ravens the best assistants in helping to apprehend poachers. Very often, when they investigate a site where ravens are circling, the poachers are still gutting their kill.

Sometimes, humans have more to offer than moose guts. George Schaller told me that in Tibet, ravens inevitably check out most remote human encampments, probably for food scraps. Similarly, Gary Clowers has reported that the seashore ravens in Baja California check all the tourist campsites in the morning. Ravens live in Inuit villages and other far-northern American settlements, and they now are moving south into the big cities, including Los Angeles, which is already inhabited by crows. All across America and Europe, ravens congregate at dumps. The food bonanzas to be found there likely have a historical antecedent in the remains of hunter-killed carcasses of mammoths, giant ground sloths, and caribou. Ravens are, and likely always have been, not just wolf-birds. They are *our* birds as well, and it is small wonder that they hold such a prominent place in our myths and legends.

Throughout the hundreds of thousands of years after humans came from Africa to pursue the vast northern herds of large antlered deer, aurochs, and mammoth, the raven, *Corvus corax,* would have almost certainly been there with them, scavenging on the kills. When we breached the American continent after the lowered level of the oceans left a land bridge to Alaska during the ice ages, we spread south and east, killing the unsuspecting and unprepared megafauna. Never in their evolutionary history had the animals encountered such hunters as humans. It was an opportunity that the raven also would not have failed to exploit. Corvids are known from fossils dating to the Miocene epoch, and raven fossils in the American southwest date back to the Pleistocene (Magish & Harris, 1976). Ravens could have been here at least hundreds of thousands of years before us, along with the wolves. As far as Raven was concerned, Man, the new predator, was probably just a surrogate wolf who also usually hunted in packs.

The wolf had been indispensable to the ravens in the north, and farther north the polar bear took its place and still does. To the south were great cats. With their sharp, shearing teeth, the predators brought down and cut open animals that were diseased, weakened, or dead. The raven's long, strong bill could pick meat from the hide and from leftover bones. The mammalian hunters might have tried to take all the meat they could, but unless they were starving or the hunting was poor, they would have left piles of entrails that the birds who tagged along or monitored most hunts would have eaten. Human hunters may even have deliberately left meat, because the presence of ravens could have been seen as an omen for a successful hunt, as they were omens for a successful raid to the Vikings.

If the hunting was good and the hunting bands were small, then hunters likely ate only part of a mammoth before going on to kill the next. Ravens and wolves would have feasted. Since ravens to this day associate with hunting bands of wolves, I suspect that in the north they might still follow people with dogs or associate with dogs near people. Why should a raven care whether it is following a pack of wolves or a pack of fur-clad human hunters? Raven would preferably follow the best. Ravens are the quintessential northern bird. It is no accident that they are also the bird most closely associated with humans in the culture and folklore of northern people, whether Norse, Inuit, or any of a large number of others.

Odin, the ruler of the Norse gods, also called "Hrafna-gwd" or Raven god, kept two wolves at his side and two ravens on his shoulders. The wolves and the ravens accompanied him on the hunt and into battle. Thus, ravens were for thousands of years associated with wolves, and with mind, men, and the gods. From the wolf-raven associations came the northern name "Wolfram," from Wolf-rhaben or "wolf-raven," once a great warrior's name.

I could not follow a Viking raid to trace origins of the old myths. But I could jump on a plane and in several hours be in the Arctic to see the raven where it has not been persecuted in recent history due to recent ignorance, and where it does not shy away from humans by hiding in the wilderness. I could be where, I hoped, it might instead still seek humans out and associate with them. I hoped to see at least traces of a long relationship, having heard vague rumors that Eskimo hunters "talk to" ravens, and vice versa.

Note: A parallel example to the one I am proposing here is that of the African greater honeyguide, *Indicator indicator,* as discussed (pp. 164–169) in Donald Griffin's book *Animal Minds.* The honeyguide is a small bird that feeds on bee grubs, wax, and honey. It communicates with the ratel or honey-badger, *Mellivora capensis,* guiding it to bee nests that the badger then opens. They then share the food. In some areas of Africa, the bird has transferred its honeyguiding behavior from ratels to humans, entering a cooperative relationship with them.

TWENTY-ONE

Tulugaq

SITTING WITH A CUP OF COFFEE
after arriving early at the Montreal Dorval airport, I waited to catch a
flight north to Iqaluit, a community of about four thousand people on
Baffin Island at the edge of Frobisher Bay, just west of Greenland and
north of Hudson Bay.

The Boeing 727-200 that I'd be flying on had only six seats of
passenger space. Most of the plane was crammed with cargo. Not a
good sign. I had imagined a life of hunters, not people eating
imported bread and potatoes. Subsistence these days does not come off
the tundra. If ravens live off people, they'd be living off the same life-
line via the airline, not from what hunters provide from the land.

As we popped through the low clouds for a brief stop at Kuujjuaq,
I was stunned by the sight. I saw pure white in all directions. The
landscape was studded with black spruce trees, willow thickets, and

Baffin Island. The Eskimo dogs on the ice are
almost always accompanied by ravens.

larch. Soon after continuing on to Iqaluit, I saw no more trees. For an hour, I saw only blinding white. Then the outlines of the tiny settlement appeared. Iqaluit at last. Stepping out into minus 21 degrees Fahrenheit in a light snowstorm, I immediately saw ravens flying in twos and threes against the white sky that blended with sea and land. Ravens stood out, crisp and clear. I could recognize their calls as raven in an instant, but they spoke a different dialect than the one I was used to. From then on, I heard sounds or nuances of calls almost every day that I had never heard before.

After Lyn Peplinski, the director of the Iqaluit Research Center, had shown me my berth for the next few days at the Center, I immediately left to walk along what I presumed to be the Frobisher Bay shoreline. A man was shoveling snow out of a boat. Jokingly, I asked him if he was going out for a ride. "Not till early July," he told me. "But the shoveling is something to do on a Sunday afternoon." We exchanged names. Kalingu Sataa, a carver of stone figurines, soon talked about seal hunting, ravens, and stone-carving. "Do they sell fur parkas here?" I asked, because I had been told they are now almost impossible to buy.

"My parents, who live in the house over there, have a caribou parka," he said. "My mother made it years ago. She might sell it."

We walked to the house just across the street, and I met a smiling elderly couple, Kalingu's parents, and Naudlak, his sister. Kalingu's father, Akaka, had been born in an igloo, and lived in Iqaluit most of his life, as had Kalingu himself, who was perhaps forty years old. Arctic explorers Freuchen and Salomsen had written that Eskimos believe ravens show them the presence of bears and caribou by their flight, and conversely, that ravens follow hunting parties, as they follow polar bear. I asked if ravens ever show hunters where animals are, so that the animals would be killed and the ravens could feed. Like most elderly Inuit and young preschool children, Akaka spoke no English, so Kalingu translated: "Yes, by dipping their wings." That was what I had long wondered about and had come to see.

Anyone who has watched ravens will have noticed them flying along, pulling in one wing, then righting themselves again. Do Inuit really *believe* that means the raven is talking to them? Kalingu was ambivalent, saying it was "just something that you learn when you grow up Inuit." Then he was quiet. Eventually, he said that these days people don't watch or notice ravens very much anymore. They watched them much more closely in the old days, when the *tuktu,* the caribou, were more scarce. In those days, they had no guns and they could get to hunt the *tuktu* only after long dog-sled travel. For 4,500 years in this land, hunting was the ultimate skill. Hunting required keen senses, guile, knowledge, strength, and endurance. When stalking caribou, hunters often took off all their clothes so that they would not rustle as they crept to within spearing distance. I presumed that was in the summer, because the March subzero temperatures were not conducive even to wearing traditional western clothing. I was freezing.

Returning to the topic of parkas, Akaka asked if I wanted to see his. His wife brought it from cold storage outside. It was a beautiful piece she had made out of light-colored caribou fur for the back and front, and dark-colored fur for the shoulders. Its hood was trimmed with wolf fur. He said he would not likely ever wear it again, and he sold it to me for $150 Canadian. That parka would make a big difference, perhaps *the* difference, just a few days later in Igloolik on an island in the Foxe Basin, the third and last settlement I would visit. I would accompany Charlie Uttak and several of his friends there on a several-day-long caribou hunt and Arctic char ice fishing expedition some 100 kilometers from the village. It would be an unforgettable ride at 50 kilometers an hour in minus 40 degrees Fahrenheit, swerving around frozen-in blocks of sea ice, hanging on to sealskin thongs on a sledge, the *qamutiik* (or *komatik*), pulled by a Yamaha snowmobile.

The ravens were all over Iqaluit. They were perched on power poles, on roofs and on back porches of residences, and on commercial buildings with people coming and going. I watched about twenty of them loitering at a day-care center where people were walking in and out with their kids. Nobody paid them any attention. Ravens were in practically every backyard where dogs were tied up. Dozens of ravens congregated on the ice with groups of Eskimo dogs. Almost anywhere, at any one moment, I could see ravens perched or flying about. Most were vocal. On many street corners, I saw lone ravens sitting on power poles, absorbed in elaborate monologues as if exercising their varied repertoires. At the same time, they expressed a lively variety of body postures. It looked like play, because there was usually no visible audience of other ravens to whom they could have been "talking." It would be stretching to speak of raven "calls," because the birds were stringing together series of very different sounds into often unique sequences. It was easy to imagine that the birds were talking to each other or to themselves.

As the visiting "raven man," I was invited for an interview at the CBC station with Gail Whitesides as host. She introduced the interview with a song about the northern raven. After the interview, the station switchboard lit up with people offering anecdotes and raven call imitations. By far the greatest number of anecdotes I heard were about ravens' play. The ravens use power lines as their toys, hanging upside down, swinging up, and somersaulting over. They sometimes hang from them with their bills. Sliding down roofs is another pastime. One woman told me of watching a group of ravens take turns rolling down her roof. As they got to the bottom edge of the roof, they either walked or flew back up to roll down again.

The next most talked-about aspect of the town's clowns, the ravens, was their feeding antics. "Ravens eat anything," which makes them disgusting to some people. A large congregation of them surrounding the sled dogs on the ice left not one dog scat in sight. If they cannot intercept enough food at the front end of the dog, they get it eventually at the other end. Feeding at the dog's front is preferred, but it is more dangerous, although ravens can reduce the risk by working in teams of two or three. Without carnivores, the ravens would here starve, even though out on the open land, I was told, ravens occasionally chase down and kill ptarmigan. Overall, the carnivores, especially humans, provide them with most of their food, and beyond the town on the tundra, ravens were scarce.

People who had been to many different towns in the north all agreed that ravens from the western towns, such as Inuvik and Yellowknife, were "totally different" from those of Iqaluit. The Iqaluit ravens still have the decorum to keep out of your way. In Inuvik and Yellowknife, they are "totally fearless," "cheeky," and "they will scream at you if you get between them and a garbage can." "They've even landed on top of my dog!" someone told me. Some people hope to scare them off by setting up plastic great horned owls, but one man who had tried this told me, "They act as if they are thinking: 'Is this a fake owl intended to scare us off? Good, it's a great perch to sit on, so there!' "

Hall Beach, the next hamlet I visited, is a mostly Inuit community of about five hundred people, where two DEW (Distant Early Warning) radar towers dominate the landscape. They were built in the 1950s to alert us to Russian nuclear-tipped missiles coming over the North Pole, but now have other uses. I'd seen an active raven nest on the DEW tower at Barrow, Alaska. Mike Wesno, who teaches at the Hall Beach school, picked me up at the tiny airport next to the towers with his snowmobile. It was dusk, and minus 30 degrees. Would I like to go for a spin? Indeed. After dropping off my gear at his house, we took a frigid evening ride out to the towers. At least a hundred ravens were perched all over the high steel girders, and the birds were not huddling together, as I expected at these low temperatures and howling wind. This site was a nocturnal communal roost, not a nesting site. Suitable sites for communal roosting must be few and far between on the tundra, and DEW towers suffice for both.

The next morning, I awoke to the Eskimo dogs' mournful howls all around the village. As in Iqaluit, the dogs were tied without shelter below the houses on the sea ice. Curled into balls with their backs to the wind, the dogs looked at me with one eye over their furry tails tucked around them as I walked by. They were all well attended by retinues of ravens. I noticed one raven, whom I called "Scraggly Tail," loitering around a particular dog pack for several days in a row.

The few ravens that foraged at the edge of this town were much shyer than those at Iqaluit. Nor did they here subsist on garbage; the dozens of plastic garbage bags along the street were not ripped open as they lay there for at least three days. Most of the ravens were getting their food elsewhere. I presumed it was beyond the ice flow edge, where tidal currents maintain open water and where there is drifting, or pack, ice. It is the seal and walrus hunting ground of the polar bears and Inuit.

Jona, one of Mike's Inuit students, graciously offered to take me on his dogsled out across the half mile of solid sea ice up to that flow edge. The dogs plodded across the blinding white snow as we left the frozen shoreline at the edge of the village. Jumbles of almost translucent turquoise ice-blocks were frozen in place, and we wound our way around them until we came to the edge. The water flowed by, silent and black, like oil. It was eerie.

The water was covered with tiny loose ice crystals. A seal popped up, looked at me, and quickly submerged. In the distance, I saw faint white ghostlike outlines of drifting ice floes. This was the world of polar bears, walrus hunters, and ravens. I realized why nobody had seemed anxious to take a newcomer out in an unsteady boat with the lethal combination of minus 30 degree temperatures, dark drifting water, and fog.

We loitered awhile, examining an igloo and the bloody trail where a seal had been dragged on the snow. After returning to the village, I felt a strange tug to *walk* back. This time out at the edge, I saw two ravens ahead of me perched on an ice block. One flew off, veered toward me, and made a wide loop back toward the flow edge where I had just seen the seal. It then rejoined its companion, again perching on the ice block it had just left. If I had been a hunter, I might have interpreted the raven's behavior as a sign beckoning me to go in the direction of the flow edge. I would have then approached this seal. But had I gone in another direction, I may also have found another one.

Both ravens lifted off from their perch when I came near them. The sun was already low on the horizon, and in liquid, sliding wing strokes they swung toward the village where the dogs lay curled on the ice, then disappeared as black dots on the horizon toward the DEW tower.

In the afternoon, I gave another presentation on Maine ravens at the school. Afterward, Mike and his wife Ina and about a dozen of the high school students and I went to the dump, where we found six ravens. There was not much raven food beyond one grotesquely burned dog sticking out of the snow and one *iqumaq* with some meat still remaining inside. *Iqumaq*s are "sausages" about four feet long and one foot thick that are made by sewing walrus hide together to enclose raw walrus meat and fat. They are buried in the ground on permafrost to be stored for many months and to ferment until they acquire the proper flavor.

Near 3:00 P.M., we saw many ravens returning to the DEW tower from the distant ice pack. When the seal and walrus hunting is good, the bears eat only the fat, leaving the rest of the carcass. The ravens and perhaps Artic foxes finish the rest. An old hunter, Noah Piugaat-tuq, whom I met in the village, explained to me that scavenging ravens are noisy, and polar bears will become used to their calls. Polar bears sometimes feed on dead marine animals that ravens find first, and the bears are attracted by the raven's calls. When hearing ravens' cries at food, bears without a kill to feed from often become distracted or attracted. Inuit hunters therefore imitate raven calls as a technique to get closer to a bear.

I wished I could have stayed longer at that sea hunter's village, but I had to catch my flight to Igloolik. While waiting for the plane in the afternoon, when the sun was still 10 degrees above the horizon, I watched the ravens sky-dance over the DEW towers. They had come back to their roost at least four hours before sundown. They played in twos and threes, repeating violent chasing flights that took them high into the sky. They dove and tumbled down over and over again. The walrus hunters had probably been successful; these ravens were well fed.

No raven had wing-tipped to me so far to indicate potential prey, as Akaka Sataa at Iqaluit had talked about. Akaka had also told me that an incantation was required by the hunter to elicit wing-tipping, and the magic words to address the raven were not given away to just anyone. In the old days, the incantation was bought from the shaman, because the magic words were very valuable. Abe Okpik, an elderly man of Iqaluit who was no longer a hunter, and whose uncle was named none other than Tulugaq (raven), later had told me that when out on the land hunting caribou, or out on the ice hunting polar bear, a hunter seeing a raven fly over used to look up to it and call its name loudly three times: "Tulugaq, tulugaq, tulugaq." Having the bird's attention, he would then yell to it, telling it to tumble out of the sky in the direction of the prey. If the raven gave its musical gong-like call three times in succession, then the hunters went in that direction and killed it. "They believed in the raven strongly, and followed it," said Okpik. "And after they killed the caribou or the polar bear, they always left the raven the choicest tidbits of meat as a reward." It seemed absurd to me that a hunter could signal to a bird, and the bird would in turn provide information asked of it. Yet I wanted to keep an open mind to the possibility of communication.

The wing-tipping behavior is unique to ravens. Anyone who watches free-flying ravens anywhere will eventually see a raven "tumbling out of the sky" and simultaneously tipping (or tucking in) just one wing that tilts the body to the side. In Yellowstone Park in late January, we saw this behavior, especially in ravens flying in pairs. I saw one raven dip its right wing five separate times, each time accompanied by a metallic, two-note gong-like call that may very well have been the call to which the Inuit referred. I have seen the wing-tipping in Maine in birds near a kill, when they were presumably well fed and "feeling their oats." The three-note gong-like call is also common, although I don't know its meaning. My friend Glenn Booma, who mimics it perfectly, says it almost invariably attracts nearby ravens. Could it indeed be an attractant call meant for other ravens that is then given to humans with the same intention? Could the wing-tilting and the call logically be connected with prey and with hunters in the Arctic?

Visibility in the Arctic landscape extends for many miles in all directions. There is little that is man- or caribou-size that could remain hidden from the view of a raven flying above the low hills. A hundred or so years ago, and thousands of years before that, the only humans walking on the Arctic landscape would have been hunters after prey, and the raven needed them to survive. Given that situation, a raven would make the connection between humans, caribou, and food in a flash. But could a raven communicate location?

A hungry or starving raven who knows that the proximity of humans with caribou means a meal would likely have felt exuberant on seeing human hunters when caribou were grazing nearby. Knowing that caribou were potential food, it might have learned that caribou in combination with people *is* food. In one plausible scenario in the evolution of communication between ravens and human hunters, one can envision a hungry raven feeling exuberant when flying over humans and seeing a herd of caribou on the horizon at the same time. As do other exuberant ravens, it may have tipped a wing as an expression of its emotion, then continued on toward the caribou to wait for the expected feast. After all, whenever it had seen humans and caribou before, it always later found entrails and other fresh meat.

The hunters could have learned as well. They would have learned that the appearance of a raven or ravens meant that large mammals were near. It would have been a small step for humans to presume that if the signaling raven flew to where they could not see, such as over a hill, that the potential prey would be there also. They would have felt the raven's signal had been intentional. Subsequently, hunters seeing a raven flying by might wonder if it had seen them, hoping the bird would again guide them. "Tulugaq, tulugaq," they would shout to get its attention. If it did not see caribou from up high, the raven may have had less cause for jubilation. Perhaps it simply flew on without giving any acknowledgment, and the hunter would not have followed it. Since a raven can live longer than a half century, there may have been specific individual ravens that learned the tricks of the trade of hunting with humans, and vice versa, and they could pass it on to others in a mutually reinforcing cycle that could end up resembling the well-known honeyguide example in Africa, where a bird routinely leads people to beehives full of honey.

What seems less logical to me is that the raven would tip its wing precisely in the direction of the game. It may have been that hunters were directed to game regardless of what direction they thought the raven had tipped its wing. Being alerted, they would start to look harder or perhaps travel farther. Caribou are often spread out in more than one direction. A hunter would have been led to game even if he went south, for example, over a rise from which he had a vantage point to see caribou towards the east. In either case, the hunter who *believed* would have been more successful than one who did not, and the price he had paid for the magic words would have been worth it. As a hunter myself, I know that conviction promotes action, and only action can produce success.

The practice of following ravens must once have been common, because it has inspired humorous Eskimo tales. I presume any truly humorous tale must have a highly serious antithesis. In "The Raven and the Hunter" tale, a raven tells a hunter who wants to settle near some seal breathing holes he has found precisely where to camp. The hunter foolishly heeds the raven and camps where directed. There in the night he is killed by a boulder falling from the mountain above. The raven then flies down and pecks out the hunter's eyes, saying, "I don't know why all these hunters believe my silly stories."

TWENTY-TWO

Caching, Cache-Raiding, and Deception

EVEN A PARTIALLY EATEN MOOSE OR elk carcass is a huge food bonanza for a raven, but that is not a guarantee for a continuous food supply. The bird may then face days or weeks without food, since any one carcass is usually ephemeral due to competitors ranging from bacteria to carnivores. There is great advantage to hoarding food for future use. Food caching by animals often looks as if it involves conscious planning, though it need not be conscious behavior. For example, bees make and store honey for use months in advance, even building elaborate receptacles of wax for this food. But these are largely programmed, inflexible responses.

What might a raven's caching behavior look like if the bird *were* consciously aware? Overall, a conscious bird should take into account various consequences of its behavior in a continually changing scenario, not just proceed according to a script. For example, it might respond to

Raven making a hole in the snow into which it will cache the meat it has temporarily placed onto the snow.

others' activity. If it found competitors at a carcass, the cognizant raven would hurry its pace. It would decide whether to feed and then cache, or cache and then feed. It would decide whether to fly off with each piece of food that it removed from the larger chunk, or to wait until several chunks were removed that could be carried all at once. With the latter strategy, there would be the question of putting pieces temporarily to the side to be picked up later before departing. This option would only be taken if no other ravens that would take it were near. If others were near, the cacher would leave immediately after tearing off a large chunk, or collect small chunks and store them safely in the gular sac of its throat. When ready to depart with the meat, the conscious bird would already have picked a general destination. If many birds were near, it would fly far away. If it saw others nearby who were uninterested in food, it could simply walk a few paces and then hide the food. If others *were* interested in its food, which it would determine from watching their behavior, then it would fly far out of their sight before caching. If after flying far away, it found another raven, it would fly off again and try to find another secluded place. After caching its packet of food, it might return to the carcass and see another bird caching. It would watch intently from hiding and wait until the cacher flew away, then fly over to that site. Before it recovered the others' cache, the original cacher (who saw the presumptive cache thief fly in the direction of its cache) would fly at the cache-robber and chase it off, provided the cache-robber was a weak bird.

What does the raven's actual, as opposed to hypothetical, caching behavior look like? Take Goliath. Like most other ravens, he would cache surplus food if he had recently experienced a shortage. But caching, to him, involved much more than just hoarding. During one observation, I saw him grab two pieces of meat and fly into the side aviary behind a big rock. He put down both pieces of meat, dug a hole in the dirt, then shoved one piece of meat in, walked a few steps to pick up leaves, and brought them back to cover the meat. He then picked up the second piece of meat, flew to another spot, and repeated the process. Meanwhile, Fuzz had been perched quietly on the dead beech stump in the other aviary some one hundred feet distant, intently watching Goliath through the wire screening. A few seconds after Goliath returned to beg for more meat from me, Fuzz rushed into the side aviary and found both pieces of Goliath's hidden meat. He ate one on the spot and recached the other for future use. That was not an isolated incident (see pp. 298–302).

On February 6, 1998, my six nine-month-old birds were all around my feet, feeding on scraps of meat. Blue, the largest male, got the biggest piece and hopped off with it to bury it deep in the snow. Then he hopped right back to get more meat. All continued to feed. Minutes later, Orange, the next-largest bird, walked toward Blue's cache, but Blue flew over to intercept, and Orange aborted his attempted theft. Blue and Orange next both flew into the other aviary, out of sight of the other birds. White, a small female, immediately walked toward the site of Blue's cached meat. She dug at the precise spot in the snow where Blue had cached the chunk of meat, pulled it out, and cached it elsewhere. A minute or so later, Orange came back, leaving Blue in the other aviary. With Blue safely out of sight, he went to Blue's cache a second time. He dug. Of course, he found nothing because White had already raided it. I have never seen a bird go back to a site from which it had removed food or seen another remove its food. It was comical to watch Orange. He looked again and again at the other's empty cache site, much as I had done looking for a radio transmitter (see Chapter 6), as if in genuine disbelief to find that Blue's food was gone when he knew Blue could not have taken it. He didn't realize that White had taken it instead.

As long as I kept giving each bird pieces of meat, there was little reason, or time, for any of them to do much cache-raiding. To see more of what was involved in this intriguing behavior, I needed to provide more motivation for it. I started by cutting two red squirrels into twenty approximately equal pieces, and I allowed only three birds into the experimental aviary: Orange, the next-to-dominant bird of the six, and Red and White, the two most subordinate ones. I showed all the pieces I had in my hands, then I started handing out the pieces one at a time, but only to Orange. Since a raven takes all it can get, especially if that means getting food away from others, he cached each piece as quickly as he could and instantly came back to me for the next piece. Orange was very dominant over Red and White, so he had less reason than a subordinate bird to hide his caches. He made the first ones in plain sight, as Red and White watched from a perch nearby. Red was brave and flew down to dig the first piece out of the snow, and she managed to get it (or Orange allowed her to get it). But then Orange chased Red twice around the aviary until she dropped the meat. Her cache steals and the subsequent chases occurred several more times, and each time she ended up empty-billed despite her thefts. Finally, whenever Red dug up a cache, she would drop the meat without even bothering to fly off with it when Orange approached. White also ventured to steal a couple of caches and was also violently chased until she dropped the stolen meats.

After Orange had made about a dozen caches, he occasionally stopped to feed. While he was bent over picking on a piece of meat, Red and White unerringly went directly to his buried caches, pulling out his cached meat, and feeding as well. When Orange noticed their thefts, he again gave chase, even while picking up and carrying along the piece he had last been feeding on. Ultimately, he could not hide or control his twenty caches all at once within the confines of the aviary, and both Red and White got to feed.

Next, I threw in new variables. I let in the other half of this raven group, Blue, Green, and Yellow, and I gave them a calf head. Blue was the most dominant bird of the six, and he fed by standing on top of the food, as is the dining etiquette among ravens. Blue focused on trying to extricate the calf's eye, while all the others except for Orange fed amply on the calf meat. The eye was extremely difficult to dislodge, and Blue worked more than ten minutes at this task. All this time, Orange sat watching at the sidelines, undoubtedly digesting a stomach full of squirrel meat. Being the next in line in the dominance hierarchy just below Blue, Orange was always the main object of Blue's aggression at food. Orange held back and waited. During the moment that Blue was pulling the eye free, when I (and probably Orange) *knew* he'd fly off to cache it, Orange flew off his perch and landed on the calf head and started tearing at the meat. As Blue spent more than two minutes walking and looking for a place to cache the eye, Orange had two minutes of feeding time. Having finally cached the eye, Blue returned to resume feeding at the calf head, and Orange left it. Several minutes later, I happened to see White, the most subordinate bird of the group, fly by me into the other aviary, carrying the calf eye in her bill. She had found Blue's cache.

This anecdote reinforced my impression that the ability of these birds to anticipate the actions of others, coupled with their good memory, are traits that can compensate in competition with larger and more dominant associates. My observations were possible only because I was so closely in their midst. My rearing them from nestlings, and daily association with them for ten months, had won me their trust, which made the expression of their fine-grained unfiltered and hence complex behavior possible in my presence. The aviary also compensated for my inability to fly. I could follow them here, while at the same time provide an experimentally crowded situation that elicited flexible and innovative behaviors that otherwise might occur only rarely in the field where the birds can more easily avoid each other if they choose.

The ravens routinely recovered food they saw me bury in the snow. They also seemed to anticipate the actions of not only other ravens, but other animals as well. Gerald Fitz, a ravenphile from Lowell, Vermont, is convinced that his tame raven quickly figured out his buddy, the Fitz's beagle. Gerald had rescued the raven as a nestling from a local granite quarry, where a raven's nest had been destroyed three years in a row by blasting. The raven had grown up with the beagle. The two played with each other, and the raven followed the dog around, apparently enjoying its company. In the winter, the beagle followed the raven when it cached food in the snow, predictably taking the hidden food. The raven responded by caching in high places that the beagle could not reach. He also cached food in vertical bolt holes that had been drilled into a sill. After caching meat in the drill holes, he capped them with pebbles that fit snugly so that the dog could not lick them or the meat out.

Terry McEneaney, who has been watching ravens for a long time, has many stories to tell. "I saw the pair of magpies near the house here [in Gardine, Montana] become terribly excited, vehemently scolding a raven. I knew something was up so I continued to watch. Soon the raven approached the magpie nest in a juniper tree. Since a magpie's nest has a nearly impenetrable dome of thick limbs, the raven couldn't just tear into it. The raven proceeded to methodically pull out one stick at a time, until it had finally made a hole. Then he reached in and pulled out a pinfeathered young, flew by the house, and cached it nearby under a bush in the open sagebrush. He flew right back, got another young, and cached it in a different direction under another sagebrush. He then repeated this with a third and a fourth young. He kept the fifth one in his bill, and flew off with it and went right by his nest in the nearby cliff, as if to show his mate, who was brooding their own young. She left the nest and joined him as he cached this fifth one, then she followed him back to the magpie nest. She now took the sixth young magpie and brought it back to their nest to feed their three raven young. He took the seventh and cached it under another sagebrush, then came back to get the eighth, the final one. This one he did not cache. Instead, he took it into a dense juniper where he had some protection from the screaming magpies, who had in the meantime recruited helpers to harass the offending ravens. There, in the safety of the branches, he dismembered the young magpie and ate the whole thing."

Lesly Woodroffe of Alna, Maine, told me a similar anecdote of a raven that raided a nestful of four young cottontail rabbits. It decapitated them, cached them one by one, and only then stopped to eat.

Ravens' policy is: "Cache while it lasts and eat later." At the calf, deer, and moose carcasses where I have watched ravens, most of the participants exhibit almost frenetic activity as they haul off meat, especially if the meat can be torn off in large chunks. On December 24, 1991, I stretched out under a spruce blind covered with snow to watch a crowd of about forty ravens disassembling a calf carcass. In one 135-minute stretch, my four individually marked birds made twenty-three, twenty-one, eleven, and nine trips, respectively. The birds worked steadily all day long, so the forty birds may have made more than four thousand caches that day. Knowing how much they can eat per day and how much they carried, I presume they could each only have *eaten* the equivalent of one to two caches in that time.

The average time per caching trip in the incident I have just described was ten minutes. This seems long. When caching in soft snow, ravens merely thrust their bill with the meat into the snow, then release it with a push of the tongue. Loose snow from around the hole falls in and covers the meat as the bird retracts its bill. Usually, the raven makes a few more shoveling motions with its bill to draw more snow in from the sides. When the snow is crusted, the raven will lay the meat down onto the crust, peck a hole through the crust, then pick up the meat and push it down into the hole. It will cover the hole with pieces of crust or shove snow over it, as appropriate. Similarly, on bare ground, a raven will thrust food into crevices or holes, or dig holes with its bill if none exist, and then will pick up such debris as sticks, leaves, or grass from nearby to cover the food. Ravens also cache up in trees. Making the cache itself rarely requires more than half a minute, but it may take many minutes to tear off enough meat for any one cache. Just before flying off with the throat bulging full of meat, and with additional meat sometimes dangling from its bill, a raven usually stops a second or two, looking in all directions and blinking its eyes. Then it flies far off in one direction with vigor and resolve, as if having decided where to go before departing. Almost every trip is in a different direction. No two caches are ever in the same place. The big question posed by these observations is: Why do they go to all of this trouble to fly long distances, and to disperse all of their caches, when they could potentially cache nearby without flying, or flying only short distances? The key is, individuals or lone pairs cache much closer to the source than do birds from crowds.

Unlike the pair that McEneaney watched and the many pairs I also have watched, ravens in *crowds* in the field almost always cache food very far from where they get it. I have seen ravens from these crowds in the wild make many thousands of caching trips, but I've seldom seen them actually make a cache. Most likely, the ravens near the food pile do not get to see others making caches, either. That's probably no accident. Ravens may have to make their caches far enough away so that other ravens don't find them. Watching from the top of a tree, as a raven might, I sometimes see the birds carrying meat fly off out of sight, over the next ridge, down the valleys. When I've found cache sites—tracks and scrapes in the snow—it has been only by walking at random in the woods. Caching close to a carcass not only invites thieving by other ravens, it also poses another danger. Such keen-scented carnivore scavengers as coyotes, fox, fisher, raccoon, and weasel, as well as other animals from bumblebees to squirrels, concentrate their search for food near where they previously have been successful. That is, if dozens of ravens put their caches in the same area, all the caches would be in greater jeopardy than if they were spread out.

All of my observations of raven-caching behavior indicated more flexibility than had ever been observed in any other animal. I wanted to get publishable results, and turned to more experimental conditions for a tighter focus on specific aspects of the caching behavior, such as memory, using fifteen individually identified wild-caught birds who were kept in one of the side areas of the aviary off the main complex. From this pool of birds, I would capture from two to four birds at a time and place them into a second side area of the aviary. Continuously well fed birds don't cache food (Gwinner, 1965), so I kept them without food for two days to induce the caching response. After the two days, they were allowed into the large experimental enclosure where a pile of chopped meat chunks had been placed on the snow. Up to an hour commonly elapsed before the birds ventured to contact the food, but after they did, they usually began caching immediately. I allowed them another half hour of feeding and caching, then chased them back to their side aviary. I waited for different durations, from one day to one month, before I would let them back into the experimental area to see if they could recover their caches. If the memory interval to be tested was more than two days, the birds had to be fed in the meantime, and then again allowed to go without food for two days before the retrieval part of the experiment, so that they would be motivated to recover caches.

I found that the birds easily remembered caches made a day or two before. They had poor ability to remember caches made two weeks earlier, and they were virtually unable to recover month-old caches. As usual, some of the details of these experiments, which I began in February 1992, were more interesting than the general results.

In one experiment, White Slash and Blue Diamond (names referring to symbols painted on wing tags, which I'll here shorten to Slash and Diamond) were allowed on February 4 to feed and to make caches. For the first two and a half hours, Slash spent most of her time flying about. She fearfully approached the bait—chopped pork lung—dozens of times, doing jumping jacks with her bill open in fright. Finally, she edged close enough to grab a piece. Diamond had perched nonchalantly during that time, but after Slash got a piece of meat, he suddenly became active. He followed her as she fed briefly. She eventually cached her piece of meat, but not until Diamond was almost but not quite out of sight at the opposite end of the aviary. Immediately after Slash cached, Diamond flew over and dug in the snow at her cache site for three minutes. The snow was deep, and he apparently had not watched closely enough to succeed in recovering Slash's cache. A few minutes later, Slash herself showed no problem recovering her own cache. She then continued feeding.

I left them without food for another day, hoping they would become more motivated to make multiple caches. In the morning session, they both acted uninterested in the pile of chopped pork lung. In the afternoon, Slash approached the meat within one minute after I put it down. She fed, and then again made a cache in the snow. As on the previous day, Diamond immediately flew there to dig. Strangely, Slash did not intervene. Instead, as Blue was preoccupied in trying to dig up her first cache, she erupted into a virtual caching frenzy, making seven caches in seven minutes while Diamond remained thoroughly preoccupied with his digging. Finally, he again gave up, then came back to watch Slash, who abruptly stopped caching, instead flying round and round the aviary with a chunk of meat in her bill, as if suddenly not knowing where to put it. She finally cached it only when Diamond once again resumed trying to dig up her *initial* cache, the only cache he watched her make. This time, Diamond was finally successful in finding the meat. After eating it, he came to the bait pile himself. When Slash was preoccupied in feeding there also, Diamond suddenly made four caches in three minutes. Afterwards, I chased both of them out of this portion of the aviary into the adjoining aviary in order to learn if they could recover their caches the next day.

On February 6, I allowed both birds back into the enclosure to find out who would recover which cache, and how fast. Although Slash had acted frenziedly the previous day while caching, she was calm and composed today. Nothing seemed to disturb her, not even me. She calmly went to one of her caches and ate the piece of meat retrieved from it. Then she flew to the next one, and so on.

Diamond, who had the day before made only four caches of his own, first went to where Slash had made three caches and probed briefly in the snow. He did not seem to know the precise locations and recovered no meat, although for Slash he was probing too close for comfort. She flew over. Diamond apparently knew Slash's intent to punish his intended trespass, because he made a submissive display before she even got there. He had never made submissive displays to her before. Slash did not attack. Instead, she walked by him and simply reached down into the snow and pulled out her cached meat to hide it elsewhere.

The observation done, I opened the gate leading to the main aviary where the raven crowd was feeding on a food pile. Slash instantly flew in to join them at their feast, grabbed a piece of meat, and came right back into the part of the aviary from which I had just chased her! Although she was a wild-caught bird, she perched with her back to me. She held the food in her bill and all the time watched Diamond instead of me, probably waiting for an opportunity to cache when her competitor was not looking. Neither paid much attention to me. Their attentions were on each other and where to cache their food so that the other could not see them.

In another experiment on February 10, I observed recoveries of caches that Yellow O and Diamond had made the day before. Diamond had made thirty-three caches, and Yellow O only two. In one hour, Diamond recovered eight of his but made ten mistakes—six unsuccessful recoveries of his own caches, and four apparent recovery attempts where there was no cache. Yellow O, in contrast, recovered both of his caches plus two more of Diamond's, and made no mistakes. Although the birds probably were capable of recovering many more caches, they had no need because they were soon satiated, so I chased them out to continue the experiment the next day.

Later in the month, on February 27, Number 106 Blue and White Blank were tested for their ability to recover the caches they made two weeks ago. White Blank had made five caches, and today went to all five. Number 106, a new bird from the same group, had made thirteen and tried to recover only three. He dug for a full eight minutes at one site, before he got his piece of meat.

I later allowed four birds to cache simultaneously, removed them from the aviary, and let them back in one month later. Only two of these birds had cached, and these two walked continually all over the enclosure giving the impression of searching. Perhaps they knew the food was in this enclosure somewhere, but they didn't seem to know exactly where. Meanwhile, the two birds who had done no caching at all sat quietly, hardly moving. They acted like birds in the control experiment.

The control consisted of forty food caches that I made in the birds' absence (while they were out of sight in a side aviary). I marked the locations of those false caches in snow with branch sprigs and twigs. Then I let the birds in. None of the forty false caches was recovered by them. I recovered them all myself after one day.

My experiments were the first to prove that any bird could remember others' cache locations, and my anecdotes consistently demonstrated that ravens also have counter-strategies that foil the considerable cache-parasitism capabilities of their fellows in a crowd. In experiments to test this three years later, in 1995, I provided Fuzz, Goliath, Houdi, and Lefty with piles of food when they were either all four together as a group, or when they were each provided food in isolation, one at a time. In the first situation, when I put down a pile of food chunks, all four instantly descended onto it, grabbed all they could, and started caching immediately. Their caching trips were short. Each bird attempted to return immediately to the food pile to get another load of meat, as if to carry it away before the others might get it. Only the bird who eventually got the very last piece of meat from the pile invariably did not cache that piece immediately. Instead, it carried the last piece about, leisurely fed from it, and eventually cached what remained of it with great deliberation. Similarly, when only one bird at a time was given access to the same amount of meat, caching proceeded at a leisurely pace. The birds therefore appeared to monopolize as much of the temporarily available food as possible, and caching was their way of choice. If that was all, they should have cached closer to the source when they were in a crowd than when alone, to increase caching speed. Instead, with others present, they went farther to cache, going into the side aviaries where they could cache in private. The results made sense if the birds also attempt to take the uncached food away from others and to hide their caches from their sight.

Ravens' caching behavior relies on their capacity to "track" objects. If they see a piece of food put into the snow, they "know" it is there and may dig six or more minutes for it. As soon as I or another bird removes it, they stop. Their minds seem to be like a chalkboard. They have the capacity to register many locations on it, but with each retrieval they "erase" that site from further consideration. I saw something even more sophisticated in one probe. In four different trials, I showed three of my ravens two halves of a cashew nut, and as they watched me from a perch, I buried the halves next to each other in a bare patch of snow. White twice found one of the nuts in less than a minute, then left. Orange showed faith that there was more to be had by continuing to dig after the one piece had been removed. The pieces of nuts were difficult to find in the loose, powdery snow, and in two trials he dug for 3:15 and 2:40 minutes, respectively, without finding them. In the third trial, Orange got one piece in 75 seconds, and both Orange and White continued to dig for another 2:50 minutes before giving up. At the fourth trial, Orange got a piece of nut in five seconds, then Red joined in and they both dug for 70 more seconds.

Although it may seem logical to suppose that the ravens have awareness of others' actions because they anticipate others' responses before they occur, not everyone agrees with that conclusion. The alternative supposition is that the behaviors are stimulus-response phenomena in an infinitely complex chain of unconscious reflex responses to stimuli, ad infinitum. Most behaviorists would be convinced of awareness if, as has been shown with some primates, ravens lie. What is a lie? I would argue that pretending to make a cache and then actually hiding the food someplace else amounts to a lie.

In deception, a signaler behaves in such a way that a receiver registers something that is, in fact, not occurring. As a result, the signaler benefits and the receiver pays a cost. That is a behavioral ecologist's definition. To a psychologist, however, lying is not only the giving of false information. Lying, from the psychological perspective, implies that the giver of information is aware of the fact that he is giving false information and that he knows this information will be interpreted by an intended receiver. I have evidence that ravens lie (pp. 300–301) when I wear my hat as a behavioral ecologist. However, as I have indicated, I attempt to go beyond that perspective in this book.

I had on numerous occasions seen ravens make false caches—bury food and then immediately recover it to bury it elsewhere—i.e., they lied. However, I had not necessarily interpreted that as conscious lying. As an ecologist, it seemed simplest for me to assume that proximally the birds were only playing or being fussy about their cache sites, with no conscious deception involved. Making false caches can function to deceive, since false caches are sometimes checked by others (see p. 301); but as much as I believed the birds were aware in some ways, I did not think this behavior, by itself, *proved* awareness. Maybe they always act as if they are being watched without having any conscious knowledge of their actions. Maybe making a false cache is beneficial and costs little to do, and they evolved complex programming to do it unconsciously.

In a complex animal, the results of any one experiment almost always encompass several realities or perspectives. To see the proverbial "elephant" described by several blind men who touch it in different places, especially one as huge as consciousness, it must be examined simultaneously at many points, and from many perspectives.

The jury is still out, but nevertheless the details of the ravens' behavior indicate to me that the birds predict and anticipate the behavior of their conspecifics. That may be as close to exhibiting consciousness as it gets. To react flexibly and in an anticipatory fashion to a wide *range* of actions (as opposed to a few specific ones) before they have occurred makes the hypothesis that the animal knows what it is doing increasingly more plausible, and the idea that all of their behavior is unconscious, unlikely.

A much-discussed current idea that holds promise in understanding the evolution of consciousness is that the most challenging problem faced by individuals is dealing with their companions. The theory of "Machiavellian Intelligence," or selection for social expertise, is the current front-runner as an explanation for the evolution of high intelligence in the hominid line. As the caching behavior suggests, the same rationale could apply to ravens.

Whitefeather, in the Maine aviary.

Goliath preening Whitefeather while she holds her head still for him.

Fuzz and Houdi, allopreening and bill-holding.

Shed in the aviary where first wild pair (now nesting near Weld village)
and then Goliath and Whitefeather nested. Goliath is perched just outside the aviary.
Picture below shows the four (standing) young of Fuzz and Houdi,
and the two younger young (reclining) of Goliath and Whitefeather.

Matt and Munster, a pair of male siblings who bonded, built a nest, and who tried to mate and then had vicious fights (but they later made up).

LEFT: *The white nictitating eye membrane is used for communication and it also serves to protect the eye.*

RIGHT: *Many of the raven's feather bases are white. Note brown eye of this fully adult bird. (Baby's eyes are blue.)*

Two baby ravens about a week old, and two unhatched eggs.

Regurgitated raven pellets look much like owl pellets. The indigestible remains of these pellets found under a raven roost contain fur, bones, eggshells, and insect exoskeletons.

Female knocking display. Note dark mouth color of an adult.

Wild ravens at a carcass near my house in Vermont.
Female is at the left, and male to the right. Note her lanceolate long throat feathers
during her bow, when both wings are elevated and her tail is flared.

Here the wild bird (perch at left) is watching one of the four hiding food, and it then dug in the snow (inset) to retrieve and steal the cache. Inside aviary in Maine.

Planning by a raven. The bird had made a groove, cutting off the tip of the bread-loaf-sized chunk of suet. Then it cut a groove to cut off an even larger piece. Note peck marks in groove and small suet chips it had left next to groove, as it concentrated on the later, larger, reward.

Young raven testing the reaction of a dog.

TWENTY-THREE

Morality, Tolerance, and Cooperation

As BEHAVIORAL ECOLOGISTS, WE
try to reveal rules of behavior as though we were discovering truths. In
reality, the word "rule" as applied to animal behavior is a verbal short-
cut. A "rule" means nothing more than a consistency of response. It is
not adherence to dictum. Animals adhere no more to rules than we do
by showing up at the beach when it's 110 degrees but not when it's 30
degrees. Rules are the sum of decisions made by individuals that are
then exhibited by crowds, not vice versa. Rules are thus a *result*. They
are the average behavior that we and many animals are programmed
with, learn, or make up as we go along.

Animals exhibit consistency of response only where it serves their
individual interest. The necessity for consistent response with respect
to treatment of others is obvious among social animals, to whom others of their group are as much a part of the environment as is temperature and other factors. In social animals, it seems almost proper to talk

The "gang of four" feeding amicably together.

of rules in the sense of dictums, since certain responses are not only evolved but also sometimes enforced.

Whether or not an animal's behavior is acceptable can depend on whether it is directed to members of its own social unit versus others. Sanctions and punishments are often applied by the group to enforce its "morals," which are the reflections of its interests. In humans, morals are even used as a weapon, especially in the political and religious arenas, to help enforce the interests of specific groups or party lines. Through extensive work spanning many years, we have discovered some "rules" of raven behavior relating to groups. We have found, for example, that on average, vagrants disperse widely and individually, forming ad hoc gangs to overpower resident pairs who defend their food.

Groups of four to a dozen birds sometimes associate in what appear to be social partnerships. Goliath and Whitefeather even tolerated juvenile vagrants at a calf carcass near their nest site in winter. I've seen six to eight birds arriving suddenly at a bait when there was only a pair for weeks before. Do ravens have groups within which, as with other animals, enforced rules of behavior are observed? What rules of behavior apply? To find out, I needed an established group that could be observed for extended periods of time. Since the foursome of Goliath, Fuzz, Lefty, and Houdi had been kept together for months and had long since stopped fighting among themselves, they were an ideal group to observe.

On February 2, 1995, I released a single wild raven into their aviary—a female in her first year. She had mottled mouth color and brown-tinged tail and wing primaries. Goliath and Houdi, then the most dominant male and female, respectively, both gave violent chase. Goliath cornered the stranger in the shed, put on his macho display— ears, erect posture, bill up—and crowded in upon her. Occasionally, he yanked her wing and tail feathers, to which she objected vociferously but without challenge. Fuzz, the subordinate male, at no time approached her. In contrast to Goliath, Houdi pursued the new female relentlessly. She crowded up to her and pulled her wing and tail feathers. The newcomer sometimes tried to ward Houdi off by reaching out with her foot, but kept retreating, all the while making rasping protests as she was being physically abused. The question of dominance and sex having been established, the four birds then ignored her completely, quite in contrast to similar human situations I was aware of.

Feeding rights were another matter. When I placed food onto the snow and all the others fed at it, the stranger was always attacked if she came close. Perching by herself and watching them feed, she then erupted in a temper tantrum. She violently hammered her perch and the walls of the shed where she was hiding. Whenever she watched the other birds feeding, she resumed her outburst. She vented her anger only when the others were at the food that she was prevented from reaching. Hours later, when they were not at the food, she ventured out of the shed and down to the ground to try to feed, but the four violently jumped at her whenever she came close to the food pile. When they jumped at her, she crouched in a submissive display with her neck pulled in, wings lowered, and tail vibrating. This is the raven's extreme entreating display saying, "Please . . ." (". . . let me feed," ". . . let me mate," ". . . don't pick on me," depending on context). The gang of four didn't relent. Theirs was an extraordinarily high degree of cohesiveness or cooperation in a group of ravens.

The wild bird eventually became even more afraid of her four conspecifics than she was of me. She perched calmly within fifteen feet of me, instead, concentrating on the others, especially when they buried caches of food in the snow. She flew down repeatedly to try to dig out these caches, but attempted the thefts only when the cacher had its back turned or was preoccupied with feeding. At first, she tried to raid a cache immediately after the cacher had finished making it and had flown back to the food pile. Each of the four cachers were soon wise to her ways, leaving the food pile to come back and chase her if she as much as flew near their latest cache site. After a few false starts, she raided earlier caches to which the cachers were less attentive. By this strategy of deception, she sometimes succeeded in recovering the others' food and getting a bite to eat before being caught raiding.

Within a couple of days, Houdi's aggression toward the new wild female subsided somewhat. Nevertheless, all four still chased her from any food pile I provided, and she still fed herself exclusively from their cached food. I presumed that in a few more days she would be tolerated at the carcass itself, and all would be well, given a superabundance of food. When I had to leave to go back to Vermont, I made sure the aviary was well stocked with food, including two large calf carcasses. Given the abundance of food, I expected tolerance to blossom, because it normally does within groups.

When I came back, I found, to my great shock and surprise, the stranger lying dead on the snow. There was congealed blood on her breast feathers, skin was torn from both her legs and wings, and there were peck and puncture wounds with hematomas all over her head and the base of her bill. She had bled, and only live animals bleed. Ravens regularly peck out the eyes of animals they intend to eat, and they have a strong social inhibition, or "rule," that prevents them from pecking the eyes of other ravens, yet this raven's eyes were punctured. In short, there was no doubt whatsoever that my four friends had killed her. She had died trying to defend herself, lying on her back in the snow, using her feet in defense. They had not eaten any of her flesh.

Perhaps my gang of four, being confined together for an unnaturally long period, had become an "exclusive club," and this had reduced their tolerance of strangers. But that was not the cause of the murder. It was only an ingredient that led to it.

The raven's murder seemed extraordinary because I had previously seen that even starving ravens do not kill the weakest of the group (p. 212). Ravens routinely threaten or mildly attack others who are in the process of robbing their caches. I had also seen several birds unite to attack another individual for reasons that were not clear to me. But this killing was a severe punishment. It went far beyond the usual behavior aimed at repelling a competitor from a cache or for showing displeasure over a mild infraction. This was censure of the severest kind.

The murder was probably punishment for repeated infractions. Feelings of outrage that are used to justify cruelty are reminiscent of our own behavior, and have been a major source of human suffering the world over. We cannot expect more from ravens.

Retaliatory acts are known among birds. Many birds harass those they have seen raid their nests. At the University of Groningen in the

Netherlands, which I once visited, a pair of magpies built a nest next to the biophysics building, while a pair of carrion crows had a nest nearby across the parking lot. Soon there were frequent magpie-versus-crow battles. In the escalating battles, the crows finally killed one of the magpies. From the crows' perspective, justice was served; from the magpies', injustice.

Individuals of the same species can also be singled out. Once in the winter woods in Vermont, I saw a very excited group of ravens make a concerted effort to pin down and attack one individual. I and others have observed crows doing the same thing. To me it had at first seemed logical to assume that the birds were attacking the other simply because it was a stranger, rather than because it had done something they disapproved of, or because it has harmed only one of its attackers. To admit to the possibility that it had harmed one and the *others* felt outraged would be to go beyond the angry raven to the potentially moral one with standards of behavior that it expects others to follow.

Most birds attack others that come near their nests, mates, food, or territories. The attacked birds are punished for presumed intentions. Few would interpret this behavior as moral—the birds are simply protecting their self-interests. We tend to think of morality as behavior that promotes not just our own direct interests, but those of the group. The distinction is not as clear as it may appear, however. For example, a crow defending its nest from a raven may in fact endanger itself and assist its mate and its helpers at the nest. Colonial breeders defend the whole colony as well as their own interests. A bird that has once caught a certain nest raider in the act will remember that species, and even that individual. For instance, at Cornell University where Kevin McGowan climbs crow nests to band the young, he is singled out among all the students and professors for attack by crows when he walks across campus. He is attacked first by the parents of the nest he is nearest, but then the neighbors join in, helping to defend the first pair's nest.

Ravens would remember an individual that consistently raided their caches, if they caught him in the act almost every time. This is possible in the confines of an aviary, where the thief cannot escape vigilant eyes. Violators are remembered, and apparently, violators are attacked. That's primitive justice in their system and ours. In such a system, the "others"—those that are identified for one reason or another, or that identify themselves as "others"—are often automatically excluded.

I presume that the murdered wild raven in the experiment above had raided the caches of all four captive ravens. All four may simply have been defending their individual interests. If so, their behavior was not moral, in the sense we use the term. On the other hand, if one of the four birds did not see any of its own caches being raided, but still lustily joined in the murder because the others' caches were raided, then academically speaking, it was a moral raven seeking the human equivalent of justice, because it defended the group's interest at a potential cost to itself.

Two subsequent experiments confirmed that group interests can drive individual volition. In one of these I introduced three yearling ravens (who had grown up together) to my group of six who had been together for two years. The six attacked these three instantly, even though none had raided caches. More tellingly, though, when one of my six attacked one of the three, its five group-mates jumped in to help in the attack. In the second case, when one pair got ready to nest, one of the mated birds at times became aggressive to one of the four other birds in the aviary. Interestingly, whenever one of the pair started an attack, first its mate and then all of the other birds also helped in attacking the same victim. These observations show that outrage is exerted, and apparently felt, along "party lines," obliterating the birds' original individual tolerances. Futhermore, the gang attacks are instigated by leaders; there are clear followers. The extreme case, the murder, was thus a social—and hence moral—act.

Intolerance is something for which ravens are well known. I previously discussed (Chapter 7) mutual intolerance as a mechanism that likely keeps nests dispersed. Yet, in our radio-tracking studies of Number 8130 (Chapter 8), the resident bird easily traversed back and forth over the territories of seven local pairs, even in the breeding season, when territorial birds are most intolerant of others of their kind. In the winter at the beginning of the breeding season, I had also sometimes seen several pairs of ravens come to feed at the same carcass, and I often had seen Goliath and Whitefeather socialize with neighboring pairs in the air (pp. 127–130). I received other intriguing reports, suggesting that ravens could be tolerant of others even at their nests.

Hans Christensen and Thomas Grünkorn, in following about 800 broods of ravens in the Schleswig-Holstein area of northern Germany from 1985 to 1996, on five different occasions (1991, 1992, 1994, 1995, and 1996) found a nest defended by three adults. The pair's helper, likely a female judging from its size, contributed to feeding the young. In Switzerland, Markus Ehrengruber and Hans-Rudolf Aeschbacher saw a third raven, likely a male, feeding an incubating female at a nest, and this "helper" was sporadically at or near the nest, but this nest and young were eventually destroyed. The cause was unknown. Lorenzo Russo, studying ravens on the island of Stromboli in Italy, told me of being mobbed by a pair of ravens at a nest as he was climbing. The pair left briefly to come back with three helpers, who joined the first two against him. Chris Walsh noted the same behavior in May near a raven nest on Raven Rock just west of the village of Moretown, Vermont. Chris and a companion were being chastised by the resident pair, who sometimes came as close as twenty feet over their heads. Chris wrote, "The display was mesmerizing, and we stayed." But "soon the pair departed, flying south down the valley until out of sight. To our amazement, some minutes later, a group of eight ravens appeared from the same direction, making a direct beeline for us. They resumed, as a group, to make the same type of agitated displays the pair had made." Similar mobbing by ravens of a golden eagle preying on a raven has been reported (Dawson, 1982).

We have no clue who the apparently very rare "helpers" at raven nests might be. They could be neighbors defending their nearby nests, grown offspring that have not dispersed, strange males seeking friendship for eventual extra-pair copulations, and/or neighbors who have joined in a coalition for mutual defense.

Carsten Hinnerichs at the University of Potsdam in Germany told me that late one winter, at a time when ravens get ready to breed, he saw six ravens at different occasions over two weeks regularly flying together as a group of three pairs. The pairs had fuzzy head feathers, made *glug-glug-glug* calls, and acted like typically courting birds. Although juveniles gambol in crowds throughout the year, nobody had ever before reported seeing ravens engage in mating displays as a group—only as isolated pairs.

The six ravens that Carsten saw as an apparent group of pairs at nesting time were an anomaly that excited me greatly, because it could shed light on the flexibility of social arrangements. Carsten found three fresh raven nests in "a little colony" in the pines nearby at the edge of a big dump. By April, each of three nests held an incubating bird. The spacing of the raven nests so close together that they were virtually a small colony seemed extraordinary. Here was concrete evidence for an amazing shift in territorial behavior from that which Grünkorn had documented in northern Germany (Chapter 7) and from what I had been used to seeing in ravens in New England. These nesting birds, like some others on an oceanic island (Noglaes, 1996), apparently were tolerant of each other. In Maine, even a small food item such as a dead snowshoe hare is fiercely defended, and other ravens coming into the vicinity are aggressively chased for miles. A larger food item such as a calf or a deer carcass is also fiercely defended; but, at least in the aviary, after the bird's bellies are full, their intolerance toward at least *familiar* individuals almost vanishes. The German ravens' extraordinary tolerance might not be so unusual at all. Perhaps it was a common response to a rare situation.

The place, the Zehlendorf dump, was a raven's Shangri-la. There was no end to good food, and it came reliably and without end, delivered by dump trucks day in and day out, all year from one year to the next. Perhaps this steady food supply was sufficient to enable communal nesting. But then all three nests that had eggs ultimately failed. Carsten found eggshells strewn on the ground below the nests, as well as torn-out nest lining. A study by Lo Liu-Chih from Taiwan, at the University of Saarbrücken in Germany, provides one hint of what might have gone wrong.

Liu-Chih's detailed analysis of many hundreds of raven nests compared nest success in different ecological sites in Germany. One of these sites was at the Rostocker heath, and another was in the area around the Grevesmühler dump. Like Carsten, Lo found a virtual colony of ravens in the pine forest within one hundred meters around the dump. Surprisingly, despite the large food supply, nesting success at the dump was much poorer than in the wild heath area. Birds started later, many nests were not finished, and three-quarters of all nesting attempts failed altogether. The "successful" nests averaged only three young, as opposed to five, typical in the nearby areas. What was even more strange, the young had thin thigh muscles, characteristic of birds that had received inadequate amounts of food. Lo thought that although food was more plentifully available at the dump, the adults living there had to spend so much time defending their nests, they had less time to forage.

But defending the nests from whom? Was there really strife among the closely nesting pairs? It seemed to me that an equally plausible hypothesis was one diametrically opposed—that the pairs cooperated. Perhaps they even defended "their" dump and their nests from the hordes of nonbreeding juveniles that are invariably present at all dumps. Near the time of the three nest failures seen by Carsten, there had been a sudden influx of about fifty ravens that took up residence at the dump. Eventually, in mid-May, the crowd had swelled to nearly five hundred. Perhaps it was they who destroyed the nests. Perhaps the decreased nesting success of pairs at the large food bonanzas that Lo and Carsten had described was not so much an indication of strife among the pairs as conflict with the ever-present nonbreeders. To try to find clues, I asked Carsten to show me the Zehlendorf dump, and my visit there in July 1997 was the highlight to my visit to Germany.

Playing detective, I sensed two strange things. First, the coincidence of the nest failures with the arrival of hundreds of juvenile vagrants; and second, the extraordinarily close placements of the nests right next to the dump, with the much more distant placement of the vagrant's sleeping roosts. I paced off the distance between the three nests next to the dump (100 to 160 paces), and closely examined the ground under each nest. In Maine, I had on four occasions (years) seen the ground littered with nest lining after a crowd of juvenile vagrants had been in the area. In spring 1998 in Vermont, when I had a large feeding crowd by the house, both local raven nests failed for the first time. As in New England, the ground under each nest at this dump was littered in all directions with tufts of nest lining. Normally, nest predators take only the nest contents. No animal other than another raven would so systematically engage in such strange behavior as tossing tufts of nest lining all around. Here was more circumstantial evidence that the strife was not among the pairs, but between the nonbreeding vagrants and resident breeders.

When I visited the Zehlendorf dump, there were still about five hundred ravens there. Opposite the pine trees on the other side of the permanently stocked feeding place, dump trucks roared in and the wind swept up a large mound of sand on top of which crowds of presumably well-fed juvenile unmated bachelor ravens gathered to play in the wind. As we watched the large queues of ravens on the crest of the sand bar, one after another would spread its wings, be lifted up, fold its wings, and be gently let down again, riding the air waves like surfers on a beach. Many others used the high thermals instead. They came flying in from miles away. With raucous cries, they circled, dove, and tumbled among the cumulus clouds. At night they gathered in a noisy raucous throng about a mile into the pine forest from the edge of the dump in a communal roost where the ground for over an acre was littered with feathers, pellets, and mutes.

Something still didn't add up. Why did five hundred ravens fly as much as a mile to roost at night, when they could have roosted directly at the dump itself? Why instead did the multiple pairs always nest directly *at* the dump, when they *could* have nested isolated from each other and this crowd, ten or more miles away from all the bustle? Why did the nests fail in one case in the egg stage already? I suspected there was a reason for the proximity of the nests to each other and their location directly at the dump. As my aviary experiments had shown, ravens' tolerance for one another increases markedly both as a function of amount of food they have available and the length of time of their mutual association. It was almost certain that the three neighboring pairs nesting at the Zehlendorf dump knew each other, opening up the possibility of a mutual alliance.

Vagrants are very often successful in overpowering pairs. Why not vice versa? Why could neighboring pairs not team up when they no longer need to compete for food? By sharing the dump, the two to three pairs would still have more food than needed, but it could make a huge difference to their nesting success if they could repel vagrants that destroy nests. As this book was going to press, Carsten informed me that the small colony of three raven pairs at the dump had now increased to nine! Had more been allowed to join the club, as more individuals had gradually gotten to know each other due to their continuous association at the permanent food.

More food means less competition and less aggression (or more, depending on circumstances), whereas less food provides more competition and leads to more aggression. A small or medium-sized carcass is worth defending, and in the forests of Maine, neighboring pairs meeting at a rabbit carcass are induced to fight over it. Fighting would destroy mutual trusts and sharing. Yet sharing would be mutually beneficial at a *large* "permanent" food pile such as a moose carcass, because meat shared by one pair with another would have a minor immediate negative impact on the sharers, but a major positive immediate impact on the recipients. If the recipients were to act in the same way at a future carcass, then all would benefit in the long run. Ravens may live for decades (Clapp et al., 1983), and they nest at the same site year after year. I wondered, therefore, if or how neighboring pairs might get to know each other, so that sharing at a future potential moose carcass could be possible for mutual benefits? Might it be in such aerial play as I had commonly seen among adjacently living pairs, which reminds me of interdepartmental touch football?

Play by Ravens

Play is notoriously difficult to define. We can all recognize it at the extremes, but we cannot fit it into an exclusive category of behavior. According to most definitions, it is behavior that is *seemingly* purposeless, but behavior may seem purposeless simply because the observer hasn't figured out what the benefit is. Benefits could be long-delayed and have many repercussions, not just one. For example, play fighting may function to develop skills for later use, establish important social relationships with peers, and build muscle tone and coordination needed to escape predators. However one may define play, young animals are not likely motivated to perform play for the rewards they ultimately may get from it. Play is often nonstop action with bizarre twists that has its own phychic reward, like hanging upside down by the feet. I have not attempted to organize my observations into such categories as "object manipula-

Raven hanging by its bill from a long, thin,
flexible branch (left), and hanging by feet from rope (right).

tion," "social play," or "play fighting." Rather, in the following I report activities as they occurred, in specific snippets of time.

On November 11, 1993, I saw Lefty dangling upside down from a thin branch of the pine tree in the aviary with her head held horizontally. Within a few seconds, she let go with both feet, turned in the air, and landed right side up on a branch below her. She regained her previous perch, flipped over backwards to hang again, then let go and flew off.

With time, she incorporated more variety into this oft-repeated stunt. Two weeks later, she hung with only one foot, grasping a piece of bark in the other. She repeated her upside-down hanging maneuver three times in about two minutes, always holding the same piece of bark either in the bill or a foot. At last she landed on the ground, where she picked up a bark chip and stuck it into a hollow plastic tube. She lifted the tube at one end till the bark slid out the other end. She repeated this maneuver three times. After she dropped the bark chip, she scraped freshly fallen snow off twigs with her bill. Within fifteen minutes, she picked up another bark chip, picking at it, carrying it, and at last caching it under a log. Goliath, seeing her hide something, came and dug in the ground nearby. So did Fuzz, but neither bird found anything of interest. Then Lefty chased Houdi, stopping often to wipe her bill on branches and singing a gurgling-gargling song with head fuzzed out and throat hackles puffed out. Lefty then resumed chasing Houdi.

In early December, I again saw Lefty hanging upside down. She was dangling from a very thin and flexible branch with her right foot while her left foot held a wood knot. While still hanging upside down, she passed the knot back and forth between her bill and the free foot. Sometimes with the knot in her bill, she hung with both feet. She flew up to the same branch three or four times in succession with the knot in her bill, then transferred it to the other foot before flipping over. Meanwhile, Houdi, Goliath, and Fuzz were dragging sticks on the ground and didn't seem to pay her any attention. For a few seconds, Houdi fluttered from a swaying perch, hanging with her bill only. Within the next half hour, Houdi had twice more tried to dangle by her bill from the same branch, whereas Lefty hung upside down by one foot, holding a small rock in the other. She picked up a pebble, and while holding it in her bill, turned over frontwards on another twig, continuing to manipulate the pebble in her bill. The pebble eventually fell or was let go. Then she let go with her feet to fall and then fly.

Two days later, I saw Houdi pick up a foot-long thin twig and carry it all over the aviary, occasionally stopping, holding it with her feet, pecking at the ends. After five minutes, she dropped this stick and snapped another off the pine tree in the aviary. This twig was two feet long. Fuzz tried to take it away from her right away, and Houdi gave it up without any resistance, then walked over to a calf carcass to feed. Lefty then picked up Houdi's first stick but dropped it in less than thirty seconds. Fuzz, too, dropped his long stick.

All birds were then without sticks. Goliath snapped off another long branch, and Houdi came over and pecked at the exposed stub where the branch had snapped off. Fuzz and Lefty meanwhile had a tug-of-war with a gray squirrel skin. Goliath had broken the stick off, then Houdi got it because Goliath dropped it. He then pecked at the place where the branch had snapped off, where Houdi had pecked before him. Goliath joined Fuzz and Lefty at the squirrel skin; the latter then both departed and Goliath quickly lost interest. Lefty resumed picking on the squirrel hide, while Houdi joined her. Lefty flew up onto a long perch that somehow loosened and fell down. All birds became frightened and flew into their sleeping shed, where they stayed for a minute. After venturing out, they refused to go near the fallen stick, and only after one hour and thirty-six minutes did they come down to the ground. I gave them a new toy, an aluminum can. Within seconds, they approached it, just gently touching it with their bills and then doing jumping jacks. Soon they jabbed it vigorously— all except Lefty, who showed no interest. The can was soon torn apart. Houdi ended up with the bottom part, and Fuzz had the top part. Both parts were shredded ever finer.

Four months later, when I gave them another beer can, Goliath got it first and hammered it for five minutes before losing interest and dropping it. Houdi grabbed it, but Lefty gave chase. For about one hour Houdi, with beer can in bill, was chased at intervals about the aviary. Finally, she dropped it. Did Lefty then swoop down to grab it? No! She let it lie. So did the others. Houdi went on to hang upside down with her right foot, while gurgling her raven song. Goliath rolled on his back on the ground, holding a branch up in his feet.

Lefty hung by both feet, then by only the right foot. After looking all around, she smoothly turned back up to the branch by going forward, assisted with wing-beats. She made her head fuzzy and called hoarsely, then did the same sequence again, this time dangling by one leg before letting go and dropping, after pecking the branch she had perched on.

One day in May, at the age of one year, Houdi repeatedly peered into an eleven-inch-long, square, brown plastic tube. After peering into one end, she picked up a green tennis ball, another long-neglected toy, and stuffed it down into the hole. It fit snugly, and she pounded it in by hammering it with her bill. When I looked into the aviary the next day, the ball was loose. Houdi retrieved it and again shoved it into the hole, stuffing dead leaves behind it. She then pulled the leaves and ball out, and repeated the process twice more. Fuzz began to take interest, pulling the leaves and ball out, pounding the ball back in, shoving debris behind it, pulling all out, and repeating the process twice. At first, Fuzz had some trouble getting the round ball to fit into the snug, square hole. He braced the tube with his foot to hold it fast while he pounded the ball in. When he was finished, Houdi came back and inspected what he had done. Fuzz came back and inspected also, putting more leaves behind the ball. Then he pulled it all out again, pounded the ball in deeper with his bill, then pulled it back out and pushed it into the other end of the tube, again stuffing leaves behind it. In the afternoon, Fuzz took the ball out of the tube and hid it under leaves nearby. By then, I had become satisfied that they were not afraid of this new object, the tube, and that they had learned it was hollow. Perhaps they had been productive. I then did tests to find out if they could put their knowledge of the nature of the tube to use, to see if they might keep track of objects within it that they could no longer see with their eyes (pp. 304–309).

Their antics were always fresh, unpredictable, lively, and constant. Nothing seemed to stop them, even when at the end of January temperatures dipped to minus 45 degrees Fahrenheit. In the morning, they were almost white-headed. Moisture from their warm, exhaled breaths instantly turned to ice crystals and condensed onto their fluffed-out head feathers. The birds started attacking a plastic milk container that had already received a thorough pounding the day before. Their only visible concession to the cold were their belly-feathers, fluffed out so that they reached down to their toes when they stopped to perch.

They often played "catch-the-stick," but the most popular activity going was snow-bathing. In its simplest form, this involved skooching down and flapping their wings, just like bathing in water. But there was more. They also slid forward on their breast and belly feathers, being assisted not only by wing-flapping and leg-flailing, but also by gravity. Houdi twice rolled completely around sideways down a short bank of snow. Goliath, Fuzz, and Lefty did only partial rolls. As always, these ten-month-old birds were vocal throughout, making nearly constant soft utterances. They pushed themselves forward in the snow on their bellies and went through all the motions used for bathing, especially after a new fluffy snow had fallen; but snow play was not restricted to fresh snow. There had been snow on the ground for well over six weeks by Christmas 1995, and on the evening of December 24, I gave Fuzz and Houdi a "present"—I made a two-foot-high snow pile in their aviary. As I had anticipated and hoped, both birds appeared to enjoy their new toy the next day. They perched on it and pulled out several loose sticks that had gotten packed in. Houdi slid down and turned one complete roll. She repeated the maneuver six consecutive times. Fuzz didn't try it, although he was by Houdi's side almost continually.

Many of our own behaviors have a conscious purpose, even if we enjoy them. Adults bathe, for example, to achieve a purpose, or at least so we rationalize, and we do not generally call it play. If the same behavior in ravens does not have a conscious purpose, I wondered, is it then play? To find out if ravens bathe for the purpose of getting clean rather than just for fun, I did a simple experiment. I tried to get my later group of six young ravens, then only a month out of the nest, dirty. Would dirty birds bathe more than when they were squeaky clean? It wasn't easy to get them dirty, because no raven is tolerant of mud-slinging. I finally outsmarted them, but only briefly, by spraying them with a thin solution of honey (which they disdain as food) through a squirt gun. I even succeeded in dumping some flour on them, gumming up their feathers. Birds with honey and flour preened more, but the honey-flour treatment did not cause them to bathe.

For a, to me, foul concoction, I set fresh cow dung to soak and rot for a day, wrung it out through a screen, then diluted it and sprayed it on them with the plant sprinkler. I managed to douse three of them. After that, none came close enough when I was holding the sprinkler for me to do any more damage. I put the rest of the smelly solution in a tin can, and succeeded in dousing two more birds by judicious dung-flinging. Then I tossed dried peat at the wet birds for good measure. After that, they wouldn't come near me even when I tried to lure them with their ultimate delectables, potato chips. My sample size wasn't exactly statistically adequate, but I'll report the results nevertheless. It's this: The birds didn't bathe after the treatments.

Bathing is a party activity. In the spring, when the first open water runs in the brooks, I had twice come on raucous aggregations of ravens who flew up when I approached. The fresh snow on the ice all up and down the edge of the brook was padded down with footprints, and there was evidence of water splashed and imprints of raven bodies rolling in the snow; but it was the ravens' sounds that had originally attracted me to the beach parties. I suspect that the participating ravens had an uproariously good time on their first water bath of the year.

The youngs' first bath of their lives, when they are days out of nest, is a memorable sight. They make the acquaintance of water cautiously with their bill, dipping it in, splashing it back and forth. Then they walk in hesitatingly, perhaps dipping their whole head in and violently shaking it back and forth. Increasingly more contact with the water is achieved as they gradually first lower their rear end down, followed by the front end. Soon they beat their wings violently.

Splashing is accompanied by numerous comfort sounds. When thoroughly soaked, the birds hop out of the water, seek a perch, shake, and begin to preen. As the birds alternately bathe and preen, they act intoxicated. Sometimes the fun wears off only after an hour or so.

Nobody watching the bathing performance of young ravens would ever get the impression that the birds were trying to remove dirt. Like kids splashing in the pool, the birds might get clean, but if they do it is strictly incidental. Bathing occurs regardless of when or if dirt is removed. Nevertheless, the birds could potentially learn with experience that this particular activity could pleasantly cool them on a hot day and/or clean their feathers. My observations so far suggest that any possible utilitarian functions are strictly secondary. A proximal reason is: They do it because they like it. The ultimate, evolutionary, question is: Why does it feel good to them, so that they do it?

I made a nearly full-scale study to try to find stimuli that induced them to bathe. All through one summer and into the winter of 1997, I took notes on which of my six ravens bathed, when they bathed, under what conditions they bathed, and how much they bathed. At intervals of a day to a week, they were given an opportunity to splash in a pan of fresh water from my well that was always at about 52 degrees Fahrenheit. The results were, to put it mildly, "senseless" and idiosyncratic. For example, as expected from the previous test, birds spattered with feces on their backs (this time from others perched above them) did not jump into the bath more than others. Neither did air temperature have much to do with it. On a sweltering, 81-degree, sunny August 10, only one bird hopped in once for a quick splash. On the other hand, on September 19, when it was overcast, windy, and 62 degrees, there was a constant queue at the water pan. Forty-five baths were taken within twenty-five minutes. All the six birds had bathed, each from four to seventeen times. Finally, I expected none to bathe on the early morning of October 29, an overcast and windy day of 33 degrees and with snow on the ground. Since it had just rained, all the birds already had recently gotten wet. I hoped to get a "No baths" entry in my notes at least once, to have a zero point on the graph. What happened? A *record* number of baths: forty-nine in thirteen minutes! Even more strange, for seventeen of the forty-nine baths they used the dirty mud puddle in the aviary that I had never seen them use before. I have no pat theories to explain their behavior, and I hesitate even to mention it, because assuredly some will think I'm exaggerating. I will try to exonerate myself in this case by giving more details, because it could help to provide additional insights into the raven's mind.

As always with ravens, the story is in the fine details. First, I had given the birds their last fresh pan of water for bathing seven days earlier. For a young raven, seven days without a bath is a very long time. The birds prefer to bathe every two to three days, and one of the most consistent variables that determine whether or not and how much they bathe is the number of days since the previous bathing session. The exception being, as previously mentioned, that when Houdi was incubating eggs and brooding young, she didn't bathe for two months.

With humans, temperature is such an important bathing variable that we could predict the number of people at a beach and call it the bathing law. To my ravens, temperature (except at 32 degrees Fahrenheit and below) seemed unimportant. With the total sample of 582 baths taken at temperatures from 33 degrees to 90 degrees Fahrenheit, temperature just wasn't an important variable. I finally did get my zero point on the graph of bathing frequency versus temperature at an air temperature of 15 degrees Fahrenheit below freezing. But in that instance, maybe the water wasn't available long enough before it froze. So they snow-bathed instead.

Perhaps the most important factors determining whether a raven will take a bath is whether it sees another bathing. As soon as one of my six started to take a bath, all of the others rushed to the water dish and tried to bathe as well. Of course, a contest ensued, with the most dominant birds being first. If a subordinate bird hopped into the bath, a dominant bird simply shoved it aside. Sometimes a subordinate then prostrated itself in extreme submissive displays, as if pleading, but that never helped. Each bird took multiple baths, whether it really wanted to or not. It went like this: When a dominant finished its bath and hopped out to preen itself, the next would hop in and start to splash. The dominant, who had just done all the bathing it wanted, then almost always stopped its preening and hopped in again, in order to displace the bather. Having splashed some more, he would hop out to preen again, and so on. After a half-dozen to a dozen such exchanges, the dominant male would finally allow himself to preen in peace, and the second in line would bathe, behaving in the same way to the next in line below. So it went on down the line to the most subordinate bird, who could bathe in peace.

Given the above, bathing in the mud puddle suddenly seemed rational. With all the birds competing at the water pan to bathe, one of the frustrated subordinates suddenly got the bright idea to use the other water, the mud puddle. The more dominant bird, at that very moment bathing in the clean, roomy water pan, immediately jumped out to chase the subordinate bird out of the mud puddle. He jumped into it himself and churned the puddle into a muddy broth. The vacated water pan with nice clean water was, of course, instantly used by another bird. The mud bather's frolic was therefore brief as, hopping rapidly, he returned to evict the opportunist from the water pan. Mix six birds into the fray, each trying to bathe and at the same time keep the others from bathing in a game of king-of-the-bathtub, and you end up with a frenzy. That's the reason there was a record of forty-nine baths in thirteen minutes. It was fun to watch, but it sure threw a curveball at my efforts to graph bathing frequency on any and all of the different variables that I had anticipated might be relevant. Even after taking notes on the context of 582 baths, I still had no idea why they bathed when they did. So I had to let it drop, as just another one of those interesting but unpublishable idiosyncratic results that might, like many others, only confuse some current academic thought.

A half year later, the results were different: The older ravens had become less enthusiastic bathers and were less influenced by the activity of their peers. But on November 15, at 48 degrees Fahrenheit, after they had not bathed for a month, the six took fifty-seven baths in seventy-eight minutes.

The most commonly seen raven play occurs high in the air. The following examples show some of their games.

Hanging games.

- In the autumn of 1983, Johanna Vienneau of New Hampshire was hiking above the tree line on Baldface Mountain with her dog, who was about twenty-five feet ahead of her. A raven came flying over and dropped a rock about an inch in diameter a foot away from the dog, a near-hit.
- Cedric Alexander, a wildlife biologist from the Vermont Department of Fish and Wildlife, told me that on December 5, 1993, an employee of his came into the office reporting that a raven had dropped a four-inch sprig of spruce while flying over him at about three hundred feet.
- Rod and Amy Adams and Eliott Swarthout, who were employed in September and October watching hawk migrations at Lipan Point on the South Rim of the Grand Canyon, wrote me: "Once we have seen a raven drop a rock in flight and successfully catch it. Once we also saw a raven drop a red object and catch it. Also dropped and caught were a coyote tail, and a vulture feather—the latter was missed by six ravens!" (Could they really catch rocks but *miss* a feather?)

Locking talons in flight.

I've seen ravens loitering for hours in the updrafts of the hills and mountains of western Maine. Again and again, they ride the air elevators and dive down in pairs or small groups. Once, on November 19, 1992, I was in a spruce tree watching groups of five to twenty birds return to a roost. Most were flying methodically. Suddenly one, who was coming back alone at high altitude, closed both wings to its sides and bolted straight down. In rapid succession, it made three 360-degree spins around its axis. Then it extended its wings, banked slowly, and descended in a graceful arc to land in the top of a pine near the roost where others were already settling for the night. Why the extra flourishes? Do the birds act out something they visualize in the brain, which other birds don't? Or do their odd behaviors just "happen" without their conscious knowledge? Could we suddenly do a back-flip without thinking about it first? If so, why should a raven be capable of it, if, as is generally presumed, they are unconscious?

At Grandfather Mountain (5,965 feet) in North Carolina, a plaque is given to anyone who can maintain an hour or more sustained flight with a hang glider. Those who earn the honor are said to belong to the Order of the Raven. "They [ravens] fly with the gliders, usually staying five to twenty-five meters off the gliders' wing-tips. Sometimes they get playful and dive past the glider, coming within a few feet. Other times they stay directly above or below, often getting pretty close," Michael E. Miller told me. What fun it must be to glide with ravens!

Tim Hall, also a former hang glider and paraglider pilot, wrote me: "I have had the pleasure of sharing the air with ravens on many occasions. I was paragliding in the mountains about eight miles east of El Cajon, California, traveling above a 3,000-foot cliff, when I came upon a group of about ten ravens. They were swooping and doing barrel rolls. After a few minutes, another raven joined the others. This raven was carrying what appeared to be a twenty foot long about two inches wide white plastic streamer, similar to surveyor's tape, in its beak. It started swooping down through the others, folding its wings and swooping back again. After doing this several times, it started handing the streamer to other ravens. Several ravens took turns catching the streamer, swooping with it, then releasing it, and the next raven took its turn. None played 'tug-of-war' with it. This activity continued for about twenty minutes. Those ravens that did not participate landed on a bare tree at the top of the cliff and appeared to watch the flying antics of the others." Miller's and Hall's information is valuable and hard to get. Not many biologists do field observations from hang gliders, although researchers watching dolphins from boats observe very similar antics, including bow riding, chasing, and object play.

Not all the raven's playthings are inanimate. Wildlife biologist George B. Schaller reported African white-necked ravens' (*Corvus albicollis*) play with gorillas at Kabara, in East Central Africa, in his book *The Year of the Gorilla:*

"The Virunga range spread glistened in the heat. I sat on a grassy knoll with my back against a bluff, my feet dangling over the depth of a dusky canyon. Ahead of me were the gorillas, and beyond them the slope swept upward, the breeze moving the green and silver leaves of the senecias until I was dizzy with the light. While eating my lunch, I spotted the ravens high above me, small black spots against the white of a cloud. When I whistled, which to the birds signified food, they descended in leisurely arcs. [The African ravens are not nearly as skittish as the northern raven, and the Schallers had tamed this wild pair.] The gorillas ducked when the gliding shadows of the birds passed over them, and the male jumped up and roared; females screamed, some looking at me, others at the ravens. Then, as if in play, the ravens swooped at the gorillas, diving low over them again and again. The male grew angrier than I had ever seen him, and the females milled about in utter confusion. The apes obviously failed to find this a game, and the ravens, well satisfied with their mischief, landed in a heather tree near me to consume the rest of my lunch. Then they headed into the valley, but an hour later they returned and again flew in unison at the gorillas."

I will not indulge in endless speculation on the ultimate benefit that these many kinds of play could have to ravens. Your guess is as good as mine. There are few data to prove or disprove almost any hypothesis we might come up with. Nevertheless, some types of play that are common in ravens do have demonstrable effects on their breadth of diet and their interactions with carnivores.

Ravens are omnivores who, with a combination of following others, curiosity, and learning, are able to find and utilize appropriate insects and fruit as these foods come into season in any one environment. This behavior of ravens differs from the more stereotyped innate responses of say, a red-eyed vireo that looks for caterpillars under the leaves of deciduous trees, or a flycatcher that hunts flying insects by sallying from a perch, or a kingfisher that dives for fish. These birds undoubtedly learn, but they are programmed to be exposed to a very narrow niche. Ravens seek wide exposure and experience, and profit from it. For that they have evolved curiosity (Chapter 5).

Ravens get significant amounts of prey by hunting, scavenging, and by association with predators. All of these potential food sources are also opened up to them by learning under the broad rubric of *getting to know* potential prey and predators. The exposure to these indirect food sources is achieved with play and curiosity, followed by learning. Object manipulation is a kind of play that results in identification of what berries, insects, and other objects might be appropriate to eat. Similarly, getting to know prey and predators is also proximally play, in that the behavior is not done initially out of motivation for immediate food reward. Even after being fully satiated, Jack, my pet raven free in the Maine woods, did not hesitate to fly in pursuit of birds or butterflies; and I once even saw him take off after a snowshoe hare, annoy a languid old dog, and taunt an aggressive big cat. He quickly learned what he could and could not get away with.

In the wild, ravens are well known to take the measure of wolves, coyotes, and eagles. Some miscalculate. For example, Jim Brandenburg once filmed "an aggressive, dominant, nearly black wolf as it grabbed and shook a raven." In that particular case, the raven got loose and escaped, but not all who miscalculate so badly are likely to live. Practice at the dangerous game of testing the limits of the predators could pay huge dividends. This strong tendency, seemingly to court danger, is another record that is engraved in their behavior, of an ancient evolutionary history with carnivores. As with human young, these natural tendencies for specific play are most obvious early, to be later blurred by learning and cultural influences.

John Sawyer, a neighbor of mine in the nearby village of Weld, had on numerous occasions seen one of my ringed ravens behind his house. This raven knew his cats. One day, hearing the raven call loudly, John saw his cat approaching out of the woods, carrying a mouse. The raven hopped right behind it onto the lawn, then erupted in loud caws. The surprised cat stopped, and in the instant that it looked back, the raven rushed in boldly, grabbed the mouse, and flew into a tree to eat it while the cat meowed in frustration. The raven had gauged all of the cat's moves perfectly.

There are innumerable reports of ravens presumably showing their bravery to impress potential mates, as they pull wolves' or eagles' tails. In four days of almost continuously watching dozens of ravens feeding along with wolves at Shubernacadie, Nova Scotia, in March 1997, I did not see the behavior once. On the other hand, I have seen all of my young ravens exhibit equivalent behavior within minutes of being exposed to a dog or cat. After observing only wild birds, I had originally theorized that, to use an analogy of Indian braves touching their enemy with a lance, ravens "count coup" with wolves and other carnivores to show "bravery" and gain status with mates. I now know I was wrong. I reject this hypothesis because of new evidence: First, the behavior is most prevalent in young birds long before there is any mate bonding. Second, birds do it in total isolation, without any potential audience; and third, it is not associated with *other* ritual behaviors used to impress potential mates. All three points are in concordance with the idea that the behavior is play, and that the play serves in education. Ravens do not, of course, try consciously to educate themselves. They act out neutral patterns that have evolved to be internally rewarding, because those individuals that are proximally rewarded in behaviors that ultimately benefit them engage in them more, survive, and leave more offspring.

Sometimes it is the seemingly senseless little things they do that make me wonder if they can be motivated by thought, or whims, rather than blindly following a programmed script. On the morning of February 19, 1998, I went to visit and socialize with my birds in the aviary, as I did almost every morning. There was an abundance of food: one cottontail rabbit, one side of calf, and two partially eaten gray squirrels. Yet, as always, the birds followed me and loitered around me like so many stray puppies. Blue and Yellow made comfort sounds and edged up to me, making eye contact. Mostly it was Green who caught my attention, as she tried to pick a three-inch rock out of ice on the ground. She worked diligently, and when she got it out she dropped it and tried to get another more challenging one that was embedded deeper. Giving up on that one, she then found three more rocks that she did dislodge from the ice. Usually as she was working on one rock, the other birds would come over to watch, help for a while, but then wander off. One rock she got loose weighed nearly a pound. I never saw her, or the others, ever again take an interest in rocks.

Ironically, throughout the animal world, a variety of play behavior is generally acknowledged to be correlated with intelligence, even though doing senseless things only for the fun of it as opposed to some purpose is thought to be stupid. I suspect that play is almost literally like intelligence. Play is an acting out of options, among which the best can then be chosen, strengthened, or facilitated in the future. The difference is that with play, the options are all played out overtly, not only in the mind. With intelligence, only the best options (and some of the worst) are played out overtly, and often much more quickly.

TWENTY-FIVE

Deliberate Acts?

IN EARLY JANUARY 1989 JOHN
Marzluff and I had caught another group of ravens from a feeding
crowd. As usual, most of these birds were pink-mouthed, immature
vagrants, but three of the crowd were dark-mouthed. We felt we could
have captured a mated pair among these three, since adults commonly
travel closely together in pairs, especially at this time of year near the
beginning of the nesting season.

We marked all the birds with wing tags, then released the pink-
mouthed birds into the main aviary and segregated the three adults
into a side aviary. Two of the three adults almost immediately preened
and made soft cooing sounds to each other. They seemed glad to be
together, and neither of them interacted with the third. Seeing these
social developments, we removed the third bird to leave the couple in

*Ravens may look at you with one or both eyes. This bird has
been digging in the snow with its bill, after sufficient exercise to
heat its bill so that snow melted and stuck to it.*

its new "territory." For two years and three months, they became key players in several studies.

This pair of wild birds did not nest that first spring, but by the second spring, they built a nest in their shed and then raised four young. In the third spring, they rebuilt the nest, but the male got out through a hole in the chicken wire. He stayed in close contact with his mate by frequent calling and by perching on the trees nearby.

After being free ten days, the male still had not abandoned his mate, and we were hopeful of luring him back inside. To try to catch him and still keep her, I cut a hole at snow level in one side of the aviary where I installed an inward-directed two-foot-long funnel of wire mesh, like a lobster trap. I placed meat at the end of the funnel and hoped that it would attract the female. The male might then be directed through the funnel to meet her and the food. I anticipated that once inside the aviary, he would fly up to his familiar perches and be caught.

I did not anticipate correctly. He did gain access to the meat and his mate inside the aviary, but he went back through the tip of the funnel and escaped again. It was now my move. I built a blind in the snow nearby, under the branches of a large fallen pine tree, and waited there until he went back into the cage. I planned to jump up and run and yell as soon as he was inside. Ravens have lightning-quick flight responses. I guessed that in the excitement, he'd lose his head and fly straight up. I tried my ruse. In response to my yells and mad rush toward him, he suspended his natural impulse to fly up. Instead, he did the more difficult thing. He attended to what was relevant in the totally new situation. He sought out the narrow entrance hole of the funnel instead.

My move again. I camouflaged the hole so that it was difficult to see from inside the aviary. The funnel had been straight, but now I angled it, then stuck twigs into the snow around the funnel entrance. The raven had no problem solving the first part of the puzzle—he got into the aviary through the obstacles, and his mate joined him on the ground. He then stayed close to the funnel entrance, as if ready to make another quick getaway through the exit. I gave them several minutes, hoping his attention would wander. Then I jumped up and yelled like a banshee. This time the raven missed the camouflaged entrance. He flew up and didn't have time to find the hole before I got there.

This incident engaged me for two reasons. First, I had no idea that a raven would so faithfully return to its mate and/or nest; and second, I was surprised at the bird's escape through a funnel, a strange new contraption that all ravens fear and would normally avoid, when before he had always flown *up* when disturbed. I doubt if any other bird would have been able to keep track of such a specific escape route, or to concentrate on precise and strictly relevant factors despite major distractions. Did the bird have insight into spatial relationships?

We all have "cognitive maps," which we use to correctly negotiate over territory. By use of our mental maps, we can return home without having to retrace the exact path we used when leaving. Ravens may also have a cognitive map of their territory, although we have no way of knowing. Their reactions to three-dimensional geometry may be easier to examine.

Each morning at dawn, Houdi and Fuzz pounded on my bedroom window and called loudly until I got up and gave them a treat or a toy. One morning, I gave them two shiny pennies. Fuzz carried both in his throat pouch, then stuffed one of them into the crack behind a two-inch-wide branch nailed onto the side of a post. Stepping to the side and using the point of his lower mandible, he hacked a sliver of bark off the branch and used this sliver to cover the penny. Houdi watched him, and several minutes later came to retrieve his hidden penny. It was too deep in the narrow groove to be grasped from the top, so she pounded it down instead, repeatedly peering down below the branch to see if it had appeared yet on the other side. Fuzz came, and shoving her aside, first looked at the top and then the bottom.

I later gave Houdi a rectangular piece of cheese on the same high platform outside my window beneath her nest then containing young. She tried to cache the one-and-a-half-by-two-inch cheese by inserting it into one of the two-inch spaces between the floor beams. At first, she held the narrow side of the cheese into the gap, where it would have fallen ten feet down to the ground if she had released it. Normally, ravens simply drop food into cracks, but this one had no bottom. She twisted her head around and inserted the broad side of the cheese, which fit snugly into the space. She walked away, then brought back a pigeon feather and covered the cheese with it.

On February 18, 1996, I asked Goliath to solve a geometrical puzzle by giving him a favorite food: corn chips. They were about two and a half inches in diameter, and I lay thirteen of them down. How would he handle them? He crunched three, one after another, and ate all the little pieces. Then he stacked four, one above the other. Lifting the four-pack, he flew off and cached it. Soon he came back, ate one chip, and again stacked four before flying off to make another cache. One corn chip was left. He came right back to eat that one.

Food is a reliable motivational tool for a raven, allowing one to try to trick the trickster. Lorrell Shields told me about an Alaskan oil pipeline worker who, for amusement, tried to frustrate a begging wild raven by throwing it *two* donuts. The raven wanted to fly off with both, of course, but a raven can grasp only one at a time in its bill—at least, so the pipeline worker thought. This raven, instead of grasping the first donut with his bill, stuck its bill through the hole in the donut. That left its bill-tip free to grasp the second donut. Then it flew off with both. Similarly, Terry McEneaney saw a raven in Yellowstone make off with a roll of toilet paper. The ends of the roll were too thick to get its bill around, but the raven carried the roll by sticking its bill into the center hole.

As already mentioned in the Preface, unlike all other Passeriformes, or perching birds, ravens occasionally use their feet in imaginative ways that could involve insight. If not, why do some of the thousands of other species with the same tools not do the same? Ravens are the only passerines who may carry objects, including eggs, in their feet in flight. Rare individuals may also carry multiple food items in their bills in surprising and equally imaginative ways. I talked with Peter Kevan, an entomologist,

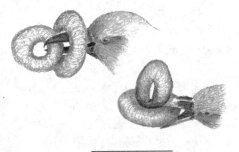

Two ways to carry
two donuts at once.

who saw a raven fly into a tree where there was a grackle nest with half-grown young. Instead of taking many trips to carry these young off individually, the raven emerged from the tree with the whole nest in its bill. Similarly, one year when the crossbills nested in abundance in the red spruces on Bald Mountain, I found a crossbill nest at the edge of Webb Lake, discarded on the snow beneath a raven's nest. Had the raven carried the whole nest back from the mountain?

It is potentially conceivable that using nests as baskets is innate. However, it seems difficult to envision evolution encoding the fine arts of cracker stacking, donut handling, and toilet paper carrying in ravens specifically, or that these skills were learned after lengthy trial-and-error behavior.

In fall 1995, I wondered if my four untrained two-and-a-half-year-old ravens at the time were as clever as the Alaskan bird was reputed to be at the donut-handling task. To find out, I left six Vermont Kof-fee Kup Bakery donuts in the open air for a week to firm them up a bit, and dumped them into a pile in the aviary with Fuzz, Goliath, Houdi, and Whitefeather. It was their first exposure to donuts. I expected the well-fed birds to cache the donuts for future use, rather than eating all six of them on the spot.

Cracker stacking.

In fifteen seconds, Fuzz stood over them, glancing repeatedly at the other three birds, who kept their distance. He bent down to feed on one donut, and Houdi, his preening partner, started to come near him. He rebuked her only mildly as she grabbed a donut and flew off with it to eat it alone. He stayed and leisurely fed in place, then picked up a stray twig from amongst the donuts, and walked four feet away to cover the twig with pine needles as if he were caching a donut. At four feet, he was still close enough to defend the donut pile from Goliath and Whitefeather, who came down to try to get their share. Seeing them approach, Fuzz chased them off and resumed casual feeding. Next, he picked up another piece of debris and cached that nearby also, while continuing to cast glances at his two fellow ravens, whom he still had managed to render donutless. Returning to the donuts after his trash cache, he took a few tiny bites, then cached a donut inside a nearby plastic drain tube. In three leisurely trips, he brought pine needles and other debris to shove in behind the donut. While he was thus partially diverted, Whitefeather managed to find an opening to fly down and grab a donut. Three down, three to go.

Having cached one donut, Fuzz casually ambled back and took his second from the pile, walking with it into the weeds about six feet away. This time, he made four separate leisurely trips with debris to cover the cached donut. Goliath, who had not yet had a taste, watched the whole time and made repeated tentative approaches. Whenever he tried to get within four to five feet of the donuts, Fuzz drove him away.

Fuzz returned to the two remaining donuts, again walking nonchalantly. This time, he finally picked two up at once. He did it in a unique way that seemed to me at least as clever as the Alaskan's. He put only his upper mandible through the donut hole, then he maneuvered the second donut to lie horizontally on his lower mandible so that the lower curve of the vertical donut was partially inserted into the hole of the horizontal one; the hole of the lower donut served as a basket for the upper. He then left to cache the two. This time, instead of staying nearby, he flew far; he no longer had to simultaneously guard any remaining donuts while caching those he held. However, he had possibly miscalculated Goliath and/or forgotten the already cached donut. As Fuzz flew off, Goliath immediately flew down, yanked the plug out of the tube, and grabbed Fuzz's first cached donut. Fuzz quickly came back after having cached his two donuts, and he chased Goliath all over the aviary until the latter dropped the stolen goods.

Maine ravens obviously knew how to hold their donuts, but if they seemed a little more idiosyncratic or creative than the Alaskan raven, it was probably because the Alaskan pipeline workers eat bigger donuts than those from the Koffee Kup Bakery. The donuts I used were a puny three inches across, but I had enlarged the hole to one inch.

When I gave the ravens four similar donuts two weeks later, Fuzz at first guarded them, pointedly watching all the other birds. He again made a fake cache nearby with a piece of debris before caching a real donut. Houdi checked the fake cache. Fuzz saw her and then recovered his just-cached donut, bringing it back with him to the pile. He next made four false caches of twigs and bark. He again went to the trouble of covering the trash he had cached with billfuls of dead pine needles. After the fifth false cache, Fuzz finally cached, in a great rush, three donuts within ten feet of the donut pile he continued to guard. I knew with near certainty that if he flew off to cache at a distance, where they couldn't see him, he would immediately sacrifice what was left. Did he? Was that why he made the false caches—to try to confuse the others, who he knew would try to recover the caches they saw him make? Was he lying, playing, or engaging in displacement activity because he didn't know what to do?

When he finally got to the last donut, he flew off with it into an adjoining aviary, where the others could not see where he would hide it. And as I expected, the others took that as the opportunity to steal the donuts he had cached nearby, while simultaneously guarding the pile. They also examined his false caches, proving that he had been in fact a credible liar, regardless of what he may have intended. He had not been cooperative in repeating his two-donut carrying trick. He had had a much different agenda than mine. He was calling the shots here.

For my third donut test, I closed Whitefeather into the adjoining aviary and left her four donuts. The three birds, Fuzz, Goliath, and

Houdi, flew and impatiently walked back and forth along the wire by
the donuts they could see but not reach. In contrast, Whitefeather
casually walked to the donuts, daintily picked at one, and walked off
with it. It took her a full six minutes to cache just that one. She
cached a second one with equal languor, then just left the other two
donuts to lie tauntingly where they were—in front of the other birds,
who frantically paced back and forth along the wire screen, until she
finally did cache them, in no great haste.

About a month later, on November 10, 1995, I confined the four
birds in half of the aviary, where they had little room to get away to
cache in privacy, and I put four donuts on the ground. As always, Fuzz
was at them immediately, keeping the other three away. He was not
hurried, picking up crumbs and only caching a few of them nearby.
He allowed his mate, Houdi, to come near, and she grabbed a half
donut and left. A little later, Whitefeather rushed in and grabbed a
whole one. Fuzz instantly mounted a vicious pursuit, and she dropped
it. He picked up the purloined donut and returned it to the pile,
putting it next to him. The grab-chase-drop-return sequence was
repeated three times in ten minutes. He continued slowly to walk
around the donuts as if guarding them, then broke one apart into two
pieces and cached the pieces nearby, one at a time. He broke the next
donut into five pieces, caching these pieces also one at a time, the
small pieces first. With only a half donut and one whole donut
remaining, he picked up both together by placing the half one onto
the hole of the horizontal one balanced on his lower bill. Twelve minutes had elapsed. Fifteen minutes later, he was still carrying the one
and a half donuts with him, while the other birds had meanwhile
uncovered the caches he had made. Perhaps he knew that, given the
confines of the aviary where he could not escape the others' eyes, he
couldn't cache what he now held, either. That's flexibility. Other
examples are legion, but not all examples are equally valid.

When I give my ravens dry bread, they sometimes dunk pieces in
their water dish. Such behavior is not hardwired. It is rare. It has been
interpreted as a deliberate strategy to soften the dry food to make it

more palatable. To me, it looks more as if the behavior is simply incidental. Birds eating dry food get thirsty. While carrying the dry bread
around, so that others won't get it, they arrive at the water dish to
drink. I've seen Whitefeather stop at the dish, hesitate, look all over
for a safe place to put the bread down, and finally just drop it into the
water where she can guard it easily while she drinks. Then she
retrieves it out of the water after her drink. Ravens drop rocks, pieces
of meat, sticks, and other toys into water as well. It is possible that a
raven could, by such seemingly random behavior, have an "Aha!"
experience and then quickly learn goal-directed bread dunking. However, they seem quite content to eat dry bread.

There are numerous reports of ravens dropping sticks, stones, and
other objects. It is usually assumed that the birds are acting deliberately for some purpose. However, most reports of object-dropping
behavior that I'm aware of are too incomplete for conclusions. Bob
Sam told me of a raven in Sitka, Alaska, who took an unshelled walnut, flew up with it several times, and dropped it on the concrete.
The shell didn't break. Finally, after the raven dropped it on the
street, he perched atop the Sitka Hotel. A car came along and ran
over the nut. Then the bird flew down and ate the nut's contents.
Hilmar Hansen, a railway worker from Montana, told me of ravens
placing deer leg bones onto rails he was inspecting, and coming back
to feed on the marrow. These acts may or may not have been deliberate. More anecdotes, and especially more details to each anecdote are
needed to come to conclusions as to what drives the raven's behavior.
Have there been a thousand unreported instances of ravens flying
with nuts and eventually dropping them at random? Would a raven
go to pick up the pieces of *any* bone or nut that was run over by a
vehicle? Had the bird previously dropped a tough walnut in frustration at random, to be rewarded with food by a passing car? Did they
learn by trial and error, and then know by insight what they had
done? Intelligent behavior may result from a combination of curiosity, exploration, persistence, patience, keen observation, learning, and
opportunism; but insight is difficult to prove because it is not necessarily a prerequisite for clever behavior.

Roger Smith of the Teton Science School wrote me: "I was band-ing young ravens on the east side of Teton Park in 1992. The nest was in a Douglas fir, about 35 feet up. While banding the third of five young, one of the adult ravens landed on a branch approximately 1.5 meters from me. This bird began vigorously tapping and rubbing its beak on the branch, vocalizing, and moving back and forth on the nest branch. I then noticed it had pulled a cone off the branch and began making low vocalizations with the cone in bill. By this time, I was banding the fifth nestling when, to my surprise I was hit in the face with a cone. I stopped and watched as the bird walked down the branch about two feet and pulled off another cone. At this point, I tried not to make eye contact with the bird but watched the behavior more closely. The bird walked back along the branch to the same distance from me, began the same beak tapping and vocalizations, then flicked the cone toward me, only this time the cone landed in the nest. The 'throwing' movement was incredibly quick. The bird flew off the branch only moments after this second encounter."

It is not possible to say with certainty if that raven knew what it was doing, although it seems unlikely the bird had learned the act of throw-ing, that cones can be pulled off for throwing, or that throwing them at nest predators might deter them. Insight could account for all three.

Mental representation of successive images that are projected in the mind like a movie is consciousness and the modus operandi of intelligence. We use mental projection so routinely in almost all activ-ity that we take it for granted. When we throw a ball, we see a trajec-tory to the intended target. We make choices to achieve very specific anticipated results. Without anticipating or projecting the results of the various alternatives to generating a sequence of actions, no intelli-gent strategy is possible. In effect, the fundamental capacity to develop strategy requires a capacity to visualize that which is out of sight, and that which has not yet happened but can happen. To ravens, that which is out of sight is, as with us, also not necessarily out of mind, as can be seen even in trivial examples.

One time, I had a white opaque plastic bag with a fresh DOR (dead on road) woodchuck. Goliath had seen me cut off a leg for him, and then put the chuck back into the bag. I walked away down a woods path. I hadn't gone far before he was behind me, begging. Odd, I thought. How could he be hungry after I had just given him a whole woodchuck leg? I left him another leg. Instead of eating from it, he hurriedly hid it as I was walking off, and then flew right behind me, again begging some more. "Okay, Goliath, you know there is more in that bag. You want the whole thing. So here it is." He does not follow me if he sees me put a piece of food into a bag, and I then retrieve it and give him the whole thing. He keeps track of objects even after he no longer sees them. Is that capacity advantageous in the wild?

Kristi Dahl published an anecdote in 1996 in *Wyoming Wildlife* that suggested the immediate usefulness of keeping mental track of objects out of sight. From her porch in Grand Teton National Park, Kristi observed Uinta ground squirrels scurrying around in the muddy sage grassland when the shadow of a raven overhead made them dash for their burrows. "We watched carefully as the adult raven landed nearby and approached a squirrel burrow. . . . The bird began to peck at the dirt, scooping loose soil with its beak. Suddenly, the raven stopped digging and gave a series of high-pitched, throaty yells. It was immediately joined by a young raven, evidently its offspring. The young bird began to beg and call as the adult continued digging. . . . Finally, about eight inches below the surface, the raven found its lunch. Pulling up a full-grown ground squirrel from the hole, it stabbed the animal several times. . . . The squirrel was torn apart and fed to the young raven before the pair flew off."

It is only a small step from seeing something in the mind when it is out of sight, such as prey or an enemy, to remembering past moves and anticipating future moves and reacting appropriately. There are many suggestive examples that are consistent with the idea that ravens are capable of all of the above, although most of the examples are not tightly enough constrained to allow unequivocal interpretations. Here are a few of them:

As reported in the Manchester (England) *Guardian* of June 25, 1995, a trapper in Prince Albert National Park in northwest Saskat-

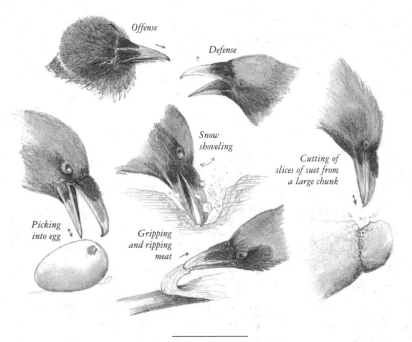

Offense

Defense

Snow shoveling

Cutting of slices of suet from a large chunk

Picking into egg

Gripping and ripping meat

Some of the varied uses of the bill.

chewan observed a raven feeding on an animal killed by a wolf. The raven occasionally interrupted its meal to lie still on its back. Eventually, the trapper noticed that this happened every time ravens flew by overhead. He conjectured that the raven played dead so that to the birds overhead the scene would look like yet another raven had died from poisoned meat bait and would give the place a wide berth. Again, we need more details to draw such conclusions. Ravens roll on their backs in play. What, precisely, were the timings of the back-rolls versus the overhead flights?

When the woodfrogs were chorusing near my aviary, I got four from a local pool. I put the first frog out of sight into a square, brown, foot-long PVC drainage tube. The birds had played with the tube before and presumably knew it was hollow, but they were shy at their first sight of a frog. They perched and watched from up in their loft, about thirty feet away. As I walked away from the tube, Houdi flew

Prying

Crushing, holding, and carrying

Combing

Caressing

Breaking limbs

down and first looked in one end, then ran around and checked the other end. The frog could not be reached from either end, but she picked one end of the tube up and the frog slid out. She grabbed it, flew back up onto her perch, ate the frog, then came down to walk on the aviary floor. When she walked past the tube, she did not peek in, apparently remembering that I had put just one frog in and she had taken that one frog out.

Several hours later, I put the second frog into the tube. Fuzz and Houdi were on full alert, intently peering down from their perches at the tube as I walked out of the way. Fuzz was the first one down this time. He bent his head down to ground level and peered into one end of the tube. He must have been able to see the frog, but he could not reach it. Was it therefore unavailable? No problem. Fuzz did not hesitate even a fraction of a second. He quickly walked around to the other end, reached in, and pulled the frog out. While he was eating it, I put the third frog into the tube. Houdi peered at the tube from her perch for a half minute before she flew down and looked in. Since the frog apparently had not moved, she simply reached in and pulled it out. After eating it, she retrieved the tennis ball she had played with previously and stuffed it back into the tube.

While the ball was solidly stuck into one end of the tube, I put the last frog in the other end, presuming it would escape deep into the tube to lodge against the ball. Houdi left her perch and looked in. The frog could not be reached. It must have already moved down the tube to the ball. So she walked to the other end of the tube and tried to remove the ball. Fuzz joined her. He first peered into the open end of the tube, then walked to the other end and tried to remove the ball. It must have been stuck solidly, because it took him a full half minute to remove it. When he finally did, he grabbed the frog. Had he "filled in the blanks"—imagined where in the tube the frog lay? Do ravens have X-ray vision or can they reconstruct spatial relationships in their mind? I speculated that the latter was more likely than the former.

Two days later, I introduced two round, white PCV drainage tubes into the aviary, four feet long and four inches in diameter. The tube was a different kind than ones I had used in previous experiments. As they always do with new objects, both birds inspected this new tubing from up on their perches, craning their heads down, twisting and turning their heads rapidly. Fuzz made deep rasping caws, a normal response to feared predators or strange things. Then he came down to make an inspection. He stood sleek and tall, and jumped nervously all around the tube as if performing a dance. He leaned down with his head close to the ground and peered into one end. He walked the four feet to the other end and peered in from that end also. Houdi meanwhile watched him from her perch. When he had finished and had gone back up to the perch, she joined him without examining the tube herself. I left the tube, being satisfied that they knew or would soon learn that it was hollow, not a white log.

In the meantime, I had introduced them to snakes. They had learned that snakes are good to eat and that they slither. Weeks later, I brought a live, foot-long green snake. Holding it by the tip of the tail,

I dangled it in front of the white PVC. Then I dropped it. The snake slithered into the tube, and I stepped back. Both ravens had been watching me from their perches thirty feet away. Fuzz came down, walked to the mouth of the tube, looked in, then hopped rapidly the whole four feet around to the other side. He reached in, pulled the snake out, crushed its head, fed from it, and cached the rest, covering it with leaves. Houdi watched. A few minutes later, she went to Fuzz's cache, retrieved the remains of his snake, and ate her fill also. Fuzz tolerated the theft because they were a pair by then, and he wasn't very hungry. Later, with another group of birds, I dropped food into the same tube, now held up vertically. In that case, when I dropped food in front of them, instead of looking first from the perch at the end where they saw me put the food, they hopped down to the ground and went on their first try to the end where the food had ended up (snug on the ground and where it could not be seen except after digging). That is, they had anticipated its movement through the tube.

I wrapped a lump of butter in paper as the ravens watched me, and put it into the tube, stuffing wads of green foliage behind it. Fuzz did not hesitate to pull out all the foliage to get the butter. I then put an egg in so far that it could not be reached from either end. Fuzz checked into the tube from both ends, walking back and forth, for a total of eight inspections, first at one end and then at the other, as if not believing that what could be seen at one end could not be reached from the other. After the eighth look, he finally picked up one end of the tube. The egg rolled out, and he ate the yolk. One explanation for him picking up the tube might have been that in frustration he would try almost anything. Another is that he knew the egg would roll out. Not all apparently deliberate acts are as ambiguous.

A pair of wild ravens on the frozen carcass in front of my spruce blind were chipping little pieces of meat off one at a time by partially opening their bills and using the pointed tips of their lower mandibles as chisels backed by the force of the momentum of their swinging heads. In contrast, they pulled off softer meat in small chunks by grasping and pulling, using the small hook at the tip of the upper bill. Chunks of meat were stacked piece by piece into a pile. Finally, the birds grabbed their whole pile and flew off with it.

Ravens in crowds always act differently than those alone or in pairs. They never stack meat, perhaps knowing that any loose piece would instantly be taken by another bird. They instead either fly off with only one large piece at a time as soon as they have detached it, or fill their throat pouch with small pieces before flying off.

A large chunk of beef suet in the woods in the back of my house could not be pulled apart into pieces like meat. The frost-hardened suet could, however, be handled by hacking into it, chipping off small, loose pieces similar to how woodpeckers, crows, blue jays, chickadees, and nuthatches invariably do it. Ravens normally feed on suet in this way as well. But one day, one raven did something imaginatively different.

The raven pair that fed there never allowed me to get close. They flew off even when they saw me near a window. I had gone to the food at the edge of the woods, as usual, to provide new food, and as I started to walk there, the raven that had been feeding flew up. I had not observed its feeding behavior directly, but it had left its tracks in the snow. Most interestingly, it had also left an intriguing record of what it had done. This bird, rather than randomly picking off small chips for immediate eating, had carved a groove around one corner of the fat (see p. 306). This groove was carved using precisely aimed blows. The object had obviously been to cut off a manageable chunk from a larger, immovable one. Many of the small pieces of fat that had been chipped off during the cut were not even ingested.

I want to indicate briefly why I was so excited about the seemingly trivial fact of a raven carving a groove in fat, when birds are capable of infinitely more complex innate behavior, such as weaving ornate nests or navigating by sun-compass, using an internal clock. That the raven had carved the groove was a fact, and given that no other birds and perhaps only a very rare raven would do such a thing, it was a highly singular fact. One robin does not make a spring, but this one fact was visible proof that the bird had forgone immediate gratification for a reward later on. That's planning. There were no secret raven trainers out there in the woods, it was not a learned plan. It was a plan derived from mental visualization. It was an invention. It made the raven's life easier but was hardly necessary. Given the constraints of the alternative interpretations, what the raven had done was news. I felt it should be splashed across the front cover of *Nature* and *Science*. Of course I knew it would not be. As expected, it was rejected several times for publication because it was "just an anecdote." Science, to be publishable, is almost defined as that which is strictly replicable, and there was no way that anyone could drop off a hunk of suet and expect a raven to slice a piece off by hacking a groove through it. Not even my aviary birds would do it. This was a raven Einstein.

The record of peck marks in the suet seemed to suggest that the raven not only had thought ahead, but also had acted on that thought and shown intelligence. Intelligence is not merely consciousness or awareness alone. It is not just complex behavior. Intelligence is not just super-detailed memory, rapid learning, complex vocal communication, play behavior, or tool use. Intelligence may or may not be related to all of these things, and some kinds of intelligence require them, but they are not what intelligence *is*. Intelligence is doing the right thing under a novel situation, precisely as this bird had done. Intelligence is understanding the world, and reacting appropriately to it, not just perceiving it. Intelligence is about awareness, and about testing responses in the head rather than the "real" world, where such activity may be time-consuming, harmful, or fatal.

Testing Raven Intelligence

Ravens have relatively large brains, and they do many things that look intelligent, but there was still no experimental proof that they *are* intelligent. I needed new ideas to get out of a rut, but really new, novel ideas, seldom come to us by forethought or design. They come by mucking around even as we try not to. I got my idea for a new approach on examining raven intelligence by leafing through a copy of *Ranger Rick* magazine, a present to my then young son, Stuart.

The magazine contained a short article on the "clever" things you can watch birds do, like pulling up food suspended by a string. I could not believe that chickadees could actually do that. If they did, it seemed to me, they would have to be trained. And if you trained them first, then insight or mental visualization needed to perform the task could follow the task, but could not precede it. I dismissed the whole

Solving the meat-on-the-string puzzle.

"intelligence" aspect of this, as I had with so many other stories I had heard.

My next thought was that if ravens were as intelligent as they sometimes seemed to be, it was possible that a rare individual among them might figure out how to pull up food on a string without having to go through lengthy trial-and-error learning. That is, insight might precede or accompany learning to produce the same behavior. The bird might evaluate the situation, play out a mental scenario in its mind, and from that insight perform the task quickly.

Testing experimentally for the existence of insight—the evaluation of different choices to make an intelligent decision without overtly trying them—may seem impossible to do because everything that an animal does includes prewired responses and learning. Neither of these can be simply excised from its brain.

Any one behavior is a combination of innate programming of blind or unconscious responses, learning, and insight. For example, our sexual preferences are largely innate, but giving flowers during courtship—preferably daisies rather than skunk cabbages—takes insight as modified by learning. There is no apparent limit to the complexity of animal behavior that can be innate or learned, provided it relates to the same conditions the animal has encountered predictably over millions of years. Insight requires consciousness, but it is more. It is the mental visualization of alternative choices that then guide judgment of new situations at the moment. It could be common in all sorts of behavior, but we can only *assign* it to play a likely role in behavior when it fulfills three criteria: First, to eliminate the innate component, it has to be extremely rare and *exclusive* of what the animal normally encounters and does. Second, it has to solve a problem. Third, it can't be a learned response.

In the wild, ravens' food is never found dangling on anything that resembles a string. String-pulling could therefore not be hard-wired into the innate behavioral repertoire by natural selection over millions of years of foraging experience. The behavior might, of course, be learned in incremented steps by tedious repetition where you reward the first step that contributes to the correct solution, then you reward the second step, and so on until the bird ultimately strings together a dozen or more steps. My captive ravens had never before even seen string. I had a unique opportunity to do a useful experiment. I doubted they could perform the task, but if they did, then innate programming or learning could both largely be excluded, leaving either insight or random chance as the prime contender for an explanation.

As I contemplated the problem of how a bird would get food dangling from a string, I saw a unique opportunity for testing insight, a mental capacity that defines at least one of several proposed kinds of intelligence (Gardner, 1998). First, food should give a strong emotional motivation to induce a bird to try to get the food the easiest, quickest way. The bird might try to sever the string, fly at the food directly, or attempt to knock the food off. Yet, given a secure string, there was only one sure and easy way to get the food—pull it up in successive steps, hanging on to each pulled-up loop of string.

Pulling up food dangled on a string involves many steps that must be executed in a precise and nonarbitrary sequence. The bird must 1) perch above the dangled string that is attached to the food, 2) reach down below the perch, 3) grasp the string with its bill, 4) pull the string up and over the perch, 5) place the pulled-up loop on the perch, 6) lift one foot, 7) step on the loop of pulled-up string with the lifted foot, 8) press down with the foot hard enough to prevent string slippage, 9) release the bill's hold on the string, but only after the foot is already firmly pressing the string down on the perch, and 10) repeat the whole sequence a variable number of times depending on the length of the string and the amount of slippage of the string.

I arbitrarily chose a string length of two and a half feet because I felt it would present the birds with a sufficiently challenging problem. The process would have to be precisely repeated at least five to six times, or perhaps many more times if the string were to slip. In all, dozens of discrete steps had to be assembled into one precise sequence needing constant updating as it progressed. Any one step by itself could of course be learned and/or innate, but the critical behavior was not any one individual step or steps. Instead, it was the placing of these steps into a unique sequence that would solve a unique problem.

It was practically implausible that the correct sequence of dozens of discrete steps would emerge by random chance.

The beauty of the string-pulling test was that 1) it greatly reduced the possibility that the solution could be arrived at by random chance, 2) there was no plausible genetic programming that could have coded this very specific unnatural behavior because there is no conceivable reason for it to have evolved in the wild, and 3) my ravens had been hand-reared in captivity in an aviary and had no experience in string-pulling. They had not even seen string, hence, I knew they could not have had prior opportunity to learn the behavior.

My ravens, I presumed, would not be able to get meat by pulling it up on the string. If they did not, so what? Nothing would have been lost. On the other hand, if they did it from the start, a lot would be won. Invoking Occam's razor, the rule in science that favors the simpler theory over the more complex one, a successful pull-up would suggest that the birds could solve a problem. That's proof of insight and if one grants it is a difficult problem, then its solution is also a measure of intelligence as regards that problem. Furthermore, solving the problem by insight presupposes consciousness. And all I needed for this was a piece of string and a slice of meat—not exactly an impressive investment. It was worth the try.

When I first thought about doing the experiment, I recognized two complications. First, ravens are highly temperamental animals. They would be wary of the string. In one test out in the field, I hung a piece of meat on a white piece of twine from a branch next to a rock-solid frozen cow carcass where more than fifty wild ravens were feeding at temperatures near minus 25 degrees Fahrenheit. They had to work long and hard to chip off tiny bits of meat from the carcass, and every dawn the crowd came and all chipped for hours. Any *loose* piece of meat was a valuable prize, to be taken instantly and much fought over. So here was suddenly a morsel of choice food dangled next to them. What happened?

I was as usual well hidden in my blind of spruce and fir. I had put up the meat while it was still dark. At dawn, the birds came and instead of quickly descending to feed as they usually did, they stayed up in the trees, making angry, rasping alarm calls such as they make to strange, frightening phenomena. Only about an hour later did one of them descend to the cow, leading the others in. They looked at the meat on the string as if it were an apparition. Not one went near it. I left it there. Two days later, it was still there.

I hoped that my aviary birds would be less spooked by such a strange thing, but I expected that they would be shy, since they had never seen or experienced food on string. If they did figure out a solution to reach the meat, they still might not demonstrate their insight because of fear of approaching the string. Ideally, I would test them one at a time, but that meant catching them and building another aviary to isolate them, which would cause much disruption, money, and time. Isolated aviary birds would likely stay away for days from something as unusual as food dangling on a string. The ravens would be much calmer if they remained together as a group. There was a negative aspect to working with a group of birds, because if one pulled up food, the others might simply copy the first, and/or chase away all others. Nevertheless, the first bird I tested had none other to copy, and I was interested in the phenomenon as such rather than in how many individuals could or could not do it. It made no difference to me whether ninety-nine percent or one percent of the population of ravens could solve the problem. If problem-solving—whether you choose to call it genius (in ravens), insight, or intelligence—proved to be present even in one bird, then it exists. Even if rare, it is no less interesting a phenomenon.

Hard salami was important for my test. I first offered pieces out of my hand to make sure they would want it; they ate it eagerly. I knew if I had used soft meat and a bird had flown at the suspended prize, grabbing it with its bill, it might have torn off a small piece. Being rewarded, the bird would be predisposed to try the same maneuver again and again; it would not find it necessary to try anything different. Hence my test would fail.

The salami was three months old, having dried in the refrigerator to a leather-like consistency. The ravens would not be able to tear it off by flying at it, grabbing it with their bills, and then dangling from it by holding on to it with their bill, behaviors I was sure they would try if they tried anything.

I notched a slice of salami and tied it onto the end of tough woven string, then tied that onto one of the horizontal perches in the aviary so that the prized food item was suspended about six feet off the ground and two and a half feet down from the perch. I rushed back into the house to my desk by the window to watch.

All of the birds looked at this new thing. The dominant pair came closer and tilted their heads and stared at the salami. They looked at the string wound round the perch, then hopped up and down on the perch doing nervous jumping jacks, as they would do in front of a feared carcass. The string and the salami jiggled. After a while, and more looking, they cautiously approached. One of them pecked at the loop of string tied around the perch and quickly jumped back. They kept craning their necks to look down. One gave the string a few tugs as if trying to rip it off the branch. The meat jiggled some more, but the string was tough and did not break. The birds then seemed to lose interest. I went back out to remove the string with the salami. I would have them try it again some other day. "Just what I expected," I thought. "No way will these birds get that meat."

When I put the meat out the second time the two birds again eyed it warily, but looked less nervous than the first time. Abruptly, one of the birds, Matt, flew up to it, and to my utter amazement performed the entire sequence with only minor fumbling. I shouted for joy and pounded on the window to startle him so he would drop the food. I wanted to make sure his reward was only mental. Not having eaten, would he go through the entire sequence again right off?

This bird knew very well what he had done. After I shooed him off, he returned within several seconds, pulling the meat up in great haste. Again and again (six times) I chased him off before he had a chance to eat. Interestingly, he never tried to fly off with the salami after pulling it up. Furthermore, he always dropped it even when I rudely startled him and later when I literally had to push him off his perch after a meat pull-up when he held the meat in his bill. In contrast, when the birds get a piece of loose meat they *always* fly off with it.

After the sixth repeat of the pull-up behavior, I knew it had not been a fluke. Matt really did know how to pull up meat. As a reward after all those "empty" repeats, I finally allowed him to eat salami. My reward was an adrenaline rush, and the chance to do a series of experiments that built on these original observations. I was hooked. I was convinced that I had blundered upon an insight for testing insight in ravens.

Eventually, I saw other naive ravens do the pull-up of the same length in as little as six minutes from the time of first presentation of the meat on string into the aviary, and after as little as thirty seconds after they first contacted the string. Given that many of the birds were shy of strings, the time required to pull up the meat successfully probably greatly underestimates the amount of time it takes them to do the task, after they see meat on string. The presence of the other birds turned out to be a big problem (which I solved in another set of observations with another group of birds) because whenever a dominant bird who knew how to pull up meat saw a subordinate near meat on string, it tried to chase the bird off.

I eventually tested five different groups of ravens and two crows. In no case did any naive birds ever show any interest in pulling up, or even approaching a string without food attached to it. In all groups of ravens, but not the crows, several individuals performed the whole sequence with little or no fumbling right from the start. Two patterns were seen. In one pattern, the "direct pull-up," the bird stayed in place to pull up successive loops carefully, stepping as the loops pulled up. In the other, the "side-step," the bird pulled string laterally onto the perch before stepping on it, then pulling up again, and so on. On the other hand, a group of four three-month-old ravens, like the two crows, were incapable of reaching the meat, although they showed no fear.

To me, the most convincing point of the experiments that showed insight (because it excluded both observational learning and trial-and-error learning) was that those birds who didn't pull up meat, but who seized meat on a string that I or another bird had pulled up, tried to fly off again and again with that meat. Of course, since it was still attached to the string, it was rudely yanked out of their bills. They didn't seem to catch on until at least six trials. Those birds who *did* pull the meat up, on the other hand, never once in thousands of trials flew off with it so long as they were left on their own. I had difficulty shooing them off the perch after they had pulled up the meat, and when I did succeed in getting them to leave, they almost always dropped their piece of meat before flying off. The ravens that had pulled up the meat acted as if they knew, from the very first time they pulled it up, that to try to fly off with the meat would involve having it ripped out of their bill in flight. The significance of the remarkable behavior of *not* flying off was that it was a *new* behavior that was acquired without any learning trials. They acted as though they had already done the trials. The simplest hypothesis is that they had—in their heads.

I next demonstrated experimentally that *one* behavior—choosing the correct string when two were side by side—could be the result of *two* different mental concepts. In the setup for this experiment, the birds that were experienced in pulling up meat were first exposed only to one string with meat. In the competition among each other they all rushed to be first at that string, and they pulled quickly. That is, I trained them to expect meat every time they pulled up string, hence looking at the string closely before pulling it up became unnecessary if not counter-productive. Then I provided two side-by-side strings, one holding a rock, the other, meat. There was again competition among the five birds to be first to pull up the meat. In this situation, there was as before much haste and some birds initially made mistakes; they rushed to the strings and yanked on the first one they came to. However, if that was the wrong one, they quickly realized their mistake. They dropped that string without pulling it up, looked again, and pulled up the one with meat. That is, I had now retrained them to look. As one might expect, after a few trials they learned to look before they gave their first yank. A choice was now necessary, and they learned to contact only the correct string. Then in the test, I crossed the two strings so that a raven perched directly above the meat and pulling on the string below its feet would now (in contrast to all its previous trials) end up pulling on the string with rock, not food. Conversely, to get meat, the raven *now* had to perch above the *rock* and pull the string below its feet, a novel setup contrary to its previous string-choice training experience.

In the test, three out of four proficient string-pulling ravens first contacted the wrong string—the string attached to the pole directly above the meat. That by itself was neither surprising nor interesting, because that choice previously had always been the correct one; but what was interesting and amazing was that now they showed no evidence of learning to correct their mistakes. In dozens of trials, they continued to yank first on the wrong string; i.e., the one over the meat, switching only after they had seen their mistake; i.e., the rock jiggled. That is, consciousness of what they thought they knew took precedence over trial-and-error learning, which was glacially slow even for this one extremely simple task.

Proof of conscious involvement was shown by one bird who was correct from the beginning. Before the test, that one bird had shown identical behavior to the other three. That is, it pulled, as trained, only on the string with meat, the one directly above the meat. In the *test,* this bird immediately and consistently did a novel thing. It pulled the string over the *rock,* the string to which the meat was attached. Rather than first contacting the "string-above-meat" as the other three had done and continued to do without correcting themselves, this bird had pulled on "string to which meat is attached" on the very first and all subsequent trials. Of course, the birds could not use words, but words are unnecessary for thinking as such. (If the ravens had evolved to communicate to their followers how to pull up string or some other useful tasks, then they would of course have had to evolve the capacity to use words, and to communicate using them they would then need the ability to think with them as well.)

In subsequent tests, birds that were proficient in string-pulling were given food on a new string of a different color, texture, and thickness, with which they had never been rewarded, versus a rock on the same string on which they had always been rewarded. If they simply had been conditioned to pull on brown twine, for example, then they would choose it above previously unfamiliar green shoestring, even when the food was attached to the shoestring, which they had never pulled up before. What did they do? They all chose the *new* string that they had never seen before, much less been rewarded from, right on their first trial.

They ignored the familiar twine that had always been associated with food. In summary, they knew the solutions to several new tasks without any overt trials. Given a choice, they attended to what is relevant in preference to what they had been trained, and what they had in mind could take priority over what they experienced. In conjunction with all the other prior observations that they can keep track in their mind of what they no longer can see, I conclude that they experience some level of consciousness, and use it for insight to make decisions. Whether that is "intelligence" is subjective; but according to most people it is.

I wrote up the data, providing thoughts on what it might mean, and submitted the manuscript to a journal that is reviewed by other scientists whose names remain secret to all but the journal editor. The editor then considers the comments of the reviewers and accepts or rejects the paper, generally asking the researcher to consider the reviewers' comments in the revision if it is to be published. Constructive comments on one's work are precious, because they often catch one's errors and oversights. I anticipated few problems. Since I described something novel, I felt sure that my paper would be received eagerly and rushed into press. However, this manuscript took a different turn.

One of the reviewers felt that I needed to examine the evolutionary precursors of the behavior, and also pointed out that "Freud had shown much mental work to be unconscious and that much of human insight is unconscious." Sure. Okay. Does that then preclude further research? What is mental work anyway? What evolutionary precursors of string-pulling? The editor rejected the paper, but offered to consider it again after a new submission, subject to another round of reviews. I rewrote and resubmitted.

Some months later, the manuscript was again rejected. A reviewer claimed that I had made a "clear dichotomy between learning and genetic programming—a dichotomy that is twenty years out of date." Of course, he was partially correct! There is no clear dichotomy. But I had *tried* to create one. That was the experiment. About twenty years earlier, I had published several papers on learning and innate behavior in bees, making a point to show the relationship between genetic programming and learning. That a reviewer would now suppose I'd think that there *is* a "clear dichotomy" in real life seemed odd. The point of this experiment was that I managed to force a small chink in the armor, to drive a wedge into a mechanism, much as an experimental physiologist might ligate a blood vessel to find out what organ it might supply. The point of the experiment was that the effects of both genetic programming and learning could be minimized, to see if any behavior remained. I felt I was being taken to task for the study's strength rather than its weaknesses. Yet another reviewer used such phrases as "incredible leap of faith" and "matters of the heart." I felt that these, and several other best-not-repeated comments, were more emotional than rational. Had I violated a taboo? It might seem I suffered from attachments to my birds, thus reading human motives into them. Perhaps. But I was instead mostly wondering the opposite: what unconscious drives would move someone to reject that which they find new and unfamiliar, a very conspicuous raven behavior. Ultimately, the paper was rejected five times. Part of the price of doing what you feel is really rewarding and novel is sometimes the necessity to endure harsh criticism.

Several years after my paper was published, I got another group of six ravens and repeated the experiments, but that time in greater detail and when the birds were only nine to ten months old. I built a new aviary in which I could insert an opaque partition to separate two sections, so I could test each bird in isolation and counter the criticism of "social learning." Since string shyness had been a big problem, I had also habituated the birds to string by tying several strands tautly between branches and onto the vertical cage walls so the birds could see string but not still pull and step. The results were essentially identical, except that five of the six birds in isolation pulled up meat much more quickly, all within four to eight minutes after contacting it. However, they first tried several alternate methods, including pecking, yanking, and twisting the string, before doing the pull-up. That is, the younger birds were overtly experimental. As in the first group, one bird only flew at the meat and never pulled string.

In the future, when I have another group of birds, I will give them the task of trying to access food that is suspended below them but that they can get only by pulling *down* on a string that is *above* the perch. I predict this counterintuitive task will *not* be performed by naive ravens without lengthy learning trials, if at all. Those who already pull string up learn this trick quickly, as expected in what psychologists call transfer learning.

Any phenomenon can potentially be explained by several alternative hypotheses at the same time. The scientist then seeks to disprove each one. If all the likely alternatives are disproved but one, that one is generally considered to be the most likely answer—until new data come along that provide a better explanation. The currently accepted answer ultimately depends on the alternatives with which one begins. If insight is denied as a possibility from the very beginning, then it can never become an alternative hypothesis, and hence it can never be the best hypothesis that remains.

It is hardly to be expected that the human animal would be qualitatively different from all others. The psychologists who have studied learning in rats and pigeons have assumed (and found) similarities across species. If that is anthropomorphizing, I'm all for it. There is no evidence to suggest that humans have some new or different mysterious vital essence that other animals lack. Indeed, the raison d'être for studying animals is the unspoken assumption that results can be extrapolated to humans. Otherwise, the agencies that award research grants would not have spent untold millions of dollars on rats.

At the most fundamental level, learning, consciousness, insight, and all such correlates as problem-solving and intelligence, are simply the firing of neurons. Neurons are components of intelligence and insight, but you can't probe specific neurons and say, "*There* it is. *Insight!!*" You can no more critically define insight by examining neurons than you can discover the structure of the Maine coastline by examining the grains of sand on the beaches with ever finer detail. The relevant patterns can sometimes best be seen by stepping *back* and looking from a new, unfamiliar vantage point.

I often see "intelligence" in my ravens in the stupid things they do. One time early in November 1992, I surprised a group of them in a fir thicket. They were noisy and raucous around a long-dry cow scapula. Gathering around a dry bone is "goofy." A chickadee wouldn't do it, or a blue jay, or a crow. These birds would not be so foolish as ravens. But then, few species of birds except ravens (and some parrots) would end up pecking airplane wings, pulling off windshield wipers, swiping golf balls, and not incidentally, getting and opening food from Dunkin' Donut dumpsters, or sealed black garbage bags, or from a string, or sliding on their bellies in the snow, or doing barrel-rolls in the air when returning alone to the roost at night.

Doing foolish things like stealing windshield wipers and dancing around a dry cow scapula is, like play, one of the costs of being bright. It's a little like the "intelligence" of the immune system. Our immune system produces thousands of different kinds of molecules that most of the time do nothing useful. It may seem like a huge waste to produce them at all. If by chance, one of these odd, seemingly senseless molecules neutralizes a specific unanticipated invading pathogen, then this one is recognized and remembered by the body, to be replicated in huge numbers. That is, the body "learns" through selection. Neural networks work in the same way, but they pretty much have to be present all the time, barring some exceptions (see p. 330). Those that are used or rewarded are activated and strengthened, taking preference over others. From our own experience, we all know that we can try out what "works" within our minds even before we try it out physically. When we want to reach an apple over our heads, we can evaluate, by mental projection of our limbs, the feasibility of trying to reach it by hand. Or we can try to reach it by jumping up and grabbing, getting a chair to stand on, bringing a ladder, swinging a stick, calling the fire department, throwing rocks at it, hurling sticks, shooting the branch off with a shotgun. There are endless possibilities that we evaluate and discard in milliseconds. We may quickly come upon one that rewards us *mentally,* and when we do, we continue to run the scenario through our mind before actually trying it. If we had a raven's mind? We'd be forced to try more possibilities overtly, and we'd have a great deal fewer and less elaborate possibilities to choose from.

TWENTY-SEVEN

Brains and Brain Volume

ALL VERTEBRATE ANIMAL BRAINS consist of fore-, mid-, and hindbrains. These divisions have different functions, with the hind- and midbrains responsible mainly for integrating and processing sensory information and organizing movement and attention. The forebrain is the locus of conscious activity, playing important roles in sensation, learning, memory, and mood. There is relatively little variation among species in hind- and midbrains, but forebrains vary greatly, and in such animals with large brains as humans, it is the forebrain that accounts for the large brain size.

In general, the greater the average brain volume of a species, the more information the animals can handle. Large animals require bigger overall brain size than small animals simply to control their bodies; and in general, brain size increases proportionally to body mass or

This raven had been caching food in snow,
and is carrying food to cache in its throat pouch.

volume. If brain size is greater than what would be predicted by body size alone, that is called the "residual" factor, and is a measure of the brains' "encephalization." Humans are some of the most encephalized animals in the world, second only to some species of dolphins.

Some birds also have high cephalization. In the 1940s, the Swiss zoologist Adolphe Portman compiled data on brain volume in birds, reporting that the corvids as a whole, which include ravens, crows, jays, magpies, and nutcrackers, had one of the highest encephalization indexes, scoring a 15. The raven scored a 19, the highest member of the corvid group and therefore the highest of any bird. All other passerine, or perching, birds ranged from 4 to 8.

The potential information-processing power of the brain is presumably related to the number of units, or neurons, it contains, and the complexity of their interactions. Brain volume is closely correlated with the number of neurons, and the complexity of interconnections of neurons is independent of brain size and of species. Encephalization is

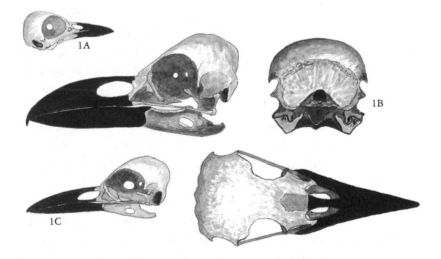

1A Bluejay.
1B Raven.
1C American Crow.

thus probably a fairly objective measure of behavioral flexibility. We intuitively infer that intelligence is correlated to brain size, and this inference is generally supported by a variety of criteria. It is also true, however, that we can't credibly claim that one species is more intelligent than another unless we specify *intelligent with respect to what,* since each animal lives in a different world of its own sensory inputs and decoding mechanisms of those inputs.

Primates live largely in a visual world, and in general have large areas of the brain devoted to processing visual information. Humans, in contrast to other primates, additionally have large brain areas committed to auditory processing, speech, and language. The large forebrains of some dolphins and killer and sperm whales are thought to be devoted to echolocation. Echolocation alone, however, does not explain their huge brains, because bats and some other whales and dolphins echolocate superbly with very small brains. Birds' encephalization could be necessary to coordinate flight, yet such insects as dragonflies manage exquisite flight (and walking) coordination of their four wings (and six legs) operating independently all with a brain smaller than a pinhead. Why would a raven need a large brain to coordinate two wings?

Brain tissue is metabolically as active and hence as expensive as muscle, and it is active day and night. Our brain accounts for only about 1.5 percent of our body weight, but it demands about 20 percent of our energy supply. At any one time, this energy is used mainly by those neurons that are active, and we can determine the regions of the brain where neural activity is concentrated using a modern technique called PET (positron emission tomography) imaging. PET scans provide pictures of the regions of the brain where neural activity is most intense, moment by moment. For example, different areas light up when we hear, see, speak, or generate words. When we see an object, a specific area of our brain indicates neural activity. When we later *think* of that object, the same area lights up. That suggests to me that thinking involves in some way the same or some of the same neurons that are involved in processing and storing incoming information, indicating a suspiciously close link to memory.

Animals have evolved to minimize energy use whenever possible. Large, metabolically expensive brains would only have developed and be maintained for very compelling reasons. As already mentioned, we can infer that sensory processing and motor coordination alone do not explain why dolphins, humans, and some birds have such large brains. One suggestion that neurobiologists have made is that the often limited stimulus load that the animal *accepts* from the environment is considerably less demanding, in terms of number of neurons required, than *what is done* with the stimuli—how large stores of memories are projected and manipulated. Different animals in effect not only see different worlds (because they have different sensors and different sensitivities), they also handle the incoming information in different ways, to *create* different worlds in their heads. For example, a bat and a dolphin both live in a world where pressure waves and vibrations of air and water, respectively, are highly important to their survival. A bat uses pressure waves as information to plot an interception with flying insects. A dolphin, however, could to use them not only to intercept prey but also to plot the ocean floor, to navigate over thousands of miles, to distinguish individuals in a herd, and maybe even to discern the moods of other dolphins (a popular theory), and to track individual dolphins for mutual interactions.

An animal may extract enormous amounts of information from the environment, organize it, and give it meaning. Before the animal encounters the patterns of vibrations or pressure waves, for example, there is no sound. The animal's sensors detect these vibrations or pressure waves, making them stimuli; the brain then interprets these stimuli to perceive them as sounds, and to manipulate or sort them to create "stories" and scenarios out of them. Since the brain creates or specifies the animal's unique world, it is difficult to apply the same intelligence tests across species. Perhaps the only objective criterion is brain volume itself.

In June 1988, on a canoe trip down the Noatak River in Alaska, my companions and I found the remains of a dead raven behind a trapper's cabin, the only human structure we saw along four hundred miles of river. At the time, I saved the bird's skull as a curiosity, or perhaps as a memento of the trip. Later that year, one of my five tame ravens unexpectedly died. I had nicknamed this bird the "cretin" because I had the subjective impression it was incredibly dumb relative to the other birds. The cretin's skull showed an injury, possibly a peck, which could have resulted in abnormal development, and possibly death. I saved the skull, and on a whim compared it with the Alaskan raven's skull. Both skulls were almost identical in length, but the Alaskan's had a strikingly larger and more rounded brain case. I next weighed a known volume of sugar and filled the brain cavities with sugar through the foramen magnum, then weighed the sugar to determine brain volume. The Noatak raven's brain volume was 18 cubic centimeters, while the cretin's was only 11.8 cc. I had never before seen two *similarly sized* skulls of the same species with such an enormous difference in brain volume. With numbers like that, you immediately wonder whether the cretin's brain was abnormally small, or the other's abnormally large.

Pursuing the possibilities, I called my friend and ornithological colleague Fran James at Florida State University in Tallahassee, who steered me to Phil Angle of the Smithsonian Institution in Washington, D.C., Phil loaned me a boxful of raven skulls from their collections. I measured brain volumes as before, determining that my Noatak raven skull was close in volume to the others from Alaska. That is, it was not abnormally large. The Alaska ravens, with an average volume of 17 cc, had a higher brain volume than the ravens from the western United States, with mean volume 13.1 cc (see Table 27.1). Brain volume of the Maine ravens overlapped both, with a mean of 15.5 cc. Alaskan ravens are larger than western ravens, and perhaps their larger brain volumes can be attributed to their larger body mass.

The northern raven's absolute brain volume of about 17 cc is twice that of an American crow and nine times that of the common or rock pigeon. Both crow and pigeon weigh about 400 grams versus the raven's 1,200 to 1,400 grams. In contrast, a domestic chicken (Rhode Island Red) weighed for comparison had a pea-sized brain volume of 3.1 cc, even though its body weighed twice that of a raven (see Table 27.2). These numbers reinforce general prejudices that pigeons and chickens don't come close to ravens in intelligence, but crows probably do.

Yet brain mass in any one individual is not constant. Recent research with some birds shows that the mass of the hippocampus, the portion of the brain devoted to the specific functions of singing and food caching, increases and decreases seasonally with use. Birds can grow and shrink brain tissue as needed, thereby avoiding expensive maintenance of tissues not in use.

There has been much discussion of why some animals have large brains. Most of this discussion has centered on human brain evolution, but the same ideas probably apply generally. One thing stands out: The large brain of hominids appeared rather suddenly. Furthermore, it did not evolve uniformly in all of the hominids. For millions of years, our ancestors, such as *Australopithecus afarensis,* walked bipedally and had an essentially modern human form, but had a brain volume just marginally larger than a chimpanzee's. For millions of years, that smaller brain sufficed. Suddenly (in evolutionary terms), a small brain wasn't good enough anymore for one small group of ancestral hominids. Something changed for them alone. What was it? The one correlation we have is that as hominids became meat-eaters, they became larger in body size, and brainier. The others who remained largely vegetarian remained small-brained. Is this change in diet a clue?

The diet connection is strong, but interpretations of it differ. The prominent anthropological argument acknowledges the diet connection and attempts to explain it by saying that meat from large animals— large amounts of high-powered concentrated food—was needed to power that metabolically costly brain, so we turned to scavenging and hunting. I think that particular explanation is backwards, because it assumes a large brain is a good thing. It isn't, necessarily. Both ravens and hawks are meat-eaters, but hawks are small-brained, and they are very effective and successful predators. Diet alone therefore does not explain raven brain evolution. Another explanation is that a protein diet *enabled* an expensive brain, and that some strong selective pressure, such as sociality, then drove the evolution of increased encephalization.

Much recent research in mammals has converged to indicate that perhaps the major driving force behind the evolution of increased brain size is social complexity. In turn, social complexity increases inordinately when individual recognition becomes possible and the animal tracks not just others, but myriad *specific* others. Ravens, like other corvids, and like dolphins and most primates, are highly social. As I have indicated (Chapter 14), they apparently recognize one another. Furthermore, not only do nonbreeding subadult ravens form coalitions against breeding adults, but adults may cooperate in pairs and perhaps in coalitions of pairs (Chapters 9 and 10). Ravens also are exposed to interactions with dangerous carnivores as they attempt to get meat from them (Chapters 19, 20, and 21). All of these interactions require instant reactions or choices that can be made much more quickly and safely in the head, rather than overtly. In short, they may require consciousness, the ability to examine, evaluate, and make mental choices before committing to action.

Table 27.1 Average Skull Measurements and Brain Case Volumes of Ravens (*Corvus corax*)

Origin	Number	Bill Length (cm)	Skull Length (cm)	Skull Width (cm)	Brain Volume (cc)	Range Brain Volume (cc)
Alaska	9	7.91	12.4	4.5	17.0	15.5–18.3
Labrador	2	7.70	12.3	4.4	16.8	16.1–17.5
Mich., Wisc., Minn.	4	7.60	11.8	4.2	15.8	15.3–16.7
France,	3	6.60	11.0	4.1	14.9	14.5–15.3
Western U.S.	8	6.73	10.8	4.0	13.1	11.7–14.8
Maine	7	7.83	12.1	4.3	15.5	13.8–17.6
Maine "cretin"	1	8.10	12.7	3.9	11.8	—

Table 27.2 Comparison of Brain and Body Mass in Different Species of Birds

Species	Body Mass	Brain Mass	Brain as % Body Mass
Golden-crowned kinglet (*Regulus satrapa*)	5g	0.34g	6.80
English sparrow (*Passer domesticus*)	22g	0.9g	4.09
European starling (*Sturnus vulgaris*)	68g	1.7g	2.50
Robin (*Turdus migratorius*)	~90g	1.3g	1.44
American crow (*Corvus brachyrhynchos*)	~400g	9.1g	2.27
Common pigeon (*Columbia livia*)	~400g	1.9g	0.48
Ruffed grouse (*Bonasa umbellellus*)	600g	2.5g	0.42
Common raven (*Corvus corax*)	1000–1500g	12–17g	1.33
Domestic chicken (*Gallus domesticus*)	2800g	3.1g	0.11

TWENTY-EIGHT

Are Ravens Conscious and Emotional?

IN THE SUMMER OF 1997, THE
Maine Fish and Wildlife Department was contacted by a woman
near Farmington, Maine, about one of my ravens. She had identified
it from the aluminum ring on its right leg and a yellow plastic tag
on its left wing. This bird had been tagged eight years earlier. She
felt it was a special bird, possibly one singled out for its "extreme
intelligence."

I have never met anyone who has known a raven who did not
think it was bright, but this woman, Diane Pickard, tipped the enthusiasm scales with superlative comments: "This bird is *so* truly amazing
that I just had to tell someone about it. Nobody is going to believe
this. But I have witnesses. It is scary how bright this raven is. He
knows what he's doing. I had *no idea* a bird could be so smart."

---

*Allopreening, the giving and receiving of caressing
and/or feather care among bonded individuals.*

In the two years that she had known the raven, he had followed a routine. In the early morning, he flew along Route 43, covering seven to ten miles and apparently scouting for roadkills, which Diane determined by following him by car. Next, his schedule included a stop at the Pickards' house to feed on suet suspended in a wire cage. He did not go on weekends when people were there, and Diane suspected that he usually did not come to get suet until she and her husband had left for work. "One day I stayed home just to watch," she said. "As soon as my husband went out the driveway, the raven came out of the woods from across the road where he'd been hiding and watching. He looked all around as if to make sure the coast was clear before coming to the suet container. When other cars came along the road, he crouched behind a tree and peeked out to one side, just enough to see. No other bird I've ever seen acts so deliberate and alert, as though he knows exactly what's up. I see all sorts of other birds here in the woods where I live, including blue jays and crows, but in comparison to this raven, they look really dumb."

Deciding to put his intelligence to the test one July day, Diane twisted ten separate strands of wire over the suet container opening, and again stayed home to watch the results. The raven came as usual, at that time of year with his mate and two young. "It was just unbelievable!" Diane said. "He started untwisting the wires. Sometimes he'd twist one the wrong way, and then he'd stop and take a break. But he always went right back and did it the right way. He went from one wire to the next, and untwisted them all. He must have had some idea of what he was doing, because he kept at it for one hour, not giving up. Occasionally, his mate would come down and just crudely

Getting the suet out of a squirrel-proof feeder.

pull on the box, then stop. He'd fluff out and act very irritated at her. When he finally got it open they both fed on the suet, but they chased their young away until they had their own fill."

I asked Diane to send me the exact suet container, because I wanted to test my own ravens. I received it via parcel post, put cheese inside, twisted the wires shut on top as she had done, and put in into my aviary. All six of my birds gathered round and pecked and pulled the wire. After twenty-seven minutes, they all gave up, and I stopped watching. A month later, the cheese I had put inside was still there. I mentioned this to Diane. She didn't think it was surprising, pointing out that mine were on "welfare." They may not have been less intelligent. Perhaps they did not have the ambition, since they'd get fed even if they failed. Additionally, the Pickards' raven was already nine years old, and mine were just over one year old. A long period of maturation could be required before ravens achieve their full reasoning power and/or persistence and patience in the face of obstacles.

Anecdotes like this are easy to dismiss. I have done so myself with numerous others, when I was not there to observe the fine details firsthand, to sift out facts from interpretations. Extraordinary cleverness can often be explained by "simpler" hypotheses, and I've always prided myself on my skepticism. But skepticism in what? With ravens I'm no longer always sure of how to distinguish a simple from a more complex hypothesis, how to know whether all of the ravens' behavior is somehow complexly preprogrammed or whether they know or learn to know what they are doing. Science normally progresses one small step, one small observation, at a time. Discarding solid isolated observations could be tantamount to discarding critical clues.

Behavioral biologists have a long tradition of outright rejection of observations that are not, like reflexes, strictly replicable. Some reject the idea that the actions of some nonhuman animals could be guided by thoughts, by conscious inner representation of the world. Only a little more than a decade ago, many scientists felt that consciousness in animals was not open to verification and experiment, and that it was therefore necessary to set it aside (Wasserman, 1985). Many still agree with that view.

As a conspicuous example of one current of academic thought I refer to the views expressed by Euan M. Macphail. In his book *Brain and Intelligence in Vertebrates* and elsewhere (see Notes) he concludes that there is little evidence of differences in general intelligence between nonhuman species. This conclusion, which is shared with many other psychologists, is mainly a projection of the fact that fish, reptiles, birds, and primates, as well as insects, show no compelling evidence of qualitative differences in learning. It ignores evidence that does not fit classical learning paradigms, and falls into the same trap of equating learning with intelligence that unsophisticates are accused of.

Why should animals show differences in learning? I would *a priori* assume they would *not* because all have nervous systems, and learning (or facilitated transmission across synapses separating neurons) is possibly as fundamental a property of interconnected neurons as is electrical conduction along the neurons themselves. Secondly, Macphail goes on to define consciousness as the capacity of any organism to feel something—anything, particularly pain or pleasure. He refers to the overworked analogy that we can design a machine that yelps when you kick it, to imply that we cannot infer pain from behavior. True, but just because a worm may writhe without feeling pain, that says *nothing* about whether you or I do. Macphail proposes that since humans can communicate our experience of pain with words, he presumes that pain, to be experienced at all, requires a cognitive self, and since words imply cognitive self, therefore only language-using animals can feel pain! I'm highly skeptical of that conclusion. I'm too influenced by the *rest* of the data set. Along with almost all modern evolutionary ecologists, I see animals as adapted to environment, with common principles applying across species. I believe, as George Schaller has said, in his classic studies of mountain gorillas, that: "Only by looking at gorillas as living, feeling beings was I able to enter into the life of the group with comprehension, instead of remaining an ignorant spectator." Donald Griffin of Rockefeller and Harvard Universities, one of my scientific heroes, similarly points out in his 1992 book, *Animal Minds,* which is devoted to exploring and reviewing possible consciousness in other animals, that dismissing the mind as trivial because one cannot personally prove it with precision by a simple test or device is like the denial of the role of inheritance before genes were elucidated. Part of the problem is that we are hampered by what Griffin calls "paralytic perfectionism." We can't define or study mind fully, in part because it will never be understood in terms of such simple units as genes, because it is an emergent property requiring and arising only out of the *complexity* of billions of interacting neurons. So we say it is unknowable, and ignore it. This is folly, because the task of science is to study precisely what we *don't* understand.

Philosophers have tried to explain consciousness in weighty books, concluding we know little. Opinions about consciousness differ enormously, as John Horgan explores in his book *The End of Neuroscience.* It is true, we know little about the precise neuroanatomical hook-ups that make "it" possible, even if we reach a consensus in how to define it. I agree with most others that consciousness, at its simplest level, implies awareness through mental visualization. It is difficult for me to envision performing any novel bodily motion without first mentally visualizing it. Indeed, I well recall my "eureka" experience of learning to do the butterfly stroke one night before going to sleep, endlessly practicing the motions of all of my limbs in my head, suddenly knowing I'd have it down pat when I stepped into the pool the next day. Every athlete knows the connection between mental and physical activity, and practices mentally for coordination. That's thinking. Every step we take over uneven ground involves consciousness of where we'll plant our feet. The longer and more unpredictable the journey, the more consciousness is required to get there in the quickest, most direct way. In the night, when we dream of making a jump, our legs may twitch as do a cat's. When we dream of climbing a tree and falling, our arm may suddenly move. Our body motions, unless thoroughly trained, emanate from consciousness, from mental representation. Consciousness is a *monitoring* of motor patterns in this case, which are neurally engraved preferred pathways in the central nervous system. Is not the cat's leg twitch a function of the same process as ours, or must we invoke vitalism for humans? I think not!

I studied such neural patterns governing flight (external behavior) and shivering (internal "behavior") in bumblebees at U.C. Davis with the late Ann E. Kammer. In bumblebees, those motor patterns were neurally nearly identical. Both were monitored in the bee itself, by other motor patterns with which they are integrated in a way that coordinated the whole animal, analogously to my concept of how consciousness works. In the bees, little neural activity exists in the absence of fully expressed muscle activity. In us, there is continual neural activity in the brain, often short-circuiting the neural commands to the muscles, although in dreams some of our consciousness may be *partially* expressed, since some of the filter between the mind and the body is relaxed.

Why do we have a monitor of muscle activity (i.e., behavior) of the brain? At the simplest level, it is necessary to note that every system monitors almost every other in a physiologically integrated animal. The monitoring is a ubiquitous and necessary property in any complex organization. In our brain most of that activity is unconscious, but why not *all* of it? I speculate it is for one specific reason only: To make choices. By far, the majority of our automated responses are those that require no flexibility. As soon as we have consciousness, we also have the possibility to make alterations in behavior. That's generally useful, but it has costs. We can screw up royally.

Different options can be different "memories"—represented by different neural pathways that we can activate or deactivate, depending on feedback from the brain's reward centers. Once we have these memories encoded in our brains, we can retrieve them from the past, rearrange them into novel configurations to provide insights, and project them forward to see, for example, the throwing motion of an arm to make predictions about hitting a target. The manipulations of symbols requires similar projections.

There is no need to assume that only we have consciousness because only we can think in symbolic language. Doing a back-flip, a raven's barrel-roll, or aiming an arrow can all involve the same fundamental process of comparing and fine-tuning scenarios.

In searching for models to try to understand the physical basis of consciousness, we might consider a social insect's decision-making processes, an analogy expressed by Lewis Thomas in *Late Night Thoughts on Listening to Mahler's Ninth Symphony*. He likened a termite colony to "an enormous brain on millions of legs," in which the individual termites are mobile neurons. In a termite colony or a beehive, the "sensory" experience of thousands of individuals' feeds, like information from sensory receptors to the brain, into the collective organism out of which an "intelligence" emerges. Various options or potential choices are encoded in circuits of neurons, compared, and weighed. The more sensory input there is into the hive, the more informed, appropriate, and "rational" the colony decision-making process becomes. A collective intelligence emerges.

In the insect colony, different decisions are tried out and an intelligent decision is made, but no consciousness as such is necessary to accomplish that task. Why? Because the colony response can be, and is, glacially slow. In effect, different courses of actions are communicated, and intermediaries within the colony evaluate and act on the various options. That is, intermediaries "eavesdrop" on, or monitor, different potential options proposed by other individuals. The response takes hours, not milliseconds. For example, bee dance followers may evaluate the recruitment dances of different individual bees, each indicating various potential food locations. In a single thinking individual, in contrast, all of the decision-making, backed up by innate responses and by learning, is tried out in the brain by populations of billions of neurons connected with all of their neighbors, rather than by populations of different "mobile neurons." Consciousness is the brain's way of quickly monitoring the many options so that a choice can be made without overtly trying the alternatives first. Once choice is no longer necessary, then consciousness becomes superfluous.

Consciousness would be particularly useful when there is a need to anticipate another's moves, blow for blow. For example, a raven feeding with wolves on a carcass could potentially learn by trial and error how to avoid getting bitten and how to get meat. But how many wolf bites can a raven afford and still survive? If the raven could consciously anticipate the wolf's moves and plot its own responses, it could avoid the first fatal mistake—the only one that really matters. Once consciousness has been selected to serve one purpose, it can be co-opted for use in other similar circumstances, in the same way that an insect's wing is not only adapted for flight but can secondarily serve as camouflage, armor, sexual signaling, and shading.

The neural basis of consciousness is unknown, although opinions of what it may be range from presumptions of the existence of mysterious properties to the idea that it is little more than elaborating on memory. As previously indicated, I tend toward the latter view, because we already know that the vividness and depth of our consciousness is a function of the vividness and depth of our memories. Neurobiologists agree that consciousness requires the standard chain reactions of impulses traveling neuron to neuron in the brain, but probably in multiple neural circuits. The neural circuits traverse a labyrinthine jungle of inconceivable micro-anatomical complexity with almost unlimited possibility for input to affect the loops, and it is in this arrangement of neurons and their circuits that consciousness resides.

We'll never find proof for the existence of consciousness by picking the animal apart, or by looking at its parts in isolation. That's like trying to understand the caching behavior of ravens by grinding them up, examining ever smaller parts down to the molecules, and studying them through the laws of physics and chemistry. That's backwards. To the biologists life is made of matter, but the nature of every living thing in the cosmos is time-bound. Every living thing is, like a book, more than ink and paper. It is a record of history spanning over two billion years. The more we dissect and look at the parts disconnected, the more we destroy what we are trying to find—the more we destroy what took millions of years to make. Mind, like life or liveness, is an emergent property that is a historical phenomenon, though also still wholly a physical one. It reveals itself far above and beyond its component parts, in this case, primarily the nerve cells with their infinite interconnections that cannot and will not ever be unraveled one and all. Consciousness is not *a* thing. It is a continuum without boundaries. We can most readily see its presence or absence in the extremes. In mussels, and in men. Is one of these extremes for birds found in ravens? Is it of sufficient magnitude for us to detect with our feeble detectors, our minds?

How can we envision a progression of steps leading from memory to consciousness? At first, the neural connections between different reverberating circuits could be loose and tenuous. Perhaps one pathway merely monitored when another was active. Then, with more interconnections, it could monitor activity more precisely, ultimately to track the other circuit ever more accurately, sensing its cycle before that cycle is committed to muscular effort (i.e., behavior). By these analogies, a graceful action such as a swan dive, or a raven's wing-dipping in flight, if not already hardwired, is first an idea. Repeated often enough, the thought becomes engraved in the nervous system. It becomes automatic, and no more decisions to guide its course are necessary beyond that of the "go" signal to execute the action. That is, it becomes unconscious, like most of our vital functions that run on "autopilot."

The physical bases of emotions and feelings are potentially more of a theory problem than of mental representation, such as that underlying conscious insight and intelligence. Nevertheless, one might legitimately ask, why do organisms have subjective experience and emotions at all? I will add little here to that topic which I discussed (pp. 109–112) in my book *A Year in the Maine Woods*. I speculated there why it is unlikely that a lobster would feel boiling water, why a worm that writhes does not necessarily feel pain, and why a raven probably feels love and a dog not, and why a moth seeking a mate would likely do so without any conscious awareness. Pain has a function. It is not an "extra," even though you can build a robot that recoils from a wall or a hot stove. Pain is a warning system that has evolved where choice is possible, that when coupled with insight or learning from experience, helps us avoid potential damage. If we feel the pain of a pinprick, then coupled with insight and/or learning, that sensation spares us the potential future damage of sitting on a spike. If we have a sensory feel of jumping off the back step and combine a memory and insight with it, then we won't kill ourselves by jumping off the roof on a whim. We'd strenuously resist any such whim. An angle worm probably doesn't have the need for a memory linking each of its behaviors with sensations, because it doesn't need to make choices. We do, and so do ravens. Of course, it makes absolutely no difference what a raven's private pleasurable sensation of eating raw skunk meat is, so long as it has one that links its memory with its reward system. The sensations may change dramatically, even in the same individuals. I recall one dramatic incident. I had always loved the taste of cranberry juice after a hot jog. Later I was forced to drink gallons (by a cranberry juice producer who sponsored me on a championship race). I drank it for 100 km, even as I felt increasing unbearable agony. I tried it on my next race, and then it tasted to me like the most disgusting fluid I'd ever imbibed. I was unable even to drink water flavored with it (dispensed from the same bottle). My body had tagged cranberry juice as a pain-inducing substance. I had a choice, and exercised it. I stopped drinking. (I dropped out of the race.) My body avoided the pain it had previously experienced in association with that particular taste. Like the color of lobster pots spread around in Muskongus Bay that identify their owners, the *specific* sensations are arbitrary by themselves.

Sensations and emotions regulate behavior. We feel attraction, love, revulsion, jealousy, fear, happiness, determination, and anxiety, all governing responses largely connected with specific behavior relating to long-term survival and reproduction. They serve to motivate and guide behavior when there is no immediate reward or logic to guide us. When functioning properly, they inform us and keep us on the right track. Who, for example, has not felt happiness on being in the woods on a spring day, listening to the birds singing, the insects buzzing on the flowers, the green grass swaying in a breeze, seeing elk graze and fishes darting along the banks of a sandy stream? We are programmed to feel happy under these circumstances because we are experiencing the stimuli that through evolutionary history have represented the sort of productive environment we might seek for survival and reproduction. A raven or a rat's neural centers would likely be wired to feel more internally stimulated by different scenarios, possibly a fetid dump or a beach littered with carcasses. Since ravens have long-term mates, I suspect they fall in love like we do, simply because some kind of internal reward is required to maintain a long-term pair bond.

Less specific emotions also help us stay on track for relatively short-term goals, again by rewarding us for intermediate steps. For example, we get no reward by turning to random numbers, say 10-8-3, on a combination lock on a chest full of money. If we *know* that the combination 10-8-3 will open the lock, allow us to reach the money, and that the money will then buy us bread, we will turn those numbers as if almost "tasting the bread." As Anthony Dickinson, a psychologist at Cambridge University, has pointed out, "Consciousness is necessary to confirm value." Our values are emotional rewards that allow us to do otherwise proximally senseless things that have no immediate reward whatsoever. The same idea applies to a raven pulling a string. The bird gets no tangible reward from all the intermediate steps before it reaches the food. It needs an emotional satisfaction to do those nonrewarding steps. And it can only get an emotional satisfaction for the senseless act of pulling on a string, as such, unless that act satisfies a thought or mental scenario of getting what the string will provide.

Some believe that consciousness and the thinking derived from it are made possible only with linguistic ability. According to that very narrow view, humans are indeed the only conscious beings. But who of us translates everything we are aware of, thinks, or feels, into words? Who thinks only with sentences? We can all project mental images of ourselves or others throwing a ball, or envision how a bicycle operates, or know how to kill a bison with a spear. To transcribe such thoughts into words takes valuable time and effort. Of course, we can and do think with words, but that's not because thinking requires words. It's instead because we're social animals who have evolved to use words to communicate, allowing us to transfer useful mental images into our helpers' and associates' minds. Admittedly language was not possible until we had achieved a critically high threshold of consciousness, which probably exceeded that of any other existing animal. After crossing that threshold, we needed even more consciousness to think using words, requiring still more consciousness, and that sped us down the road of cultural evolution. It was words, then, that enormously amplified our capacity to store and transfer knowledge, to build culture, to form alliances, to wage wars. The word became the weapon. I suspect the great gulf, or *dis*continuity, that exists between us and all other animals (as the one between those of us now manning a console on a moon orbiter versus those of us wielding spears) is ultimately less a matter of consciousness than of culture.

Back to the Wild

RAVENS HAVE LIMITED CULTURES and are closely and intimately tied to their external physical and biological environment, unlike us, who are culture-bound and only indirectly connected to the external environment. Throughout my studies, the goal has been to understand the birds' life in their natural environment, which provides context to almost everything about them. Almost nothing about them makes sense unless seen from that perspective, in the same way that very little about us makes sense unless seen from the perspective of our culture. Nevertheless, to study details of their intimate lives is practically difficult or impossible in the complex, multidimensional wilds encompassing hundreds of square miles. At times, I detained groups of wild birds in aviaries for close observations. These aviaries were as large and as natural as I could make them. They did fulfill the birds' needs—hand-reared and wild birds alike successfully built nests and reared young in them—but when it was time to release the birds, I was glad for them to resume their full, but much more risky, wild lives.

*Wild ravens near my Maine cabin playing and strutting,
photographed from the cabin.*

You will recall Goliath and Whitefeather, who were my particular friends. Goliath was hand-reared by me from a nestling, and his mate Whitefeather was a female he had met in a crowd of temporary captives. For reasons that are totally inexplicable to me, they immediately struck up a relationship. Whitefeather and Goliath built a nest in captivity, laid eggs, incubated and raised their own and another pair's young. The couple stayed on my hill after I tore open one side of their aviary, and the surrounding area became their territory. After their first successful breeding attempt (inside the aviary) they returned on their own into the aviary the next year to build a partial nest, but they then aborted that year's breeding attempt, possibly saving their energies for the next.

Having Goliath and Whitefeather as permanent residents on my Maine hill seemed like a dream come true. I could observe them closely on their terms, in their environment. What more could one want than to be surrounded by interesting friends who made no demands, who provided constant entertainment, and who were teachers. In the spring of 1997, the pair even provided me with new data I used for a paper on caching behavior in a prestigious international journal. I was high on these birds. Whenever I came walking up the hill to the clearing by the cabin after an absence of one to several weeks, I hollered "Golii-ath"—and I would soon hear him answer from the forest. Then he would circle over me, spiral down, and land on the dead birch tree by the picnic table. I'd pull a dead mouse or some other tidbits out of my pocket, and he'd drop down and swagger over in his slow, deliberate walk. Rachel tells me the moment she knew she was in love with me was one morning as she watched me sharing my oatmeal with Goliath in the yard. Whitefeather, who was still somewhat shy of me, would call from the nearby pine woods. She begged from Goliath, who shared the food that he got from me.

After mid-May in 1997 and through the summer and fall, I came up to a silent hill. No raven answered my call, and I felt a loss. Had something happened to them? I talked with the neighbor living in the shack at the bottom of the hill. Yes, he had seen them, he said. They had come several times and "made a racket" in the big pine tree next to his hut. But he couldn't be sure of dates, and I couldn't be sure if he had seen my ravens, or some others, although he said he could by then recognize Goliath's distinctive voice. The pair seemed to be gone. I still harbored a faint hope that they might occasionally be stopping by when I wasn't around, and would perhaps resume their residence later on. I was absent from Raven Hill camp for most of the fall and early winter, and when I returned with thirteen students to teach a field ecology course in late December 1997 and early January 1998, I saw no sign of them. I gave up hope of ever seeing them again.

On the evening of January 10, just as we were about to leave and as I was walking out of the woods to the log cabin, I heard a raven excitedly making a loud commotion. Looking up, I saw to my great surprise a pair of ravens flying over with a third bird following, all heading directly to the aviary in back of the cabin. I stayed away from the aviary to avoid disturbing the trio. After an hour or so, I heard a female's knocking call and continuous *rap-rap-rap* calls. The pair, or at least one of them, was back, and they were excited, possibly thinking about nesting. The next dawn, their calling resumed, and I felt confident that they were back to stay. In the first reds, blues, yellows, and turquoise of a glorious sunrise, all the trees shone like silver from their heavy loads of ice after the infamous five-day ice storm that had severely damaged the New England woods. Nevertheless, I felt happy.

As it was just barely getting light, I heard agitation calls from the aviary. At least one bird was extremely upset. I had not wanted to go near the aviary at that critical time for fear of discouraging their nesting, but these calls demanded attention. What in the world could be going on? To find out, I approached cautiously through the woods, at last seeing three birds fly up from or out of the aviary. One of them saw me, backpedaled in the air, and rapidly flew off. The second followed it. The third flew toward me and landed next to me. *Goliath!* He seemed nervous at first, perhaps because we hadn't had contact since May, eight months earlier. He stayed near me, but seemed subdued and uninterested in me, even though I talked to him. I checked inside the aviary and found fresh raven tracks in the snow by the old nest. There could now no longer be any doubt. They were back! They were getting ready to nest.

I heard at least two ravens twice more at the aviary later in the day. In the evening, the calls came from near the edge of the pine grove just north of the cabin where Goliath and Whitefeather had often slept before. We had to leave that very day, and I had to be absent from Raven Hill camp for the rest of the late winter and early spring, due to teaching and other commitments in Vermont, and planned experiments with my six aviary birds there. Glenn Booma stayed at the cabin in mid-March and called me with exciting news; he had heard ravens in the aviary. He did not go near them for fear of disturbing them, but it was the proper time for ravens to start building new nests or refurbishing their old ones. I was confident that the pair's old nest—the one that Goliath had built but that was never lined by Whitefeather—would now finally be lined with deer hair and ash and cedar bark, and would soon hold their eggs.

When I finally returned on April 29, I rushed into the aviary almost immediately to find an unpleasant surprise: The nest was just as it had been left a year ago. Hardly a twig had been altered. No lining had been put in, and it was now long past the time for the ravens to have eggs. I heard no ravens calling. Dejected, I walked to the pine grove in which they had roosted and which I had thinned out to favor the largest trees. If I were a raven, I thought, I'd build my nest *here*. I looked up. There, that dark mass ahead of me—a raven nest! As soon as I saw it, a raven flew to it with nest lining in its bill.

They were nesting now in the wild, and my spirits soared like never before. At the same time, I immediately contemplated the mystery of this extraordinary late nesting. *All* the other raven nests I had ever found at the end of April had young, many ready to fledge! This was the latest raven nesting attempt I had ever seen, out of nearly a hundred in this area. Might a disagreement among the pair have been the cause of the long delay? Perhaps Goliath, having grown up in the aviary, wanted to rebuild in the aviary, but Whitefeather, who grew up wild, did not agree. Last year, Goliath had built a nest-frame in the aviary, but she apparently had not agreed on the site, because she did not cooperate to finish building. She didn't do her part, to line it with fur to receive the eggs. As a result, they had failed to breed. This year, they had checked out the site again. There had probably again been a long disagreement, but she had won this time. They built on a pine tree in the tradition of other local ravens. Much valuable time had been lost. Apparently, there are advantages for ravens to have long-term mate relationships, especially if the birds come from different traditional backgrounds and with perhaps different expectations. In northern Germany, for example, most ravens nest on beech trees; in Maine, most nest in white pines; in Vermont, on cliffs.

On May 8, when I returned for a full week, I immediately went to the nest in the pine tree. Even from some seventy-five yards away, I could see Whitefeather's dark head above the nest rim as she was incubating. She slipped off quietly on folded wings, diving down the valley toward Alder Stream. I saw only two birds near the nest at any time. Neither came near me, and neither showed alarm. Both were quiet.

I marveled at how quiet the birds were. But two days later, at 3:30 P.M., there was big excitement near the nest—lots of vocalizing. A heavy rain had just stopped, a cloud ceiling was lifting, and there was a slight wind. Were they excited about taking a flight? I rushed out of the cabin, sprinted down the path, and looked up. Sure enough, ravens were soaring in the breeze. But there were four of them together, and they were flying along amicably in a formation, as if on a mutual friendly romp in the sky. There were no chases, no aggressive calls as when intruders come or are being held off.

One pair split off and flew over the hill to the north, where less than two miles farther on I'd just seen four large fully feathered young perched on the branches around the Braun Road nest. Goliath and Whitefeather flew back toward their own nest. As fast as I could, I ran to the top of the ridge and climbed a tall spruce from which I watched the still vocal Goliath and Whitefeather flying closely together, circling, wheeling, diving, circling some more. It was great flying weather, and I could enjoy their play vicariously. I saw the pea green patches of freshly unfurled poplar trees in the valley below. Their light washes of brilliant green in the forest tapestry contrasted with the yellowish- and reddish-brown of red maples springing to life. Thick swaths of dark green red spruces and pines showed on the hills against a pale blue sky splashed with drifting white cumulus. A thought struck me: Maybe they were excited because the young had just hatched. I almost tumbled out of my tree. I had to try to climb up to the nest, but large pines are not easy to ascend, especially without climbing irons, and I got exhausted in the attempt, finally giving up. By 7:30 P.M., the birds had long since returned to the nest, but I still heard sporadic bouts of *rap-rap-rap* calls. Whatever it was that had excited these ravens, it had made a big enough impression to last for at least four hours.

On the next day at 6:30 A.M., a red-tailed hawk flew by the cabin in the valley just below and toward the raven nest. The raven's *kek-kek-kek* alarm calls erupted immediately, but lasted only the several seconds until the hawk had flown on by. Later, seven turkey vultures flew over a calf carcass I had left. Both ravens flew up, intercepted, and chased them away. They again made the *kek-kek-kek* calls, and again became totally quiet within seconds after the intruders had left.

I was still bothered by not knowing the reason for yesterday's great excitement. It looked as if the cause was the other pair's visit. But why did they come visiting? Why did Goliath and Whitefeather's emotions continue to be high for four hours after they left? I decided I had to see inside that nest. Maybe their young had indeed hatched, and the other pair's visit was incidental.

The tree was swaying in the wind, but it didn't matter. I was determined. I clung tightly and inched my way up anyway, knocking off brittle dead branches as I climbed. After I reached the first solid branch that I could grab, it was easy. Within minutes, I was on live branches, directly under the great stick nest. Goliath and Whitefeather had left, silent. I swung around under the framework of broken-off poplar twigs as thick as my thumbs. I peered into the nest cup with great anticipation.

I have peered into many raven nests, but to be physically there always leaves me awed, especially after a risky, sweaty, hard climb. Four eggs! They were the most beautiful eggs I had ever seen. Their ground-color was green, like that of the then unfurling leaves of the birches, mixed with the blue of the clearest spring sky or that of the azure "little blue," a butterfly then fluttering over the forest floor. They were blotched with dark olive green, like that of the spruce, and the gray-

brown of pine bark. There were
a few flecks of purple and a lot of
gray and black. One, unlike the
others, was most heavily blotched
on the pointed rather than the
round end of the egg. The four
lay in a deep soft cup lined
with tufts of white hair from a
deer's tail, shredded cedar and ash
bark, chunks of pea green moss,
and small tufts of black bear fur.

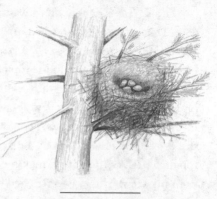

*Goliath and Whitefeather's
nest in a pine tree.*

As I photographed the nest
and eggs, I thought again of the
chicken egg experiments I had
done before. There was no need to repeat that experiment here,
because I already knew that ravens incubate chicken eggs in their
nests, but immediately eat them outside. I wondered instead how
quickly the ravens would eat a chicken egg left on the ground below
the nest.

Later that day, I left two white eggs on two different pine
stumps, one below the nest tree and one some thirty feet south of it. I
could think of all sorts of logical reasons why nothing would happen.
I knew, if it were a crow, the bird coming back would instantly see
the white eggs and fly down to eat them. It would not matter if the
eggs were resting on the ground or floating a foot above it. But a
raven would notice something "off," even if they were lying on a
stump. My ravens were afraid of a branch that appeared to move by
itself, they quickly approached round things, yet they hid in fright
for a day and a half from a helium balloon floating above a string. An
egg on a pine stump below the nest? Ravens have expectations. Eggs
magically appearing on a pine stump would not be one of theirs. To
them they might appear as to us oranges floating in midair: spooky.
And just calling it "neophobia" would not explain it, that much I
knew. I sprinted away to climb my spruce and watch from the nearby
ridge.

In just two minutes, a bird was back, making long rasping calls, odd ones. Then I heard high-pitched, upward-inflected calls that sounded like a question, as we might say, "Wh*aaat?*" I had never heard these calls before. By the time I got to the top of my tree, the pair was flying toward Gammon Ridge, all the while continuing the strange new calls. The birds acted spooked. They flew about a mile to the ridge. Straining with my high-powered binoculars, I could just barely make out their two black shapes perched closely side by side on a poplar. At 5:45 P.M., thirty-one minutes later, both flew off, making honking calls, and they disappeared from my view. This was getting to be stranger than I had expected.

An irrational fear seized me: Might they desert the nest because of those white chicken eggs?! You can imagine the anxiety I felt the next morning just before I had to return to Vermont. Would a raven be on the nest? At 5:30 A.M., I walked toward it to find out. On the way, I went past the calf carcass I had left for them on the path. A turkey vulture flew up—so one had finally sneaked in at dawn. Almost instantly, a raven came from the nest area, and with loud *kek-kek-kek* calls chased the vulture down the valley.

I drove back to Maine the very next weekend to check on the ravens. Surely, the young would have hatched, and I could forget all of that funny business about scary eggs. I drove all Saturday morning, May 23, 1998, and got to the cabin near noon, walking right to the nest. Silence at the nest area! The silence hit me like a sledgehammer. The two chicken eggs were still on the stumps exactly where I'd put them. Now I had to climb the tree. Not a raven sound anywhere. I started climbing. As I got close to the nest, I suddenly heard a raven's call, then two birds called, and the pair came flying high above me, scolding me mildly. I was again hopeful of seeing the pink just-hatched young in a few seconds. When I looked over the nest edge, I felt a shock. I saw neither young, nor eggs. There was only nest lining. Oddly, the nest cup was shallow, and the nest lining seemed loose. Had the nest been abandoned and then robbed of its eggs? Absentmindedly, I pulled at the deer hair and shredded bark, and there beneath a blanket of nest lining lay the four eggs! They had been deliberately covered to hide them, or to delay their cooling while being left during a break from incubation duties. I felt them—they were cool. I had never before seen anything like this behavior. But it was a warm day, and perhaps Whitefeather could afford to take a break. Did she have notions about heating and cooling, seeing and hiding? The twenty-one day incubation term was about over. The young would be extraordinarily late, but for Goliath and White-feather's very first nesting attempt in the wild, it was not a bad start—although this nesting attempt would also end in failure.

We had already shared a long, rich history. I wondered what new adventures would be in store for all three of us. As I sometimes do after a hard ascent, I savored a few moments to look around and reflect. The "climb" to the ravens has sometimes been hard but the results have been deeply satisfying. I feel that I've won a wider vantage point for seeing some of the raven's ironies and seeming contradictions, much as I then had a vantage point for viewing the forest. There is no end to the forest, and there is no end to the mind. Indeed, the greater the complexity, the more it is mind, as the more trees there are, the more it is a forest. It can never be encompassed fully.

More important, I have come to touch the world and the travails of a totally different yet kindred being that makes me feel less alone. I've also seen lots of morning stars and sunsets, felt alive in the snow and rain, sensed the cycling of pulsating life and silent death, found new human friendships, forgotten old traumas, and felt passion and peace.

Raven reducing forward-flight momentum when landing.

Afterword

In my previous book, *Ravens in Winter,* I explored the process of trying to solve a specific problem; do some ravens attract strangers and share their food with them, and if so how and why? Food-sharing behavior was well-known, but only for group-living animals. Ravens were known as notoriously aggressive territorial birds, and sharing if it occurred would be based on a novel mechanism. So it turned out to be.

This book continues where the first left off, but with some important differences. In the original study, the ravens' behavior was treated from the standpoint of behavioral ecology, a discipline that seeks to reveal evolved behaviors that solve specific problems that relate to the population. The behaviors are assumed to have been mindlessly selected by evolution through millions of years of history. The intentions of individuals are seldom an issue. In this book, in contrast, I have tried to give a detailed picture of specific ravens I have known, and I explore the possibility that some of their behavior is derived from conscious choice. My emphasis and primary reason for writing the book, however, was to record original observations and the adventures that made them possible. My interpretations are personal opinions and they will undoubtedly be modified. But facts are permanent.

The more facts I learned about ravens, the more apparent contradictions I saw. Sometimes ravens went out of their way to pull the tails of wild wolves and eagles. At other times I saw ravens afraid of a mouse, and even a moth, a moving twig, and a turtle, and a pile of Cheerios. I saw territorial ravens spend over an hour trying to chase as many as eight others at a time away from their food. I also saw the same individuals later welcome others and share the same food.

Sometimes ravens nested in defended territories of tens of square miles, and at other times they nest within a hundred paces of one another, even though they have the option to nest apart. Individuals performed feats of insight and intelligence, while also behaving in nonsensical ways. I found ravens forming exclusive pair bonds years before breeding, and staying paired despite conflicts. Yet extra-pair copulations occur in some populations of ravens. The birds expended incredible amounts of energy establishing dominance, yet dominant birds commonly fed last and the largest strongest bird showed the most fear. They were the shyest and most alert birds I had ever known who shunned humans in New England, yet they actively follow humans in some other areas. They are as tame as sparrows and pigeons in some cities, while retreating into the wilderness in other areas. I received reports of them warning people of an imminent attack by a predator, yet they readily feed on corpses. They are one of nature's premier scavengers, yet they feared animal carcasses, and that fear was not learned. It was innate. They gobbled individual cheetos with gusto, yet recoiled from a pile of them. To get food from feared carcasses they stole it from those birds who would go near (eagles or magpies) or they recruited others and went in as a crowd. Once a crowd was at a carcass they all suddenly seemed to try to cache as much meat as possible, yet although they could have stayed near the carcass and stored food nearby, they usually expended much time and energy to fly miles to hide each morsel singly, separately and in isolation. These and other contradictions were not readily explained by theories of behavioral ecology, yet they could not be ignored just because they were not understood. I felt there must be some underlying adaptive pattern that would explain the uniqueness of these birds, as well as shed light on their prominent esteem in almost all northern cultures, spanning America, Europe, and Asia.

The key to solving the many apparent conflicts in their behavior would, I knew, reside in their natural history and an understanding of the selective pressures they face and have faced. I eventually deduced that ravens have evolved in close association with intelligent and potentially dangerous carnivores: first mainly wolves, then transferring their allegiance also to prehuman and human aboriginal hunters. To see traces of the ancient connections I traveled to the north (Baffin island in the arctic, Nova Scotia, Yellowstone Park) where ravens still, or again, closely associate with their mammalian hunters and where they rely on them for almost all their food in winter and where in turn the ravens may signal the hunters about the availability and/or location of prey. The social hunters, both canids and humans, are both dangerous and intelligent, and intelligence is required by the ravens as well, to benefit by and to remain in the hunting symbiosis.

My previous work with bumblebee foraging behavior and bees' symbiotic relationship with the plant community through pollination predisposed me to the notion of a symbiosis between ravens, wolves and ultimately early humans. I therefore became all the more intrigued by the famous Odin myth of the Ancient Norsemen.

In a biological symbiosis one organism typically shores up some weakness or deficiency of the other(s). As in such a symbiosis, Odin the father of all humans and gods, though in human form was imperfect by himself. As a separate entity he lacked depth perception (being one-eyed) and he was apparently also uninformed and forgetful. But his weaknesses were compensated by his ravens, Hugin (mind) and Munin (memory) who were a part of him. They perched on his shoulders and reconnoitered to the ends of the earth each day to return in the evening and tell him the news. He also had two wolves at his side, and the man/god-raven-wolf association was like one single organism in which the ravens were the eyes, mind, and memory, and the wolves the providers of meat or nourishment. As god, Odin was the ethereal part—he only drank wine and spoke only in poetry. I wondered if the Odin myth was a metaphor that playfully and poetically encapsulates ancient knowledge of our prehistoric past as hunters in association with two allies to produce a powerful hunting alliance. It would reflect a past that we have long forgotten and whose meaning has become obscured and badly frayed as we abandoned our hunting cultures to become herders and agriculturists, to whom ravens act as competitors.

I didn't start out with the intention of exploring or explaining the possibility of an ancient symbiotic relationship between man, wolves (dog) and ravens. As others, I had previously assumed that the well-known association between ravens and wolves and sometimes people, was simply a reflection of the raven's opportunism. However, almost by accident I found innate hard-wired responses in ravens that not only cleared up most of the contradictions I had seen, but that also pointed to a relationship with hunters that is ancient and evolved. It turned out, to my great surprise, that ravens don't feed next to wolves because they have to, they do it because they want to.

Ravens are, however, indeed highly flexible and they of course exist in many areas where wolves no longer roam. In northern Maine during the seasonal moose hunt, ravens appeared to associate with human hunters instead. At my main study area in western Maine the ravens showed great fear of all cattle carcasses I provided for them, but they overcame their fear by recruiting even strangers, and feeding in the safety of crowds. In contrast, in eastern Oregon where there are also no wolves, ravens used intermediaries to get food. Magpies and eagles fed unhesitatingly at a deer and carried off meat to cache. The ravens harassed the magpies and eagles until they dropped their meat.

Ravens are forced to associate and get along with not only dangerous intelligent carnivores at carcasses, they are also thrown into direct competition with each other. Studies in my aviary and outside it indicated that ravens recognize individuals, not only their own kind, but also of other species. The birds have specific play behavior that appears to have been selected by evolution to gauge the responses of potentially dangerous and specific carnivores. Thus, living in a world of individuals of their own and other species, rather than only one composed of different kinds or categories, the birds' world is inordinately more complex than had been previously assumed, and their intelligence relates to that social environment.

In the latter part of the book I try to confront what intelligence is and how it relates to learning, innate responses and insight. I report observations and tests that were conducted, and describe how they have been received by psychologists and others. I conclude that ravens are able to manipulate mental images for solving problems. They are aware of some aspects of their private reality, seeing with their minds at least some of what they have seen with their eyes.

Notes and References

PREFACE

General references on ravens:

Angell, T. *Ravens, Crows, Magpies, and Jays.* (Seattle: University of Washington Press, 1978).

Boarman, W. I., and B. Heinrich. "The Common Raven." In *The Birds of North America,* A. Poole (ed.). (Washington, D.C.: Academy of Natural Sciences, 1999).

Glutz von Blotzheim, U. N. "*Corvus corax* Linneus" (pp. 1947–2022). In *Handbuch der Vögel Mitteleuropas,* Band 13/III (Wiesbaden: AULA-Verlag, 1993).

Goodwin, D. *Crows of the World,* 2nd ed. (UK: St. Edmundsbury Press Ltd., 1986).

Heinrich, B. *Ravens in Winter.* (New York: Simon & Schuster, 1989).

Ratcliffe, D. *The Raven.* (London: T & AS Poyser, 1997).

Sources of raven mythology:

Goodchild, P. *Raven Tales.* (Chicago: Chicago Review Press, 1991).

Guerher, H. A. *Myths of Northern Lands.* (New York: American Book Co., 1895).

Turville-Petre, E. O. G. *Myths and Religion of the North.* (New York: Holt, Rinehart & Winston, 1964).

Recommended animal behavior texts:

Alcock, J. *Animal Behavior: An Evolutionary Approach,* 6th ed. (Sunderland, MA: Sinauer Assoc., 1998).

Sherman, P. W., and J. Alcock. *Exploring Animal Behavior,* 2nd ed. (Sunderland, MA: Sinauer Assoc., 1998).

For the work on bird orientation by G. Kramer, S. T. Emlen, and many others:

Schmidt-Koenig, K. *Avian Orientation and Navigation.* (London: Academic Press, 1979).

Specific animal studies referred to:

Cheney D. L., and R. M. Seyfarth. *How Monkeys See the World: Inside the Mind of Another Species.* (Chicago: Chicago Univ. Press, 1990).

de Waal, F. *Good Natured: The Origin of Right and Wrong in Humans and Other Animals.* (Cambridge, MA: Harvard Univ. Press, 1986).

Dethier, V. G. *The Hungry Fly.* (Cambridge, MA: Harvard Univ. Press, 1976).

Goodall, J. *The Chimpanzees of Gombe: Patterns of Behavior.* (Cambridge, MA: Harvard Univ. Press, 1986).

Griffin, D. R. *Listening in the Dark.* (New Haven, CT: Yale Univ. Press, 1958).

Heinrich, B. "The effect of leaf geometry on the feeding behavior of the caterpillar of *Manduca sexta* (Sphingidae)." *Animal Behaviour* 19, 119–124 (1971).

Heinrich B., and A. Kammer. "Activation of the fibrillar muscles in the bumblebee during warm-up, stabilization of thoracic temperatures and flight." *J. Exp. Biol.* 58: 677–688 (1973).

Hunt, G. "Manufacture and use of hook-tools by New Caledonian crows." *Nature* 379: 249–251 (1996).

Herrenstein, R. J., D. H. Loveland, and C. Cable. "Natural concept in pigeons." *J. Exptl. Phychol: Animal Behavior Processes* 2: 285–302 (1976).

Pepperberg, I. "Cognition in the African Grey parrot (*Psittacus erithacus*): Further evidence of comprehension of categories and labels." *J. Comp. Psychol.* 104, 41–52 (1990).

Payne K., and R. Payne. "Large scale changes over nineteen years in songs of humpback whales in Bermuda." *Z. Tierpsychol.* 68: 89–114 (1985).

Seely, T. D. *Honeybee Ecology.* (Princeton, NJ: Princeton Univ. Press, 1985).

Smolker, R., A. Richards, R. Connor, J. Mann, and P. Bergeron. "Sponge carrying by dolphins (Delphinidae, *Tursiops* sp): Foraging specialization involving tool use?" *Ethology* 103, 454–465 (1997).

von Frisch, K. *The Dance Language and Orientation of Bees.* (Cambridge, MA: Harvard Univ. Press, 1967).

Some of the published anecdotes of the strange things ravens do:

Bailey, D. "Snow bathing of ravens." *Ont. Birds* 11: 1 (1993).

Barnes, J. "Raven rolling on ground to avoid peregrine." *Brit. Birds* 79: 252 (1986).

Bradley, C. C. "Play behavior in northern ravens." *Passenger Pigeon* 40: 493–495 (1978).

Connor, R. N., Chamberlain, D. R., and Lucid, V. J. "Some aerial flight maneuvers of the common ravens in Virginia." *The Raven* 44: 49 (1973).

Davis, T. A. W. "Raven covering eggs." *Nature, Wales* 14: 142 (1975).

Elliot, R. D. "Hanging behavior in common ravens." *The Auk* 94: 777–778 (1977).

Evershed, S. "Ravens flying upside-down." *Nature* 126: 956–957 (1930).

Ewins, P. J. "Ravens foot-paddling." *Brit. Birds* 82: 31 (1989).

Gwinner, E. "Beobachtungen über Nestbau und Brutpflege des Kolkraben (*Corvus corax*) in Gefangenschaft." *J. Orn.* 103: 146–177 (1965).

Hewson, R. "Social flying in ravens." *Brit. Birds* 50: 432–434 (1963).

Hooper, D. F. "Ravens snow bathing!" *Blue Jay* 44(2): 124 (1986).

Hopkins, D. A. "Snow bathing of common ravens." *Conn. Warbler* 7: 13 (1987).

Jaeger, E. C. "Aerial bathing of ravens." *Condor* 54: 246 (1963).

Janes, S. W. "The apparent use of rocks by a raven in nest defense." *Condor* 78: 409 (1976).

Jefferson, B. "Evidence of pair bonding between common raven (*Corvus corax*) and American crow (*Corvus brachyrhynchos*)." *Ont. Birds* 9: 45–48 (1991).

Moffett, A. T. "Ravens sliding in snow." *Brit. Birds* 77: 321–322 (1984).

Montevecchi, W. A. "Corvids using objects to displace gulls from nests." *Condor* 80: 349 (1978).

Owen, J. H. "Raven carrying food in foot." *Brit. Birds* 43: 55–56 (1950).

Stoj, M. "A case of ravens carrying their nestlings away from the nest." *Notatk. Ornitol.* 30: 106–107 (1989).

Tåning, A. V. "Ravens flying upside-down." *Nature* 127: 856 (1931).

Van Vuren, D. "Aerobatic rolls by ravens on Santa Cruz Island, California." *The Auk* 101: 620–621 (1984).

CHAPTER I

For a study about begging behavior:

Rodriguez-Girones, M. A., P. A. Cotton, and A. Kacelnik. "The evolution of begging: signalling and sibling competition." *Proc. Nat. Acad. Sci.* U.S. 93: 14637–14641 (1996).

For a study about counting by ravens:

Koeler, Otto. "Zahl-versuche an einem Kolkraben und Vergleichversuche an Menschen." *Zeitschrift für Tierpsychologie* 5: 575–712 (1943).

CHAPTER 2

Publications exploring raven recruitment behavior at the Maine study site:

Heinrich, B. "Foodsharing in the Raven *Corvus corax.*" In *The Ecology of Social Behavior,* C. N. Slobodchikoff, ed. (San Diego: Academic Press, 1988).

Heinrich, B. "Winter foraging at carcasses by three sympatric corvids, with emphasis on recruitment by the raven, *Corvus corax.*" *Behav. Ecol.* & *Scoiobiol.* 23: 141–156 (1988).

Heinrich, B. *Ravens in Winter.* (New York: Simon & Schuster, 1989).

Heinrich, B. "Does the early raven get (and show) the meat." *The Auk* 111: 764–769 (1994).

Heinrich, B., and J. M. Marzluff. "Do common ravens yell because they want to attract others?" *Behav. Ecol.* & *Sociobiol.* 28: 13–21 (1991).

Heinrich, B., J. M. Marzluff, and C. S. Marzluff. "Ravens are attracted to the appeasement calls of discoverers when they are attacked at defended food." *The Auk* 110: 247–254 (1993).

Heinrich, B., and J. M. Marzluff. "How ravens share." *American Scientist* 83: 342–349 (1995).

Marzluff, J. M., and B. Heinrich. "Foraging by common ravens in the presence and absence of territory holders: an experimental analysis of social foraging." *Anim. Behav.* 42: 755–770 (1991).

Marzluff, J. M., B. Heinrich, and C. S. Marzluff. "Roosts are mobile information centers." *Animal Behaviour* 51: 89–103 (1996).

Parker, P. G., T. A. Waite, B. Heinrich, and J. M. Marzluff. "Do common ravens share food bonanzas with kin? DNA fingerprinting evidence." *Animal Behaviour* 48: 1085–1093 (1994).

CHAPTERS 4 AND 5

For reports of early and late nesting by ravens:

Brückmann, U. "Hohe und späte Kolkrabenbrut in Graubünden." *Orn. Beob.* 74: 209 (1977).

Hauri, R. "Horstbau beim Kolkraben im Herbst." *Orn. Beob.* 65: 28–29 (1968).

Hauri, R. "Aussergewöhnlich frühe Brut des Kolkraben." *Orn. Beob.* 67: 63 (1970).

Hyatt, J. H. "Ravens nest in October." *Brit. Birds* 39: 83–84 (1946).

Mearns, R., and B. Mearns. "Successful autumn nesting of ravens." *Scott. Birds* 15: 179 (1989).

Sieber, H. "Frübruten des Kolkraben." *Falke* 15: 31 (1968).

For a summary of food and feeding habits of ravens:
Ratcliffe, R. D. *The Raven* (pp. 75–96). (London: T & AS Poyser, 1997).

General references:
Avery, M. L., et al. "Aversive conditioning to reduce raven predation on California Least tern eggs." *Colonial Waterbirds* 18: 131–138 (1977).
Heinrich, B. "Neophilia and exploration in juvenile common ravens, *Corvus corax*." *Anim. Behav.* 50: 695–704 (1995).
Sutton, G. M. *Birds in the Wilderness*. (New York: MacMillan, 1936).

Note:Konrad Lorenz's raven Roa appears in his classic book, *King Solomon's Ring* (New York: T. Crowell Co., 1952).

CHAPTERS 6 AND 7
For an extensive review of raven movements, and territorialism and population regulation:
Ratcliffe, D. *The Raven* (pp. 118–126, and pp. 196–216). (London: T & AD Poyser Ltd. 1997).

For studies on raven population in Schleswig-Holstein:
Grünkorn, T. "Untersuchungen zum Bestand zur Bestandsentwicklung und zur Habitatswahl des Kolkraben in Schleswig-Holstein." Thesis, Institut Haustierkunde, Univ. Kiel, 1991.
Looft, V. "Zur Ökologie und Siedlungsdichte des Kolkraben." *Corax* 1: 1–9 (1965).
Looft, V. "Die Bestandsentwicklung des Kolkraben in Schleswig-Holstein." *Corax* 9: 227–232 (1983).

General references:
Bruggers, D. J. "The Behavior and Ecology of the Common Raven in Northeastern Minnesota." Thesis, Univ. of Minnesota, 1988.
Ewins, P. J., J. N. Dymond, and M. Marquiss. "The distribution, breeding and diet of Raven *Corvus corax* in Shetland." *Birds Study* 33: 110–116 (1986).
Gothe, J. "Zur Ausbreitung und zum Fortpflanzungsverhalten des Kolkraben (*Corvus corax* L.) unter besonderer Berücksichtigurg der Verhältnisse in Mecklenburg" (pp. 63–129). In *Beiträge zur Kenntnis Deutscher Vögel*, H. Schildmacher. (Jena, Fisher, 1961).
Gwinner, E. "Untersuchungen über das Ausdrucks-und Sozialverhalten des Kolkraben (*Corvus corax* L.)." *Z. Tierpsychol.* 21: 657–748 (1964).

Lister, R. "Unusual winter movements of common ravens and Clark's nut-crackers." *Canadian Field Naturalist* 97: 325–326 (1973).

Lorenz, K. "Die Paarbildung beim Kolkraben." *Z. Tierpsychol.* 3: 278–292 (1940).

CHAPTERS 10 AND 11

For review of cooperation among animals:

Dugatkin, Lee A. *Cooperation Among Animals: An Evolutionary Perspective.* (Oxford: Oxford Univ. Press, 1997).

Predation by ravens:

Allan, T., and S. Allan. "Common raven attacks ruffed grouse." *Jack Pine Warbler* 64: 43–44 (1986).

Elkins, N. "Raven catching rockdove in the air." *Brit. Birds* 57: 302 (1964).

Heinrich, B. *Ravens in Winter* (pp. 48–57). (New York: Simon & Schuster, 1989).

Hewson, R. "Scavenging and predation upon sheep and lambs in west Scotland." *Journ. Appl. Ecol.* 21: 843–868 (1964).

Jensen, J. K. "Ravens attack fulmars in the air." *Dan. Ornithol. Foren. Tidsskr.* 85: 181 (1991).

Klickla, J., and K. Winker. "Observations of ravens preying on adult kittiwakes." *Condor* 93: 755–757 (1991).

Lydersen, C., and T. G. Smith. "Avian predation on ringed seal, *Phoca hispida,* pups." *Polar Biol.* 9: 489–490.

Maccarone, A. D. "Predation by common ravens on cliff-nesting black-legged kittiwakes on Baccalieu Island, Newfoundland." *Colonial Waterbirds* 15: 253–256 (1992).

Maser, C. "Predation by common ravens on feral rock doves." *Wilson Bulletin* 87: 552–553 (1975).

Montevecchi, W. A. "Predator-prey interactions between ravens and kitti-wakes." *Z. Tierpsychol.* 49: 136–141 (1979).

Ostbye, R. "Raven attacking reindeer." *Fauna,* Oslo. 22: 265–266 (1969).

Parmelee, D. F., and J. M. Parmelee. "Ravens observed killing roosting kittiwakes." *Condor* 90: 952 (1998).

Sanders, B. A. "Some observations of the common raven as a predator." *Chat* 40: 96–97 (1976).

Schmidt-Koenig, K., and R. Prinzinger. "Raven *Corvus corax* strikes flying pigeon." *Vogelwelt* 133: 98 (1992).

Tella, J. L., I. Torre, and T. Ballesteros. "High consumption rate of black-legged kittiwakes by common ravens in a Norwegian seabird colony." *Colonial Waterbirds* 18: 231–233.

CHAPTER 12

For review of brood parasitism:
Rothstein, S. I. "A model system for coevolution: avian brood parasitism." *Annu. Rev. Ecol. & Syst.* 21: 481–508 (1990).

For review of recognition systems:
Sherman, P. W., H. K. Reeve, and D. W. Pfennig. "Recognition Systems" (pp. 69–96). In *Behavioral Ecology*, 4th ed., J. R. Krebs and N. B. Davies, eds. (Malden, MA: Blackwell Pub., 1997).

CHAPTER 13

Possible olfaction in ravens:
Harriman, A. E., and R. H. Berger. "Olfactory acuity in the common raven (*Corvus corax*)." *Phys. Behav.* 36: 257–262 (1986).
Taylor, W. P. "Note on the sense of smell in the golden eagle and certain other birds." *Condor* 25: 28 (1923).

CHAPTER 14

Farrell, R. K., and S. D. Johnston. "Identification of laboratory animals: freeze-marking." *Lab. Anim. Sci.* 23: 107–110 (1973).
Sherman, P. W., H. K. Roeve, and D. W. Pfennig. "Recognition Systems" (pp. 69–96). In *Behavioral Ecology*, 4th ed., J. R. Krebs and N. B. Davies, eds. (Malden, MA: Blackwell Pub., 1997).

CHAPTER 15

For reports of ravens nesting near eagles and falcons:
Ratcliffe D. *The Raven* (pp. 127–138). (London: T & AD Poyser, 1997).

For studies on ravens nesting in urban environments:
Campbell, B., and G. E. S. Turner. "Ravens breeding on city buildings." *Brit. Birds* 69: 229–230 (1976).
Hauri, R. "Der Kolrabe als Brutvogel am Berner Bundesshaus." *Orn. Beob.* 90: 299–301 (1993).
Jefferson, B. "Observations of common raven in metropolitan Toronto." *Ontario Birds* 7: 15–20 (1989).

Knight, R. L. "Responses of nesting ravens to people in areas of different human densities." *Condor* 86: 345–346 (1986).

Meek, E. R. "Raven breeding on city buildings." *Brit. Birds* 69: 316 (1976).

Ortlieb, R. "Stadtbrut des Kolkraben (*Corvus corax*) in Ravensburg." *Anz. Orn. Ges. Bayern.* 10: 186–187 1971).

Ulrich, H. "Brutversuch des Kolkraben (*Corvus corax*) im Stadtebeit von Berlin (West)." *Orn. Ber. Berlin (West).* 8: 167 (1983).

White, C. M., and Tanner-White, M. "Use of interstate highway overpasses and billboards for nesting by the common raven (*Corvus corax*)." *Great Basin Nat.* 48: 64–67 (1988).

General references:

Kilham, L. "Sustained robbing of American crows by common ravens at a feeding station." *J. Field Ornithol.* 56; 425–426 (1985).

Marzluff, J. M. "Foraging relationships between corvids and golden eagles: mutual parasitism?" *Journal Raptor Research* 28: 60 (1994).

Nero, R. W. "Red fox-common raven interaction." *Blue Jay* 51: 177–178 (1993).

Williamson, F. S. L., and Rausch, R. "Interspecific relations between goshawks and ravens." *Condor* 58: 165 (1956).

CHAPTER 16

For an overview of vocal communication in birds other than ravens:

Kroodsma, D. E., and E. H. Miller. *Ecology and Communication in Birds.* (Ithaca, NY: Cornell Univ. Press, 1996).

For our own work on communication in ravens:

Heinrich, B. "Winter foraging at carcasses by three sympatric corvids, with emphasis on recruitment by the raven, *Corvus corax.*" *Behav. Ecol. & Scoiobiol.* 23: 141–156 (1988).

Heinrich, B. "Does the early raven get (and show) the meat?" *The Auk* 111(3): 764–769 (1994).

Heinrich, B., and J. M. Marzluff. "Do common ravens yell because they want to attract others?" *Behav. Ecol. & Sociobiol.* 28: 13–21 (1991).

Heinrich, B., J. M. Marzluff, and C. S. Marzluff. "Ravens are attracted to the appeasement calls of discoverers when they are attacked at defended food." *The Auk* 110: 247–254 (1993).

Marzluff, J. B., and B. Heinrich. "Foraging by common ravens in the presence and absence of territory holders: an experimental analysis of social foraging." *Anim. Behav.* 42: 755–770 (1991).

General reference:

Zahavi, A., and Zahavi, A. *The Handicap Principle.* (New York: Oxford University Press).

For various previous publications on raven communication:

B. Heinrich. *Ravens in Winter* (pp. 246–252). (New York: Simon & Schuster, 1989).

CHAPTER 17

For mouth color as related to maturation and dominance in ravens:

Heinrich, B., and J. M. Marzluff. "Age and mouth color in common ravens, *Corvus corax.*" *The Condor* 94: 549–550 (1992).

Heinrich, B. "When is the common raven black?" *Wilson Bulletin* 106: 571–572 (1994).

For status, stress, aggressiveness, and hormones in birds:

Hegner, R. E., and J. C. Wingfield. "Social status and circulating level of hormones in flocks of house sparrows," *Passer domesticus. Ethology* 76: 1–14 (1987).

Schlinger, B. A. "Plasma androgens and aggressiveness in captive winter white-throated sparrows (*Zonotrichia albicallis*)." *Hormones and Behavior* 21: 203–210 (1987).

Wingfield, J. C. "Modulation of the adrenocortical response to stress in birds." In *Perspectives in Comparative Endocrinology,* K. G. Davey, R. E. Peter, S. S. Tobe, eds. (Ottawa: Nat. Res. Council, 1994).

For review of dominance and feeding behavior:

Heinrich, B. "Dominance and weight changes in the common raven, *Corvus corax.*" *Anim. Behav.* 48: 1463–1465 (1994).

For review of dominance displays:

Gwinner, E. "Untersuchungen über den Ausdrucks-und Sozialverhalten des Kolkraben (*Corvus corax* L.)." *Z. Tierpsychol.* 21: 657–748 (1964).

CHAPTER 18

General references:

Dathe, H. "Kolkrabe (*Corvus corax*) auf Schweinen." *Orn. Mitt.* 16: 16 (1964).

Heinrich, B. "Neophilia and exploration in juvenile common ravens, *Corvus corax.*" *Anim. Behav.* 50: 695–704 (1995).

Heinrich, B. "Why do ravens fear their food?" *The Condor* 90; 950–952 (1988).

Heinrich, B., J. M. Marzluff, and W. Adams. "Fear and food recognition in naive common ravens." *The Auk* 112: 499–503 (1996).

Steinbacher, J. "Weitere Beobachtung von Kolkraben auf Schweinen." *Orn. Mitt.* 16: 147–148 (1964).

Wüst, W. "Kolkrabe auf dem Rücken eines weidenden Pferdes." *Orn. Mitt.* 10: 32 (1958).

CHAPTER 19

For a detailed report and background information on the Yellowstone Wolf Project:

Smith, D. W. "Yellowstone Wolf Project: Annual Report, 1997." (Wyoming, YCR-NR-98–2, 1998).

CHAPTER 20

General references:

Allen, D. L. A. *Wolves of Minong: Their Vital Role in a Wild Community.* (Boston: Houghton-Mufflin, 1979).

Brandenburg, J. *Brother Wolf: A Forgotten Promise.* (Minocqua, WI: North World Press, 1993).

Harrington, F. H. "Ravens attracted to wolf howling." *Condor* 80: 236–237 (1978).

Heinrich, B. *Ravens in Winter* (pp. 55–56). (New York: Simon & Schuster, 1989).

Lopez, B. H. *Of Wolves and Men.* (New York: Charles Scribner's Sons, 1978).

Magish, D. P., and A. H. Harris. "Fossil ravens from the Plestocene of Dry Cave, Eddy County, New Mexico." *Condor* 78: 399–404 (1976).

Mech, L. D. *The Wolf: the Ecology and Behavior of an Endangered Species* (New York: Natural History Press, 1970).

Peacock, D. *Grizzly Years* (p. 106). (New York: Henry Holt, 1990)

Peterson, R. O. *The Wolves of Isle Royale: A Broken Silence.* (Minocqua, WI: Willow Creek Press, 1995).

Ratcliffe, D. *The Raven* (pp. 7–26). (London: T & AS Poyser, 1997).

CHAPTER 21

General reference:

Heinrich, B. *Ravens in Winter* (pp. 251–252). (New York: Simon & Schuster, 1989).

Freuchen, P., and F. Salomonsen. *The Arctic Year.* (London: Jonathan Cape, 1960).

CHAPTER 22

For general results of the raven-caching experiments and references to previous work on other corvids:

Heinrich B., and J. W. Pepper. "Influence of competitors on caching behavior in the common raven, *Corvus corax.*" *Anim. Behav.* 56: 1083–1090 (1998).

Recent work indicating complex memory:

Clayton, M. S., and A. Dickinson. "Episodic-like memory during cache recovery by scrub jays." *Nature* 398: 272–274 (1998).

Recent work showing other corvids who also remember conspecifics' cache sites:

Bednekoff, P. A., and R. P. Balda. "Observational spatial memory in Clerk's nutcrackers and Mexican jays." *Anim. Behav.* 52: 233–239 (1996).

For reviews of the field of Machiavellian Intelligence by many contributors:

Whiten, A., and R. W. Bryne (eds). *Machiavellian Intelligence II: Evolutions and Extensions.* (Cambridge: Cambridge Univ. Press, 1998).

General reference:

Gwinner, E. "Über den Einfluss des Hungers und anderer Faktoren auf die Versteckaktivität des Kolkraben (*Corvus corax*)." *Vogelwarte* 23: 1–4 (1965).

CHAPTER 23

For an examination of morality in animals:

de Waal, F. *Good Natured: The Origin of Right and Wrong in Humans and Other Animals.* (Cambridge, MA: Harvard Univ. Press, 1986).

General references:

Liu-Chih, L. "Bruterfolg als Funktion von Ökosystemtyp, Flächennutzung und Konkurrenz bei: *Corvus corax.*" Dissertation, Universitat Saarlaudes, Saarbücken, Germany, 1997.

Christensen, H., and T. Grünkorn. "Nesthilfe beim Kolkraben (*Corvus corax* L.) nachgewiesen." *Corax* 17: 66–67 (1997).

Ehrengruber, M. U., and H. R. Aeschbacher. "Fütternder Helfer an einem Nest des Kolkraben *Corvus corax.*" *Orn. Beob.* 90: 301–303 (1993).

For discussion of helpers at the nest:

Alcock, J. *Animal Behavior,* 6th ed. (pp. 569–575). (Sunderland, MA: Sinauer, 1998).

For discussion of bird mobbing behavior:

Dawson, J. W. "Golden eagle mobbed while preying on common raven." *Raptor Research* 16: 136 (1982).

Heinrich, B. *One Man's Owl* (Appendix). (Princeton, NJ: Princeton Univ. Press, 1987).

Other references

Clapp, R. B., M. K. Klimkiewicz, and A. W. Futcher. "Longevity records of North American birds: Columbidae through Paridae." *J. Field Orn.* 54: 123–137 (1993).

Nogales, M. "High density and distribution patterns of a raven *Corvus corax* population on an oceanic island (El Hierro, Canary Islands)." *Journal of Avian Biol.* 25: 80–84 (1994).

Ryves, B. H. *Bird Life in Cornwall.* (London: Collins, 1948).

CHAPTER 24

For studies on animal play:

Beckoff, M., and J. A. Byers (eds.). *Animal Play: Evolutionary, Comparative, and Ecological Aspects.* (Cambridge: Cambridge Univ. Press, 1998).

Ficken, M. S. 1977. "Avian play." *The Auk* 94: 537–582 (1997).

For studies on raven play:

Drack, G. "Akitivitätsmuster und Spiel Freilebender Kolkraben." (Dissertation, Univ. Salzburg, Austria, 1994).

Gwinner, E. "Über einige Bewegungsspiale des Kolkraben (*Corvus corax* L.)" *Z. Tierpsychol.* 23: 28–36 (1996).

Heinrich, B., and R. Smolker. "Raven Play." In *Animal Play: Evolutionary, Comparative, and Ecological Aspects.* (Cambridge: Cambridge Univ. Press, 1998).

CHAPTER 25

General reference:

Heinrich, B. "Planning to facilitate caching: Possible suet cutting by a raven." *Wilson Bull.* (In press).

CHAPTER 26

Note: String-pulling as an experimental method for examining insight has a long history. The first results (with small songbirds) were thought to suggest insight, but subsequent studies disproved these claims. As a result, the whole area of inquiry was abandoned before animals that *could* solve the puzzle were examined. The following are related publications:

Altevogt, R. "Über das 'Schöpfen' einiger Vogelarten." *Behavior* 6: 147–152 (1953).

Bierens de Haan, J. A. "Der Stieglitz als Schöpfer." *J. Ornithol.* 1: 22 (1933).

Dücker, G., and B. Rensch. "The solution of patterned string problems by birds." *Behavior* 62: 164–173 (1977).

Heinrich, B. "An experimental investigation of insight in common ravens (*Corvus corax*)." *The Auk* 112: 994–1003 (1995).

Heinrich, B. "Detecting insight in common ravens *Corvus corax*." In C. Heyes and L. Huber (eds.), *Evolution of Cognition*. Boston: MIT Press (in press).

Thorpe, W. H. "A type of insight learning in birds." *Br. Birds* 37: 29–31 91943).

Tolman, E. C. "The acquisition of string-pulling by rats —Conditioned response or sign-gestalt?" *Psychol. Rev.* 44: 195–211 (1937).

Vince, M. A. "String-pulling in birds. II. Differences related to age in Green-finches, Chaffinches and canaries." *Anim. Behav.* 6: 53–59 (1958).

Vince, M. A. " 'String-pulling' in birds. III. The successful response in Greenfinches and canaries." *Behaviour* 17: 103–129 (1961).

CHAPTER 27

General references:
Portman, Adolphe. "Etudes sur la cérébralization chez les oiseaux." *Alauda* 14:2–20 (1946) and *Alauda* 15: 1–15 and 15: 161–171 (1947).

Rehkämper, G., H. D. Frahm, and K. Zillen. "Quantitative development of brain and brain structures in birds (Galliformes and Passeriformes) compared to that of mammals (Insectivora and Primates)." *Brain Behav. Evol.* 37: 125–143 (1991).

For review of forebrain size in birds correlated to feeding innovations:
Lefebvre, L., P. Whittle, E. Lascaris, and A. Finkelstein. "Feeding innovations and forebrain size in birds." *Animal Behaviour* 53: 549–560 (1997).

For reviews on avian hippocampus growth and development in relation to food-storing experience:

Clayton, N. S. "Memory and hippocampus in food-storing birds: A comparative approach." *Neuropharmacology* 37: 441–452 (1998).

Krebs, J. R., N. S. Clayton, S. D. Healy, D.A. Cristol, S. W. Patel, and A. R. Jolliffe. "The ecology of the brain: Food—storing and the hippocampus." *Ibis* 138: 34–46 (1996).

General references:

Armstrong, E. "Relative brain size and metabolism in mammals." *Science* 220: 1302–1304 (1983).

Clutton-Brock, T. H., and P. H. Harvey. "Primates, brains and ecology." *J. Zool. Lond.* 190: 309–323 (1980).

Gittleman, J. L. "Carnivore brain size, behavioral ecology, phylogeny." *J. Mammology* 67: 23–36 (1986).

Gould, J. L., and C. G. Gould. *The Animal Mind.* (New York and Oxford: W. H. Freeman & Co., 1994).

Harvey, P. H., and J. R. Krebs. "Comparing brains." *Science* 249: 140–146 (1992).

Pearson, R. *The Avian Brain.* (London: Academic Press, 1972).

Ridgway, S. H. "Physiological observation on Dolphin Brains" (pp. 31–59). In *Dolphins, Cognition and Behavior: A Comparative Approach,* R. J. Schuterman, J. A. Thomas, and F. G. Woods, eds. (Hillsdale, NJ: Lawrence Erlbaum Assoc., 1986).

Roberts, W. A. *Principles of Animal Cognition.* (New York: McGraw-Hill, 1998).

CHAPTER 28

Notes: *The End of Neuroscience* by John Horgan (Reading, MA: Addison-Wesley, 1996) presents various divergent perspectives of consciousness: consciousness as an illusion, something explained by laws of physics, a phenomenon independent of physical substrate, as a manifestation of short-term memory, oscillation of neural firing, and other theories.

Special issues of *Scientific American* magazine have been devoted to consciousness, and in them are references to some forty books on the topic. The issues are titled "Mind and Brain" (September 1992), "The Puzzle of Consciousness" (December 1995), and "Exploring Intelligence" (Winter 1998).

For added perspectives, see E. O. Wilson in *Conciliance* (New York: Knopf, 1998), where the topic is discussed and a list of some twenty additional books is give, and Donald R. Griffin, "From cognition to consciousness," a review in *Anim. Cogn.* 1: 3–16 (1998).

General references:

Balda, R. P., I. M. Pepperberg, and A. C. Kamil (eds.) *Animal Cognition in Nature*. (New York: Academic Press, 1998).

Boars, B. J. *In the Theater of Consciousness: The Workplace of the Mind*. (New York: Oxford Univ. Press, 1997).

Calvin, W. H. *How Brains Think: Evolving Intelligence, Then and Now*. (New York: HarperCollins, 1996).

Chalmers, D. J. *The Conscious Mind: In Search of a Fundamental Theory*. (New York: Oxford Univ. Press, 1996).

Dennett, D. C. *Consciousness Explained*. (Boston: Little, Brown, 1991).

Dickinson A., and B. Ballentine. "Motivational control of goal-directed action." *Animal Learning and Behavior* 22: 1–18 (1994).

Dukas, R. (ed.). *Cognitive Ecology*. (Chicago: University of Chicago Press, 1998).

Gardiner, H. *Frames of Mind: The Theory of Multiple Intelligences*. (New York: HarperCollins, 1983).

MacPhail, E. M. *Brain and Intelligence in Vertebrates*. (Oxford: Claredon Press, 1982).

MacPhail, E. M. "The search for a mental Rubicon." In C. Heyes and L. Huber (eds.), *Evolution of Cognition*. Boston: MIT Press (in press).

Shettleworth, S. J. *Cognition, Evolution, and Behavior*. (New York: Oxford University Press, 1998).

Wasserman, E. A. "Comments on animal thinking." *Amer. Scientist* 73: 6 (1985).

Note: The opinions I have expressed in this book undoubtedly were influenced by many of the above sources.

Index

Page numbers of illustrations appear in italics.

About the author

About the book

Read on

Insights,
Interviews
& More . . .

A Conversation with Bernd Heinrich

Where you were born, Bernd?

In Poland. (I was long confused and always thought it was Germany, not knowing about all of the border changes and history—as explained in my memoir *The Snoring Bird.*)

How did you happen to move to the States?

My father was a naturalist and ichneumon taxonomist, whose friend and colleague, Dr. Henry Townes at the University of Michigan in Ann Arbor, suggested he would have an easier time publishing his work here rather than in war-torn Germany, where we had been driven during the war.

Did Bernd the boy show hints of Bernd the zoological man?

More than hints. We lived isolated in a forest, and I did not know any other (male) person who was not obsessed with nature. I knew nothing else and could envision nothing else.

Did you have many pets as a child?

Before I was ten I had a crow. It made a great impression on me. I then also raised a wild pigeon and a jay. All lived free. Growing up in Maine I had the means to make birdhouses and a cage. I had wild birds up close, and again a pet crow. I then also kept snakes on occasion, a bat, and raised caterpillars that I loved to find in the woods and fields. My mother also had a raccoon raised from a baby my father found in the woods, and a pet monkey that she brought back from Africa. A neighbor kid had a skunk (unaltered—but *not* stinky by the way) that I "shared."

What did your parents do?

My father was for far too many years of his life a soldier. He was also a museum collector, in the style of the Victorian naturalists. When we

> 66 Before I was ten I had a crow. It made a great impression on me. 99

came to the U.S., both he and my mother picked apples, worked in small factories, and then took up logging in the woods. But soon they again took up collecting birds and small mammals for Yale's Peabody Museum, the University of Kansas museum, the Chicago Field Museum, and the American Museum of Natural History in New York. In between expeditions he identified ichneumons for the Canada and Florida Departments of Agriculture as well as the Smithsonian.

Did biology inspire you to become an artist, or did art inspire you to become a biologist? When did you begin drawing?

Biology came first. But it also goes both ways. I began drawing as a kid—maybe nine years old?—when I got my first piece of paper and colored pencils and wanted to "preserve" a beetle or a bird I had seen. But my drawing has always been very sporadic, mostly because I feel very deficient at it. I didn't focus on it until I started writing books. I could not afford to pay for illustrations—and the publishers would use what I provided.

Describe your educational background. What were your interests prior to college?

I went to grammar and high school at the Good Will Hinckley School, Farm, Home in Hinckley, Maine—except my first three years of grammar school were in a village in Germany. My main interest was to be outside the school—in the woods. (My real interest was biology, but there was nothing like it being taught then.)

What was your breakout moment as a scientist?

I think it all developed naturally from my background, and one thing just "naturally" led to another. First, a "working" lab biologist, Dick Cook at the University of Maine in Orono, showed me that research was an exciting option. I think of "breakout moments" only in terms of specific projects. ▶

Meet Bernd Heinrich

© Rachel Smolker

BERND HEINRICH is the author of numerous books, including the bestselling *Winter World, The Trees in My* Forest, *A Year in the Maine Woods, Ravens in Winter, One Man's Owl,* and *Bumblebee Economics,* which was nominated for the National Book Award. He is professor emeritus of biology at the University of Vermont. Heinrich divides his time between Vermont and the forests of western Maine. ❧

A Conversation with Bernd Heinrich *(continued)*

For example, my physiology passion went from a spark generated in my cell physiology projects to a flame when I did the experiments that proved that sphinx moths regulate their thoracic temperature in flight by means of redistributing "excess" heat of their flight exercise using their blood circulation. My most memorable moment with caterpillars came in finding a fresh leaf on the ground that had feeding damage, and that had obviously been chewed through at the petiole, because I recognized what it could mean. With bumblebees it was seeing the different behaviors at different flowers. With ravens it was hearing the crowd at a moose carcass.

You once held the world record for long-distance running. If I'm not mistaken, you weren't exactly a spring chicken when you decided to take up this sport, were you?

There are many records—I believe I still hold world records for running fifty miles (maybe also 100 kilometers and 100 miles?)—for the over-forty age group. (I have not recently checked Ultra Running records.) I held the "open" (i.e., any age or sex) American records for 100 km and 100 miles on the road, and most miles run in twenty-four hours (on the track), and all were set after age forty.

How did you develop this passion for ultramarathons?

I ran several marathons—winning the San Francisco marathon, and I was the first "master" (in the over-forty age category, just like they have a special category for women, and any others considered handicapped!) at Boston the day after my fortieth birthday, and did most of my passing— all of young guys—on the home stretch. I decided I had potential to set records at longer distances, if I buckled down. So I did, maybe because I had had injuries which had made running seem almost impossible. To be able to run seemed like a precious gift—I didn't think I had any—that couldn't be wasted.

You wrote a book about running: Why We Run: A Natural History. *Was this book more enjoyable to write than your others?*

I think so—running ultra-marathons fast has been the toughest thing I ever did, and to get it down on paper felt like I was reliving and preserving some of that effort and passion that combined my academic, personal, and scientific interests. Actually, the same goes for the other books.

What has been your most gratifying experience as a teacher?

Being out in the woods around a fire in the snow with my Winter Ecology group and seeing them excited and talking over something they had not seen before.

*Do students, friends, and family shower you with raven-themed gifts—
for example, T-shirts, wood sculptures, refrigerator magnets?*

They do.

A 1989 New York Times *review of* Ravens in Winter *complimented
your field work, noting that the book's "action takes place largely in
the Maine woods, in a makeshift laboratory that resembles a butcher's
meat locker more than an ivory tower." Have you done anything to
modernize your cabin?*

Yikes, they obviously have not been to my cabin. Somebody has an active
imagination. My cabin resembles a cozy corner, and I keep it that way.
Of course that means having a good fire in the stove in the winter.

*Does your cabin have a small library? If so, what kind of books do
you keep on hand for your Maine getaways?*

Very few books—nobody goes there to read. We go to be outside.

*Which five nature books would you lug into the woods on a year-long
sabbatical?*

I'd like to reread Thoreau's *Maine Woods,* Peter Matthiessen's book on
Africa, something of Edward Abbey's, E. B. White. I'd also have a book
with blank pages to scribble in.

*You have fallen through ice during your zoological exertions and,
presumably, suffered many other mishaps. What's the worst, most
hairy thing that ever happened to you in the name of Science?*

I had two car accidents on travel to and from my Maine study site in
winter, hitting ice. Both were head-on collisions and I was alone. The
vehicles did not survive. I think my hairiest experience was once climbing
up a big pine tree to a raven's nest and getting weak and starting to lose my
grip once I got too near the top. It looked like a very long way down—and
it took a long time to get there. Of course, being chased by a spitting cobra,
a rampaging cape buffalo, and a rhino in the forest of Mount Meru during
my year in the bush in Tanzania were memorable as well.

*You've written about a host of creatures—from bumblebees to owls
to antelope. Do you favor any animal above the others?*

No. The beauty comes from the diversity. I think that I was fully enamored
with each.

Which wild environment would you most like to visit someday?

Australia? East Africa—vast open vistas with scattered trees, lots of birds,
wildlife—and no car in sight and freedom to go where I please; I'd go ▶

nuts shut up in a tourist van. Or the Arctic tundra. The main ingredient is someplace vast that shows no imprint of man. On the other hand, I am also content digging deeper into the Maine woods. The other places are just as important to me, though, in the imagination, so long as they are there in real life. I get almost as much satisfaction just thinking about it. Maybe I am incapable of imagining a world that isn't, but when I do I shut it off as a trivial distraction from what demands full attention.

What are you working on now?

Summer World.

What do you do with your free time?

I can't remember ever having any.

What do you value most?

Nature in the sense of vast ecosystems. Pristine if possible, but man-assisted if need be. We need them to be human.

How would you like to be remembered?

Speaking for the above, and with feeling and reason, but not from sentimentality.

Whom do you most admire?

John F. Kennedy, Jr. comes to mind as one example of many for farsighted vision, acumen, energy, activism, energy, and guts. ∾

A Raven Update
New and Exciting Findings

MUCH OF WHAT I FEEL toward ravens is colored by my affection for Goliath, one large raven whom I reared from a chick, and his mate Whitefeather, who was a wild bird he met in my aviary. I begin writing this postscript at Raven Hill Camp in Maine, where their legacy lives on.

The Goliath-Whitefeather pair has raised their annual clutch of young every spring here for the last ten years. After the fledging and independence of their young, the pair leaves for a month or two and then returns each fall to sleep at their nest site in the grove of now tall white pine trees that I have nurtured near my cabin.

I have on various occasions tried to be a matchmaker for raven pairs, and except for the above exception, it has almost never worked out. From day one, on April 26, 1995, when they were first together, they bonded, and the following spring they built a nest in their aviary, laid eggs, and raised their young. In early May, 1996, I had to find out if they might accept the young from a pair in another aviary who had abandoned their own.

I was at that time corresponding with Rachel Smolker, a Fellow at the Bunting Institute at Harvard-Radcliffe, where she was writing a book about her work in Australia with wild dolphins. I invited her to come and witness at close range what I expected might be interesting behavior in the birds reputed to be the brains of the bird world.

She did come. Now more than ten years later as we sat reminiscing about the occasion while our kids slept, we again drank our coffee on the cabin door stoop facing the hills that were starting to flush out in glorious fall colors. We heard the pair make their usual ▶

> 66 I have on various occasions tried to be a matchmaker for raven pairs, and except for [one] exception, it has almost never worked out. 99

7

morning wake-up calls over by the pines, before resuming their daily
rounds. "I remember coming up the Hill," Rachel told me. She hesitated,
then continued: "I stopped in the middle of the trail in the woods, and
stood still for a few minutes, thinking I felt as though I was entering a new
world." When she reached the cabin she found four hungry pin-feathered
ravens screaming for food, and me half asleep next to them. I had walked
up the hill at midnight juggling a heavy box containing the ravens and,
simultaneously, dragging a huge beaver roadkill carcass behind me. I then
climbed up to the nest in the aviary to remove the two babies of Goliath-
Whitefeather in the dark of night and replace them with the four I had
brought. In the pre-dawn, long before Rachel had arrived, I had already
been at the aviary observing results for about an hour. The adults were at
the nest, quizzically examining the doubled clutch of new young, but not
making much of a fuss.

Soon the four strangers were
begging for food, and by noon Goliath
and Whitefeather had alternately cawed
in alarm and then also fed the four
young indiscriminately along with
their own young. I left the birds free
to forage, hoping the freshly opened
beaver carcass near the aviary would
keep them around until they got
oriented. But almost immediately
another pair of ravens arrived and my
favorite birds flew out and joined them.
To my great shock, the apparently
friendly foursome left into the distance.
Would my two find their way back?

It was a long wait. To my great
surprise we heard raven calls at the nest
early the next morning. Soon we saw
them at the beaver carcass outside the aviary, ripping off meat and feeding
the six young. After this rocky start the couple fully accepted the new
young. We helped our ravens throughout the summer by coming back
at almost weekly intervals to drop off roadkill, calves, and kitchen scraps.
Whitefeather remained reticently but noisy in the nearby woods, but
Goliath would then fly to her to bring her our treats. The young were
at first not afraid of us, but as the summer progressed they strayed ever-
further away. We had many happy hours with the ravens. We missed
them after they all left in late summer.

By the next spring, in 1997, Goliath again flew in and out of the aviary
and built a stick nest in the same shed where their nest was the year before.
However, Whitefeather did not do her part, the lining it with deer fur. The

nesting attempt was then aborted, but subsequent ones went (almost) without a hitch, and continue to this day.

I have often been asked various questions that center on "what ravens do." I have worked very hard to try to find out, as have other capable colleagues, and I am now often tempted to reply: "*What* raven, at what age, under precisely what circumstances are you talking about?" My counter question is not meant as an evasion. Ravens have personalities. There are the bold, the meek, the curious, the calm, and the nervous birds. I suspect the ravens' often unpredictable behavior resides in what we call "mind"—something independent of programmed reactions to the immediate. However, in the absence of language, it is notoriously difficult to determine what sequences of images may be projected on the consciousness that may then give insight and affect behavior. In one of my first groups of tame ravens from the 1980s I had observed and then studied behavior of ravens pulling up meat suspended on a meter-long string. Unfortunately, I had not performed the proper controls because the observations had originated from opportunistic chance. However, given the then common notion that birds are near automatons, my results were so unexpected and surprising that they seemed worth reporting.

It was not until February 1998, when I had another group of naïve ravens of suitable age (at least ten months), that I could again attempt the experiments under conditions that would exclude imitation and interference. These birds were tested one at a time and in isolation of each other. The results fully corroborated my previous conclusions and I was ready to publish again. However, in the meantime, another objection was raised to my insight hypothesis. It was suggested that the string-pulling results could be due to "rapid" learning—learning so quick that it would seem like instant knowledge. That seemed like a long shot, since it would involve learning to perform a long and precise sequence—reach, grasp, pull, step on string, apply pressure with foot, release bill, etc.—and then repeating the same sequence at least six more times in precisely the same sequence, all in several minutes. However, there are seldom enough tests that satisfy everyone, and given extraordinary findings, there shouldn't be.

There was one additional test that should do the trick of answering whether or not quick trial and error learning was the explanation of the insightful behavior. It came up in a discussion with my late friend and colleague Donald Griffin. Griffin had been enthusiastic about my results purporting to show insight intelligence using a piece of string and a slice of salami. He casually told me, "You should try to find out if the ravens will try to get meat where they have to do something nonsensical, like having to pull *down* on string to have the meat come up." The crux of the experiment that Griffin suggested is that a nonsensical multi-step solution could only be solved by trial and error and would take very ▶

A Raven Update *(continued)*

long to learn, in contrast to when the animal "sees through" the various steps and uses the insight that I had suggested. I was itching to try the experiment, but since my 1998 birds were no longer naïve to string pulling I needed to wait several years until I had another group of young birds.

In 2002 Thomas Bugnyar from the University of Vienna and the Konrad Lorenz Research Station in Gruenau, Austria, came to Vermont to work with me on a series of tests of ravens' food caching behavior. We secured a new group of six young birds from two nests, and in February 2003 they were ready to serve as subjects for the reverse pull-up experiment. We looped a string from a perch over wire above the perch, and the string hung down on the other side of a wide mesh wire screen below, only slightly to the side of where the raven would perch. If the bird pulled the string down, then the meat would come up and be available directly at the wire, by the perch, where it could be grabbed. To make the results comparable to the control part of the experiment performed in 1998, these naïve and same-aged birds were also tested one at a time and in isolation of each other.

All the ravens showed interest in the meat on string—they flew to the perch and looked down at the meat. However, the critical and telling result was that *none* pulled the string down, whereas the similarly aged and similarly inexperienced ravens of 1998 had solved the pull-*up* puzzle on the same perch in six minutes on average. We conclude, therefore, that the sequence of steps the ravens make to pull up food is not random and arbitrary. It is not attributable solely to "quick learning" (although learning occurs). It is far more plausible that the task is executed with the aid of an understanding of the steps required to accomplish a goal.

The string-pulling experiments, with these and other variations, revealed a spatial intelligence that was beyond what I expected. Intelligence is often task-specific, and the ravens' capabilities seemed almost superfluous since there is no apparent need for them to access food that is suspended. Therefore, I suspect the birds' apparent insight (i.e., the ability to form a mental scenario in this task) is an incidental byproduct that has evolved in the service of some other function.

The ravens' main food, carrion, occurs in the competitive environmental challenge with each other, and often in the presence of dangerous carnivores. Northern ravens are often confronted with rock-hard frozen carcasses in the winter. Their strong bills are superbly adapted to chip off the meat one small piece at a time. As I had suggested but not proved in *Mind of the Raven*, ravens can forego an immediate reward for a later larger one, as shown by their surgically aimed pecking in lines to carve deep grooves to dislodge chunks from cold-hard suet to carry off and eat later. Craig Comstock of Maine has recently completed a detailed study of this amazing and unique behavior on a wild pair of ravens with whom he has by now had a ten-year relationship.

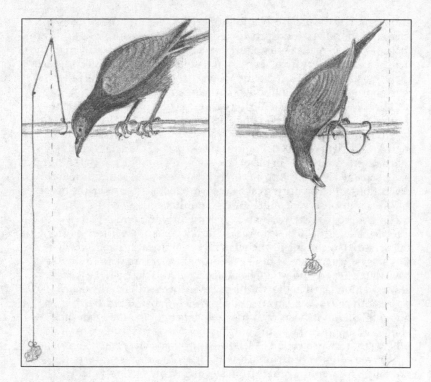

Frozen muscle meat that cannot be easily carved into chunks is handled differently than sliceable suet. Single ravens or mated pairs working at a frozen carcass stack numerous small pieces of meat next to them as they dislodge them, and when they have accumulated a sufficiently large pile they grab it in one bill-full and fly off with it. More commonly, however, there are crowds of ravens at any one carcass and no bird then leaves small pieces of meat lying around. Each bird hauls off as much meat as it can, and the amount of any one load is limited by the capacity of the gular (throat) pouch and the bill. Most of these loads of meat are cached for later retrieval and consumption. In my aviary complex the ravens went to apparent great lengths to hide their meat out of view from others (but opportunities to do so are very limited in an aviary).

The apparently costly behavior of flying far in the field or avoiding others and delaying caching in their presence began to make sense to me in 1992 when I observed, to my great surprise (in a study of ravens' cache memory), that these birds not only recover their own buried meat, they ▶

A Raven Update *(continued)*

also memorize the precise locations of caches they see others make, and then steal the others' food, provided they get an opportunity. (No control caches—those I made with no ravens present—were ever recovered, hence cache recovery involves memory, not scent.) In turn, the original cache makers vigorously defended their caches from potential pilferers who approached them, as though knowing their intentions.

The social environment during cache-making proved to be fertile ground for experiments, and Bugnyar spent two years working with the ravens in my aviary in Vermont and received the coveted Niko Tinbergen award for his work in 2006. Bugnyar set up part of the aviary complex so that a raven could cache in the privacy of its own room, but that room had a screened window into another room that could contain an observer raven. Subsequently, during cache recovery the cache-maker, the observer, and/or a non-observer could be given simultaneous access into the caching compartment. Responses depend on who a bird is paired with. Cache-makers react to a "knower" (a bird who had visual access through the window) by chasing it when it comes near its cache location where it was likely to attempt to raid. They ignore another bird who could not have watched and who would be no threat to find the cache (caches are always camouflaged in that the birds cover them with nearby debris). Apparently the ravens can attribute knowledge to one another. In the literature on similar tests with apes, it is generally assumed that such ability presupposes the actors putting themselves in the place of the other (i.e., it is a projection of self into the other's place).

A raven's ability to attribute knowledge to another is important not only in preventing others from stealing its hidden food caches. This ability could also be important in deceiving rivals or manipulating their behavior. To test for these possibilities, Bugnyar performed a series of tests that involved allowing birds that had witnessed a cache-maker to retrieve the other's cache either in private or in the presence of two categories of others. When the witnessing bird was paired in the retrieval trial with the cache-maker, it delayed its approach to the other's cache (in contrast to raiding it immediately when retrieving it in private). More significantly, it "searched" at sites distant from the actual cache. Was it trying to mislead the cache owner, and/or withhold information of its knowledge to avoid the aggression of cache defense? Either way, the other's behavior was altered to benefit the actor.

Bugnyar proceeded with still another step in the experiment: He paired the witnessing bird with *another* witness, but one who was subordinate to the first. In this case, in the absence of a cache-maker who would have defended his cache, there was a contest between two birds attempting to raid the cache; now the subordinate bird went directly to the cache to try to eat it before the dominant could reach it. These results again suggest that the birds know each other individually and anticipate their likely actions. They *deceive* by being able to put themselves in the other's place.

Bugnyar and colleagues capitalized on the raven's knowledge-attribution to examine the function of one form of ravens' play. Like most intelligent animals, young ravens are renowned for their varied and conspicuous play behavior. Since caching behavior has great importance for them as adults, the young begin to practice play-caching almost immediately after fledging.

Many ravens start to manipulate objects with their bill while they are still in the nest, and as soon as they fledge they tuck both edible and inedible items up against large objects. After about two weeks this "sticking-up-against" behavior changes to "sticking-into" crevices, and by two months of age they pick up or scrape nearby debris to cover and camouflage their caches. Up to the age of two months the birds make little or no attempt either to hide or defend their caches from others. But from then on they increasingly differentiate between their object (play) and food caches. The latter (and not their object caches) are defended and placed at a distance from other birds and behind barriers. For a test to find out if object or "play" caching indeed results in learning, Bugnyar let young ravens experience both consistent object-caching thieves (a human who was a stranger to them and who raided every play cache) vs. a consistent non-thief (a different human who examined all caches as they were made but never retrieved the objects). Our data suggest that the play (or object) caching is instinctive, while defensive food caching is learned.

The young ravens routinely inspected their own caches after the respective humans had either examined or pilfered them, as though assessing the potential pilferers' intentions. Whether they had indeed learned the individual's pilfer dispositions was determined in tests with "real" (i.e., food) caches. In these tests the human observer (or thief) merely stood nearby without approaching the caches. The results were clear: In the presence of the "thief" of play caches, the ravens retrieved the food they had hidden and/or attempted to hide it anew behind an obstacle, and in the presence of the benign observer of play caches they did not retrieve their food cache in the designated test period. These results indicate that the *play* caching had allowed the birds to perfect their food cache defenses which then reduced cache-raiding. Clearly, the birds also easily distinguish individuals not only of their own species but of another (humans).

Almost nothing of the ravens' observed social behavior makes sense without the capacity of individual recognition among their own kind. But ravens, because they gain much of their food in close proximity of dangerous carnivores, benefit from knowing these animals' intentions and capabilities as well. As in their caching skills, such knowledge is also ▶

derived from play. As I had indicated in *Mind of the Raven*, young ravens are highly neophobic (fearful of the new), being shy even of dead mice. Nevertheless, when they are a little older they have a strong innate drive to engage in what looks like foolish risk-taking with dangerous predators such as wolves, eagles, etc. That behavior usually starts by approaching the animal and touching or pulling its tail, then escalates to more risky maneuvers. It has the effect that the birds quickly learn the object's responses and capabilities. Shyness of wolves, for example, might and usually (but not always) does keep ravens safe from these carnivores, but the ability to anticipate their behavior is necessary to get close enough for procuring food from them.

The association of ravens with wolves has long been noted and it is idealized in ancient Norse and Native American legends. However, the relevant behaviors were not well known. In *Mind of the Raven* I was particularly keen to explore the raven-carnivore (and hence raven-human hunter) connection, and I traveled to Nova Scotia and Nunavut to chase human-raven legends, and I went to Yellowstone Park to make preliminary observations of ravens with wolves. Since *Mind of the Raven* was published, the study I envisioned from my initial observations has been done and is published. It confirms that wild ravens have a strong behavioral connection to wolves. Ravens *prefer* to feed alongside wolves (and other ravens) and are very reluctant to approach food in their absence. The strong behavioral connection suggests not only proximate conditioning but also an evolutionary history of mutualism.

The study of ravens' relationship with wolves is not exhausted by any measure. Indeed, the core issue is still up for grabs, and it has two separate components. First, how do the birds get away with living practically and often literally "underfoot" with wolves? Possibly ravens taste bad or their distinctive smell is unpleasant to wolves. However, wolves are not picky eaters, and a friend of mine who has cooked and eaten a raven informs me that it tasted just fine. I suspect that the ravens' distinctive scent is more of an identifying signal than a feeding deterrent, and the birds may earn near immunity to predation by gradually ingratiating themselves to the wolves. In effect, they become their "pets." Admittedly this idea sounds farfetched, but a general phenomenon is involved.

We, too, are carnivores, but we hesitate to eat the dogs, cats, birds, fish, and other animals that we adopt (or that adopt us) after getting to know them intimately. Such "artificial" animal-animal tolerances and connections are forged even between dogs and cats, and as I have experienced, can even arise between archenemies such as a great horned owl and crows. The central factor is not that ravens are often partnered with "wolves" as such, but rather that they consistently get close to efficient carnivores to get food, and a mutual tolerance eventually develops and may persist in part by culture. Some tens of thousands of years ago, ravens likely depended on another carnivore: human hunters. Although the evolutionary antecedents of a raven-human connection may now be hidden, the following example shows what is possible, and what might have been.

Dadre Traughber, a medical doctor and an outdoors woman, dropped some treats to a raven that she met while hiking in a remote area in California. The bird accepted her offering, probably because the ravens in her area do not have the strong culture of staying shy of humans (as they do in Maine and other parts of the country due to persecution in the past century). She returned to the same place on subsequent visits and also met the bird's mate. Hundreds of regular visits later over the course of two years, Traughber met their offspring, and later also small flocks of their associates. Eventually, the ravens became so tame that she succeeded in feeding them as they gathered directly around her, and one even took food from her hand. These ravens now spend time "hanging out" with her even when she is temporarily without food, and when she leaves they follow her, walking behind or flying ahead of her. They recognize her among other people and fly to her, do aerobatics as they approach and land by her feet. Sometimes they even fly alongside her car. In other words, these ravens behave near a human much as ravens do in the presence of wolves in some of the wild places where wolves still live. Traughber has re-established a local tradition with these birds that was, at one time, likely widespread with aboriginal hunters, who provided treats far larger than a few handfuls of peanuts.

What is next? Aside from unraveling the unsolved mystery of how the birds recognize each other as individuals, I expect that the next major questions to ask the ravens will concern how the social relationships amongst themselves and with wolves are established and maintained.

My science with ravens has often come close to the personal. In my latest book titled *The Snoring Bird*, I was strongly influenced by what transpired after publishing *Mind of the Raven*. As detailed in my memoir, were it not for ravens I would never have found a treasure trove of correspondence in Berlin of my long deceased father. These letters provided fresh and invaluable insights into Papa's amazing life and times as a naturalist spanning two world wars, and they greatly informed that book. Writing *The Snoring Bird* story received additional spark when, with renewed interest, I pulled his gold ring out of my drawer and took my first close look at it. I knew that it had a small black onyx inset inscribed with a tiny family coat-of-arms. The latter had never held any interest to me. Putting it under a magnifying glass I was shocked to decipher a raven perched on a person's arm. ⌒

An Excerpt from Bernd Heinrich's New Book, *The Snoring Bird*

In this remarkable memoir, Bernd Heinrich shares the ways in which his relationship with his indomitable father, combined with his very unique childhood, molded him into the scientist and the man he is today. The Snoring Bird takes readers on a multi-generation, cross-continental journey, from Bernd's father's days as a soldier in both world wars, to his family's daring escape from the Red Army in 1945, from the secluded hut in the Northern German forest where they lived for five years, to the rustic Maine farm (water from a well, outhouse) that they eventually called home. Bernd also details the incredible African expedition in which he and his father worked side by side as well as his years as a graduate student when he found his own calling as a biologist. Bernd relates this all in his trademark style, making science accessible, awe-inspiring, fascinating, and also moving.

"Visit Home"

MAMUSHA IS JUST SETTLING DOWN on her bed to watch the evening news when I arrive. Two cans of Coors, which she has opened with the point of a pair of scissors, are on the table next to her, along with a box of German chocolates. She used to make her own beer, but now, in her mid-eighties, it is Coors from a can, and because her gnarled hands are too weak; she cannot pull off the tabs. Duke, the huge shepherd-hound that she rescued from the pound, is at her feet, and a one-legged chicken lies cradled in her lap. She is mildly irritated at me for arriving unannounced (I have a tendency to either just show up or to come an hour later than I've promised, which annoys her also), but soon I have placated her and she offers me a beer.

Mamusha is the Polish word for mama or mommy. She was born and raised in what is now Poland, and despite her willingness to consume Coors, she remains, in her memories and her ways, a product of the Old World. My visits to her are usually spent listening to stories about the past. We are sitting in the low-ceilinged brick room that Papa built decades earlier for the purpose of protecting his precious wasp collections from fire; the rest of the house might be consumed, but his ichneumons would be safe. When we moved to this house near Wilton, Maine, in 1951, it was a simple salt box-style farmhouse with six rooms. Since Papa's death, Mamusha has added on haphazardly, so that the house is now a collage of thirteen rooms. The walls are decorated with pictures of flowers that she has purchased, although one wall sports a portrait of George and Laura Bush that she received free in the mail. This small brick room is her main habitation, which she shares with her house chickens. I was met in the entryway by three hens, perched on the dresser. In the living room, one drawer sits partway open to accommodate a setting hen that has made her nest there. "My chickens outsmarted me again," Mamusha says. "Yesterday I found one upstairs in a corner of the bedroom. I have too many. Next time bring your shotgun and at least help me to get rid of a few roosters. They are all so pretty—brown, black, speckled, some with feathers on their toes, some without. I can't decide which ones to get rid of."

Mamusha keeps chickens in her barn and chicken house, but her house chickens are often the ill ones. Once in a while, for some reason, some of the newly hatched chicks have trouble with their legs. They splay out to the side and the chicks can barely stand, much less walk. Mamusha has discovered that if you cradle the chick in your arms for a week or two, and sleep with it cuddled up next to you in bed, it will eventually improve and become a fully functional house pet. By the warmth of the wood stove, with feed scattered across the floor, the chickens are quite comfortable. This is Mamusha's twist on the "survival of the fittest" concept. The most afflicted chickens receive the best care, and with the warm fire nearby they breed year round, producing more and more afflicted youngsters for Mamusha to look after. Their eggs do have the deepest-yellow yolks, which Mamusha brags about. . . . 〜

Suggested Reading
Recent New Studies of the Raven Mind

Bugnyar, T., B. Heinrich. (2006). Pilfering ravens, *Corvus corax*, adjust their behavior to social context and identity of competitors. *Animal Cognition* (in press).

Bugnyar, T., B. Heinrich. (2005). Food-storing ravens differentiate between knowledgeable and ignorant competitors. *Proceedings Royal Society London Series B* 272: 1641–1646.

Bugnyar, T., Kijne, M. and K. Kotrschal (2001). Food calling in ravens: are yells referential signals? *Animal Behaviour* 61: 949–958.

Bugnyar, T. and K. Kotrschal. (2004). Observational learning and the raiding of food caches in ravens, *Corvus corax:* Is it "tactical deception"? *Animal Behaviour* 64: 185–195.

Bugnyar, T., K. Kotrschal. (2004). Leading a conspecific away from food in ravens, *Corvus corax. Animal Cognition* 7: 69–76.

Bugnyar, T., C. Schwab, C. C. Schloegl, K. Kotrschal, B. Heinrich. Ravens judge other's pilfer dispositions through play-caching (in press).

Bugnyar, T., M. Stowe, B. Heinrich. (2006). The ontogeny of caching in ravens, *Corvus corax. Animal Behaviour* (in press).

Bugnyar, T., M. Stowe, B. Heinrich. (2004). Ravens, *Corvus corax*, follow gaze direction of humans around obstacles. *Proceedings Royal Society London Series B* 271: 1331–1336.

Comstock, C. (2007). Suet carving to maximize foraging efficiency in common ravens. *Wilson Journal of Ornithology* 119 (in press).

Heinrich, B., T. Bugnyar. (2005). Testing problem solving in ravens: String-pulling to reach food . *Ethology* 111: 962–976.

Range, F., T. Bugnyar, K. Kotrschal. Simple task discrimination in ravens, *Corvus corax:* a preliminary study. *Acta Ethologica*.

Range, F., T. Bugnyar, C. Schloegl, C. Pribersky-Schwab, K. Kotrschal. Individual learning ability and coping styles in ravens. *Behavioral Processes* (in press).

Schloegl, C., K. Kotrschal and T. Bugnyar. (2006). The ontogeny of gaze following in common ravens, *Corvus corax. Animal Behaviour* (in press).

Stahler, D., B. Heinrich, D.W. Smith. (2002). The raven's behavioral association with wolves. *Animal Behaviour:* 64: 283–290.

Stowe, M., T. Bugnyar, M. C. Loretto, C. Schloegl, F. Range, K. Kotrschal. (2006). Novel object exploration in ravens, *Corvus corax:* Effects of social relationships. *Behavioral Processes* 73: 68–75.

Stowe, M., T. Bugnyar, B. Heinrich, K. Kotrschal. (2006). Effects of group size on approach to novel objects in ravens, *Corvus corax. Ethology* 112: 1079–1088.

Stowe, M., T. Bugnyar, B. Heinrich, B. Spielauer, E. Mostl, K. Kotrschal. Individual corticosterone excretion relates to exploratory behavior in ravens, *Corvus corax*.

Have You Read?
More by Bernd Heinrich

WINTER WORLD

From flying squirrels to grizzly bears, torpid turtles to insects with antifreeze, the animal kingdom relies on some staggering evolutionary innovations to survive winter. Unlike their human counterparts, who must alter their environment to accommodate our physical limitations, animals are adaptable to an amazing range of conditions—i.e., radical changes in a creature's physiology take place to match the demands of the environment. Winter provides an especially remarkable situation, because of how drastically it affects the most elemental component of all life: water. Examining everything from food sources in the extremely barren winter landscape to the chemical composition that allows certain creatures to survive, Heinrich's national bestseller *Winter World* awakens the largely undiscovered mysteries by which nature sustains herself through the harsh, cruel exigencies of winter.

"Heinrich has a rare ability to embed dense scientific explications within graceful, lightfooted nature writing." —*New York Times Book Review*

WHY WE RUN: A NATURAL HISTORY

In *Why We Run*, Bernd Heinrich explores a new perspective on human evolution by examining the phenomenon of ultraendurance and makes surprising discoveries about the physical, spiritual—and primal—drive to win. At once lyrical and scientific, *Why We Run* shows Heinrich's signature blend of biology, anthropology, psychology, and philosophy, infused with his passion to discover how and why we can achieve superhuman abilities.

"A stunningly original book. It blends personal experience in world class distance running with a firsthand account of the biology of running by one of its leading authorities." —E. O. Wilson

THE GEESE OF BEAVER BOG

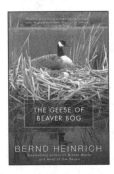

With a scientist's training and a nature lover's boundless curiosity and enthusiasm, Bernd Heinrich set out to observe and understand the travails and triumphs of the Canada geese, or honkers, living in the beaver bog adjacent to his rural Vermont home. Heated battles over territory, mysterious nest raids, jealousy over a lover's inattention—all are recounted here in an engaging, anecdotal narrative that sheds light on how geese live and why they behave as they do.

Heinrich takes his readers through mud, icy waters, and overgrown sedge hummocks into a seemingly impenetrable world. He does so with deft insight, respectful modesty, and infectious good humor, accompanied by beautiful four-color photographs and the author's trademark sketches.

"Heinrich's lyric writing and attentive observations make goose world come alive.... [A] pure joy." —*Los Angeles Times*

THE TREES IN MY FOREST

Bernd Heinrich takes readers on an eye-opening journey through the hidden life of a forest.

"This lyrical testament to the stunning complexity of the natural world also documents one man's bid to make a difference on his own little patch of land. In 1975 Heinrich ... bought three hundred acres of logged-over Maine woods and set out to restore its ecological diversity.... In his ultimate goal of creating a forest, a place of 'habitat complexity' vastly different from the sterile monocultures planted by paper companies in the name of sustainable forestry, he succeeds admirably." —*Kirkus Reviews*

"These passionate observations of a place 'where the subtle matters and the spectacular distracts' superbly mix memoir and science." —*New York Times*

Don't miss the next book by your favorite author. Sign up now for AuthorTracker by visiting www.AuthorTracker.com.